Muscle Contraction and Cell Motility

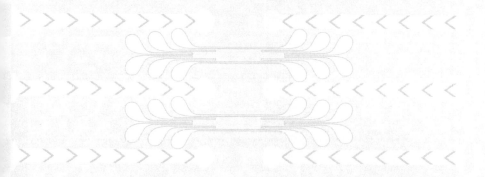

Muscle Contraction and Cell Motility

Fundamentals and Developments

edited by
Haruo Sugi

PAN STANFORD PUBLISHING

Published by

Pan Stanford Publishing Pte. Ltd.
Penthouse Level, Suntec Tower 3
8 Temasek Boulevard
Singapore 038988

Email: editorial@panstanford.com
Web: www.panstanford.com

British Library Cataloguing-in-Publication Data
A catalogue record for this book is available from the British Library.

Muscle Contraction and Cell Motility: Fundamentals and Developments

ISBN 978-981-4745-16-1 (Hardcover)
ISBN 978-981-4745-17-8 (eBook)

Printed and bound in Great Britain by
TJ International Ltd, Padstow, Cornwall

Contents

9. Effect of DTT on Force and Stiffness during Recovery from Fatigue in Mouse Muscle Fibres 235

Barbara Colombini, Marta Nocella, Joseph D. Bruton, Maria Angela Bagni, and Giovanni Cecchi

PART II: CARDIAC AND SMOOTH MUSCLE

10. ATP Utilization in Skeletal and Cardiac Muscle: Economy and Efficiency 249

G. J. M. Stienen

PART III: CELL MOTILITY

Preface

This volume provides a comprehensive overview of the current progress in research on muscle contraction and cell motility, not only for investigators in these research fields but also for general readers who are interested in these topics. One of the most attractive features of living organisms is their ability to move. In vertebrate animals, including humans, their body movement is produced by skeletal muscles, which are also called striated muscle due to their cross-striations. In the mid-1950s, H. E. Huxley and Hanson made a monumental discovery that the skeletal muscle consists of hexagonal lattice of two different myofilaments, i.e., actin and myosin filaments, and that muscle contraction results from relative sliding between actin and myosin filaments. Concerning the mechanism of the myofilament sliding, Huxley put forward a hypothesis that myosin heads (cross-bridges) extending from myosin filaments first attach to actin, undergo conformational changes to cause unitary filament sliding, and then detach from actin, each attachment-detachment cycle being coupled with ATP hydrolysis. Despite extensive studies having been carried out up to the present time, the myosin head movement still remains to be a matter of debate and speculation.

The text is organized into three parts. Part I contains nine chapters on the current progress in contraction characteristics and mechanical properties of the skeletal muscle. In Chapter 1, Sugi et al. describe their recent success in the electron microscopic recording of the myosin head movement coupled with ATP hydrolysis by using the gas environmental chamber, which enables the study of dynamic structural changes of living biomolecules related to their function. In Chapter 2, Squire and Knupp summarize the results obtained by using the time-resolved X-ray diffraction technique, detecting structural changes of myofilaments in contracting muscle in a non-invasive manner, and point out problems in interpreting the results. In Chapter 3, Bartels emphasizes the essential role of ions in muscle contraction, a topic generally ignored by muscle investigators. In Chapter 4, Sugi et al. point out that the results obtained from in vitro motility assay systems, in which actin filaments are made to interact with myosin heads detached from myosin filaments, may bear no direct relation to myofilament sliding in muscle. In Chapter 5, Ranatunga discusses the mechanism of force generation in the muscle based on his temperature-jump experiments. The 3D myofilament-lattice structure is known to be maintained by a network of a large protein, titin. Cecchi et al. in Chapter 6 and Rassier et al. in Chapter 7 show that Ca^{2+}-dependent stiffness changes of titin play an important role in muscle mechanical performance. For a full understanding of skeletal muscle performance in humans, it is useful to measure stiffness of the contracting human muscle by means of supersonic shear imaging (SSI). In Chapter 8, Sasaki and Ishii explain the theoretical background of SSI together with the results obtained from the contracting human skeletal muscle. In Chapter 9, Colombini et al. discuss the mechanism underlying muscle fatigue. Readers of this volume may become aware of discrepancies between what are stated in some chapters in this part and what are generally stated in many textbooks. We emphasize that these discrepancies reflect the general features of science in progress. Well-established dogmas in a scientific field can be denied by an unexpected discovery.

Part II consists of three chapters on the cardiac muscle and two chapters on the smooth muscle. The cardiac muscle also exhibits cross-striations and plays an essential role in blood circulation in the animal body. In Chapter 10, Stienen gives an extensive overview on various factors affecting the rate of ATP utilization of skeletal and cardiac muscles in a variety of animals, including humans. In Chapter 11, Morano also gives an extensive overview on the role of myosin essential light chain in regulating myosin function in skeletal, cardiac, and smooth muscles, based on the crystallographic structure of myosin molecule. In Chapter 12 by Minamisawa deals with proteins involved in Ca^{2+} cycling in cardiac muscle by citing vast literature in this clinically important research field. Smooth muscles do not show striations because of irregular arrangement of myofilaments, though their contraction mechanism is believed to be similar to that in skeletal and cardiac muscles. In Chapter 13, Kobayashi discusses factors that affect vascular smooth muscle diseases, including his recent interesting finding. Chapter 14, by Galler, is concerned with the so-called catch mechanism in the molluscan somatic smooth muscle, which is highly specialized to maintain tension over a long period with little energy expenditure.

Finally, Part III contains three chapters on cell motility. In Chapter 15, Shingyoji presents a comprehensive overview of the factors that influence the oscillatory movement of cilia and flagella caused by sliding between dynein and microtubule. In Chapter 16, Iwadate discusses crawling cell migration, which is caused by actin polymerization as well as actin–myosin interaction and is involved in a variety of biological phenomena, including wound healing and immune system function. In Chapter 17, Uyeda gives an extensive survey of the research on the role of actin filaments and actin-binding proteins in producing a wide range of cell activities.

This book constitutes a fascinating collection of overviews on muscle contraction and cell motility written by first-class investigators and not only provides information for general readers about the current progress and controversies in each research field but also stimulate young investigators to start challenging remaining mysteries in these exciting research fields.

Haruo Sugi
Tokyo, September 2016

PART I
SKELETAL MUSCLE

Chapter 1

Electron Microscopic Visualization and Recording of ATP-Induced Myosin Head Power Stroke Producing Muscle Contraction Using the Gas Environmental Chamber

Haruo Sugi,[a] **Tsuyoshi Akimoto,**[a] **Shigeru Chaen,**[b] **Takuya Miyakawa,**[c] **Masaru Tanokura,**[c] **and Hiroki Minoda**[d]

[a]*Department of Physiology, Teikyo University School of Medicine, Tokyo, Japan*
[b]*Department of Integrated Sciences in Physics and Biology,*
College of Humanities and Sciences, Nihon University, Tokyo, Japan
[c]*Graduate School of Agricultural and Life Sciences,*
University of Tokyo, Tokyo, Japan
[d]*Department of Applied Physics, Tokyo University of Agriculture and Technology,*
Tokyo, Japan

sugi@kyf.biglobe.ne.jp

Since the monumental discovery of sliding filament mechanism in muscle contraction, the mechanism of attachment-detachment cycle between myosin heads extending from myosin filaments and actin filaments has been the central object in research field of muscle contraction. As early as the 1980s,

Muscle Contraction and Cell Motility: Fundamentals and Developments
Edited by Haruo Sugi
Copyright © 2017 Pan Stanford Publishing Pte. Ltd.
ISBN 978-981-4745-16-1 (Hardcover), 978-981-4745-17-8 (eBook)
www.panstanford.com

we started to challenge electron microscopic visualization and recording of myosin head power stroke coupled with ATP hydrolysis, using the gas environmental chamber (EC), which enables us to observe myofilaments in wet, living state. In this chapter, we first explain historical background of our work, and then describe our experimental methods together with our findings, which can be summarized as follows: (1) the time-averaged position in individual myosin head does not change with time, indicating that they fluctuate around a definite equilibrium position; (2) In the absence of actin filaments, ATP-activated individual myosin heads move by ~7 nm at both distal and proximal regions in the direction away from the center of myosin filaments, indicating that myosin heads can perform recovery stroke without being guided by actin filaments; and (3) In the presence of actin filaments, ATP-activated individual myosin heads exhibit power stroke in nearly isometric condition, with the amplitude ~3.3 nm at distal region and ~2.5 nm at proximal region; (4) At low ionic strength, the amplitude of power stroke increases to >4 nm at both distal and proximal regions, being consistent with the report that the force generated by individual myosin heads increases approximately twofold at low ionic strength. Advantages of our methods over in vitro motility assay methods are discussed.

1.1 Historical Background

As illustrated in Fig. 1.1, skeletal muscle consists of muscle fibers (diameter, 10–100 μm), which in turn contains a number of myofibrils (diameter, 1–2 μm). Under a light microscope, both muscle fibers and myofibrils exhibit cross-striation, composed of alternate protein-dense A-band and less dense I band. H-zone and Z-line are located at the center of A-and I-bands, respectively. Although the striation pattern of skeletal muscle has been observed since nineteenth century, its functional significance was not clear. In 1954, Huxley and Hanson made a monumental discovery by using phase-contrast microscope and electron microscope. Their findings are summarized as follows: (1) Two main proteins constituting muscle, actin and myosin, exist in muscle in the form of two independent filaments, i.e., actin and myosin filaments. (2) When muscle contracts to shorten, actin and myosin filaments slide past each other without changing their lengths. (3) Myosin filaments are located in the A-band, while actin filaments originate from Z-line, located at the center of the I-band and extend in between myosin filaments in the A-band. (4) The central region in the A-band where actin filaments are absent is called H-zone.

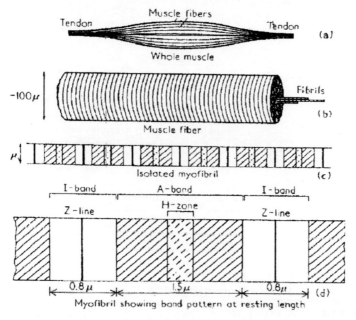

Figure 1.1   Structure and cross-striation of skeletal muscle. (a) Skeletal muscle consisting of muscle fibers. (b) Muscle fiber containing myofibrils. (c) Cross-striations of myofibrils. (d) Details of cross-striation. For explanation, see text. From Huxley (1960).

With respect to the sliding filament mechanism, a question arises: What makes the filaments slide? Considerable progress has been made concerning the structure of muscle and muscle filaments (for a review, see Sugi, 1992). Figure 1.2 summarizes the results obtained up to the present time. Myosin molecule (MW, ~500,000) consists of heavy meromyosin (HMM) with two heads (myosin subfragment-1, S-1) and a short rod (myosin subfragment-2, S-2), and light meromyosin (LMM) forming a long rod (Fig. 1.2a). When myosin molecules aggregate to form myosin filament, LMM constitutes myosin filament backbone, while HMM extends laterally from myosin filament with an interval of 14.3 nm (Fig. 1.2b). On the other hand, actin filament consists of double helical strands (pitch, 35.5 nm) of globular actin molecules (MW, ~50,000).

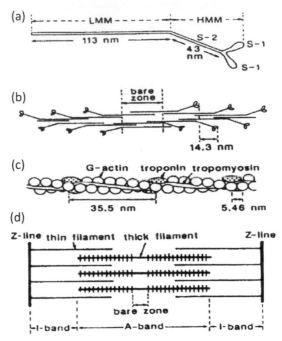

Figure 1.2   Ultrastructure of thick and thin filaments and their arrangement within a sarcomere. (a) Diagram of a myosin molecule. (b) Longitudinal arrangement of myosin molecules to form myosin filament. (c) Structure of actin filament. (d) Longitudinal arrangement of actin and myosin filaments within a single sarcomere. For further explanations, see text. From Sugi (1992).

Actin filament also contains two regulatory proteins, troponin and tropomyosin (Fig. 1.2c). In muscle, actin and myosin filaments constitute hexagonal filament-lattice structure, in which each myosin filament is surrounded by six actin filaments. The region of filament-lattice between two adjacent Z-lines is called the sarcomere, which is regarded as structural and functional unit of muscle (Fig. 1.2d).

Muscle is a kind of engine utilizing ATP as fuel. It converts chemical energy derived from ATP hydrolysis into mechanical work. In accordance with this, both actin-binding and ATPase sites are located in myosin heads. Up to the present time, therefore, myosin heads are believed to play a major role in chemo-mechanical energy conversion responsible for muscle contraction. Figure 1.3 illustrates a hypothetical attachment-detachment cycle between myosin heads and corresponding myosin-binding sites on actin filaments, put forward by Huxley (1969). First, a myosin head extending from myosin filament attaches to a myosin-binding site on actin filament (upper diagram), changes its configuration in such a way to produce unitary sliding between actin and myosin filaments (middle diagram), and then detaches from actin filament (lower diagram). The configuration change of myosin head is generally called the power stroke.

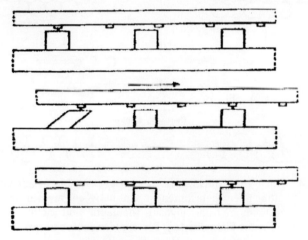

Figure 1.3   Diagrams showing attachment-detachment cycle between myosin heads extending from myosin filaments and corresponding sites on actin filaments. For explanations, see text. From Huxley (1969).

The above hypothesis has been supported by biochemical studies on ATPase reaction steps of actin and myosin in solution (Lymn and Taylor, 1971). Figure 1.4 shows a most plausible attachment-detachment cycle between myosin head (M) extending from myosin filament and actin (A) in actin filament. First, M

attaches to A in the form of M·ADP·Pi, as M·ADP·Pi + A →
A·M·ADP·Pi, and then performs a power stroke associated with
release of ADP and Pi, as A·M·ADP·Pi → A·M + ADP + Pi (from A to
B). After completion of power stroke, M remains attached to A until
next ATP comes to bind it (B). M detaches from A on binding with next
ATP (C), and performs a recovery stroke associated with reaction,
M + APT → M·ATP → M·ADP·Pi (from C to D). M·ADP·Pi, and
again attaches to A to repeat attachment-detachment cycle (from
D to A). In order to repeat the attachment-detachment cycle, the
amplitude of power stroke should be the same in amplitude as,
and opposite in direction to, that of recovery stroke.

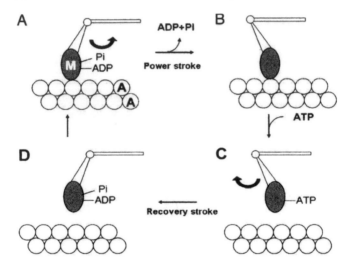

Figure 1.4 Diagrams illustrating a most plausible explanation of
attachment-detachment cycle between myosin heads (M)
extending from myosin filaments and actin monomers
in actin filaments (A). For explanations, see text. From Sugi
et al. (2008).

Despite extensive studies including muscle mechanics,
chemical probe experiments, time-resolved X-ray diffraction
experiments and in vitro motility assay experiments, the power
and recovery strokes of myosin heads in contracting muscle
still remains to be a mystery (Sugi and Tsuchiya, 1998; Geeves
and Holmes, 1999). As early as the 1980s, we have been aware
that a most straightforward way to measure the amplitude of
myosin head movement, coupled with ATP hydrolysis, was to

perform experiments using the gas environmental chamber (EC), which enables us to observe hydrated, living actin and myosin filaments under an electron microscope.

1.2 Materials and Methods

1.2.1 The Gas Environmental Chamber

The gas environmental chamber (EC, or hydration chamber) has been developed to record physical and chemical reactions in water under an electron microscope (Butler and Hale, 1981). Although the EC has been widely used by materials scientists, its use in the research field of life sciences was unsuccessful. As early as the 1980s, we attempted to visualize and record ATP-induced structural changes of muscle proteins with research group of the late Dr. Akira Fukami, who developed carbon-sealing film suitable for our work (Fukami and Adachi, 1965). Figure 1.5 is a diagram showing the gas environmental chamber (EC), used by us.

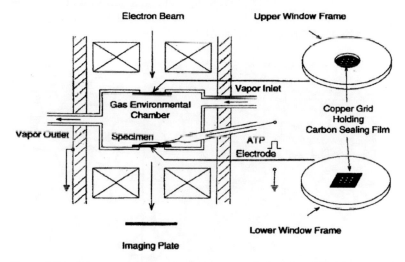

Figure 1.5 Diagram of the gas environmental chamber (EC). Upper and lower windows are covered with carbon sealing film held on copper grids with 9 apertures. Interior of the EC is constantly circulated with water vapor. The EC contains ATP-containing glass microelectrode to apply ATP to the specimen. From Sugi et al. (1997).

The EC consists of a metal compartment (diameter, 35 mm, depth, 0.8 mm) with upper and lower window frames (copper grids) to pass electron beam. Each window frames has nine apertures (diameter, 0.9 mm), and covered with carbon sealing film, which insulates the interior of the EC from high vacuum of electron microscope. The specimen is placed on the surface of lower sealing film, and covered with thin layer of experimental solution (thickness, ~200 nm), which is in equilibrium with water vapor circulated at a rate of 0.1–0.2 ml/min through the EC. To obtain clear specimen image, internal pressure of the EC was made 60–80 torr. The EC contains a glass capillary microelectrode filled with 100 mM ATP to apply ATP iontophoretically to the specimen. The EC is attached to a 200 kV transmission electron microscope (JEM 2000EX, JEOL).

1.2.2 Carbon Sealing Film

Both spatial resolution and contrast of the specimen image taken by the EC increase with decreasing sealing film thickness. Preliminary experiments made in Fukami's laboratory indicate that to obtain a spatial resolution <1 nm, sealing film thickness should be 15–20 nm.

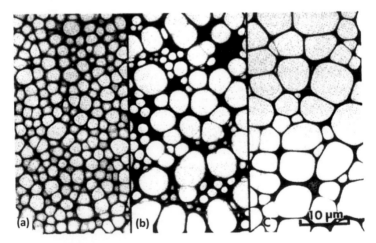

Figure 1.6 Light micrographs of carbon sealing film. Carbon sealing films with too small holes (a), with holes of irregular size (b), and with holes of appropriate size. From Fukushima (1987).

On the other hand, resistivity of sealing film against pressure difference decreases sharply with increasing its area. Fukami and Adachi (1965) solved this technical problem by using plastic microgrids made from cellulose acetobutylate, supporting carbon sealing film. Electron micrographs of the microgrids are shown in Fig. 1.6. Microgrids with fairly uniform holes of 5–8 nm diameter are suitable for electron microscopic observation of the specimen mounted in the EC. Methods to prepare carbon sealing film (thickness, ~20 nm) supported by microgrids are described elsewhere (Fukami et al., 1991; Sugi, 2012).

1.2.3 Iontophoretic Application of ATP

A most characteristic feature of our EC system is that ATP can be applied at any desired time to the specimen, so that dynamic structural changes of the specimen related to its physiological function can be visualized and recorded under an electron microscope with high magnifications (up to 10,000×). A positive DC current is constantly applied to the ATP-containing microelectrode to inhibit diffusion of ATP out of the electrode. At any desired time, a negative going current pulse (intensity, 10 nA; duration, 1 s) is applied to the electrode through a current clamp circuit (Oiwa et al., 1991, 1993), so that ATP ($\sim 10^{-14}$ mol) is released out of the electrode, and then reaches to the specimen by diffusion. Assuming the volume of experimental solution covering the specimen to be $\sim 10^{-6}$ ml, the ATP concentration around the specimen is 5~10 μM (Sugi et al., 1997, 2008).

1.2.4 Determination of the Critical Electron Dose Not to Impair Physiological Function of the Specimen

To study dynamic structural changes of the specimen, it is essential to first determine the critical incident electron dose not to impair its physiological function. For this purpose, Fukami and his coworkers (Suda et al., 1992) observed a number of muscle myofibrils mounted in the EC (magnification, 2,500×), and made them to contract by applying solutions containing 4 mM ATP at various times after the beginning of electron beam irradiation. As shown in Fig. 1.7, all the myofibrils within a microscopic field contracted in response to ATP, when the total incident electron dose was $< 5 \times 10^{-4}$ C/cm^2. If, however, the total incident electron dose was further increased, the ATP-induced myofibrillar contraction disappeared in a nearly all-or-none manner. Based on the above results, electron microscopic observation and recording of the specimen placed in the EC was made with a total incident electron dose $< 5 \times 10^{-4}$ C/cm^2, being well below the electron dose to impair physiological function of the specimen.

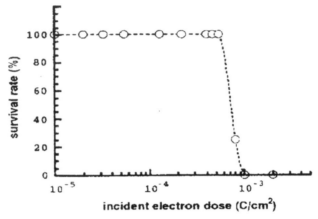

Figure 1.7 Relation between the total incident electron dose and the survival rate of isolated myofibrils placed in the EC. The survival rate was estimated from proportion of myofibrils shortened in response to applied ATP within a microscopic field. From Suda et al. (1992).

1.2.5 Position-Marking of Myosin Heads with Site-Directed Antibodies

Since the hydrated specimens are observed in unstained conditions, it is absolutely necessary to position-mark individual myosin head in myosin filaments by an appropriate means. For this purpose, we used three different site-directed antibodies (antibodies 1–3) to myosin head (Sutoh et al., 1989; Minoda et al., 2011). Antibody 1 (anti-CAD antibody) attaches to junctional peptide between 50 and 20 k segments of myosin heavy chain in myosin head catalytic domain (CAD), antibody 2 (anti-RLR antibody) attaches to peptides around reactive lysine residue (Lys83) in myosin head converter domain (COD), and antibody 3 (anti-

LD antibody) attaches to two regulatory light chains in myosin head lever arm domain (LD). Figure 1.8 is a ribbon diagram showing structure of myosin head, in which the sites of attachment of antibodies 1, 2, and 3 are indicated by numbers 1, 2 and 3 and 3', respectively, while Fig. 1.9 shows electron micrographs of rotary shadowed myosin molecules with antibodies (IgG) attached (Sutoh et al., 1989; Minoda et al., 2011).

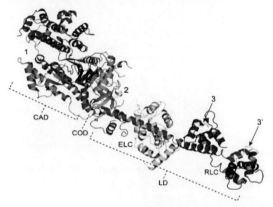

Figure 1.8   Ribbon diagram of myosin head showing approximate attachment regions of antibodies 1, 2, and 3, indicated by numbers 1, 2, and 3 and 3', respectively. Catalytic domain consists of 25 K (green), 50 K (red), and part of 20 K (dark blue) fragments of myosin heavy chain, while lever arm domain (LD) consists of the rest of 20 K fragment and essential (ELC, light blue) and regulatory (RLC, magenta) light chains. CAD and LD are connected via small converter domain (COD). Location of peptides around Lys83, and that of two peptides (Met58~Ala70, and Leu106 ~Phe120 in LD are colored yellow. From Minoda et al. (2011).

Individual myosin heads were position-marked with colloidal gold particles (diameter, 20 nm; coated with protein A, working as a paste to connect proteins, (EY laboratories) via one of the three antibodies. Since native myosin filaments are too thin and tend to curl and aggregate, we used synthetic myosin filaments (myosin-myosin rod cofilaments), prepared by polymerizing myosin and myosin rod molecules at low ionic strength (Sugi et al., 1997). As shown in Fig. 1.10, spindle-shaped synthetic myosin filaments, with a number of gold particles attached to individual myosin heads, can be recorded with the imaging plate (IP).

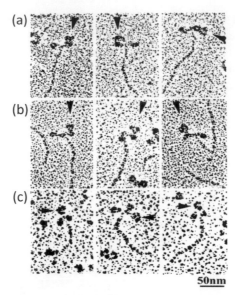

Figure 1.9   A gallery of electron micrographs of antibody 1,2 or 3(IgG)-myosin head complexes. IgG molecules are indicated by arrowheads. From Minoda et al. (2011).

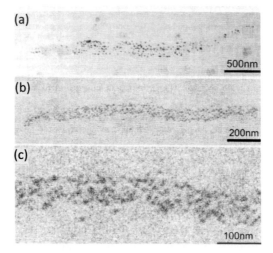

Figure 1.10   (a, b) Typical imaging plate (IP) records of bipolar myosin-myosin rod cofilaments with a number of gold particles attached to individual myosin heads. (c) Enlarged view of the filaments shown in b. From Sugi et al. (2008).

1.2.6 Recording of Specimen Image and Data Analysis

Under a magnification of 10,000×, the average number of electrons during the exposure time (0.1 s) was ~10. Reflecting this electron statistics, the image of each gold particle on the IP consisted of 20–50 dark pixels with different gradation. The center of mass position for each particle was determined as the coordinates (two significant figures; accuracy, 0.6 nm) within a single pixel where the center of mass position was located. These coordinates representing the particle position was taken as the position of myosin head. The change in position of myosin head was compared between the two IP records of the same myosin filament (Sugi et al., 1997, 2008).

1.3 Myosin Head Movement Coupled with ATP Hydrolysis in Living Myosin Filaments in the Absence of Actin Filaments

1.3.1 Stability in Position of Individual Myosin Heads in the Absence of ATP

At the beginning of our work using the EC, we examined whether the position of individual myosin heads changes with time or not, by comparing two IP records of the same myosin filament taken at intervals of several min. Figure 1.11a shows part of a myosin filament, on which a circle of 20 nm diameter is drawn around the center of mass position of each gold particle obtained from two IP records. The circles representing center of mass positions of the same particle overlap almost completely, indicating that the position of each myosin head remains almost unchanged with time. Figure 1.11b is a histogram showing distribution of change in the center of mass position of gold particles, i.e., myosin heads, between the two IP records of the same filaments. Change in position of myosin heads were mostly < 2.5 nm (Sugi et al., 2008).

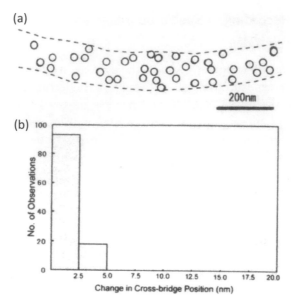

(a)

200nm

(b)

Figure 1.11 Stability of time-averaged myosin head position in the absence of ATP. (a) comparison of myosin head position between the two IP records of the same filament. In this and subsequent Figures 1.12 and 1.13 open and filled circles (diameter, 20 nm) are drawn around center of mass position of each particle in the first and the second IP records, respectively, and broken lines indicate approximate contour of myosin filament. Note that filled circles are barely visible because of almost complete overlap of open and filled circles. (b) Histogram showing distribution of distance between the two center of mass positions of particles in the two records. From Sugi et al. (2008).

1.3.2 Amplitude of ATP-Induced Myosin Head Movement in Hydrated Myosin Filaments

The stability in position of individual myosin heads made it possible to visualize and record myosin head movement in response to iontophoretically applied ATP. We compared the position of individual myosin heads between the two IP records of the same filament, one taken before and the other taken at 40–60 s after the onset of current pulse to ATP electrode, taking diffusion of ATP into consideration. In Fig. 1.12a, open and filled circles are drawn around the center of mass positions of individual myosin heads before and after application of ATP, respectively. In most cases, individual myosin heads were observed to move nearly in parallel with filament long axis in one direction. Figure 1.12b shows amplitude distribution of the ATP-induced myosin head movement, which exhibited a peak at 5–7.5 nm. The average amplitude of ATP-induced myosin head movement (excluding values < 2.5 nm) was 6.5 ±3.7 nm (mean ±SD, *n* = 1210) (Sugi et al., 2008).

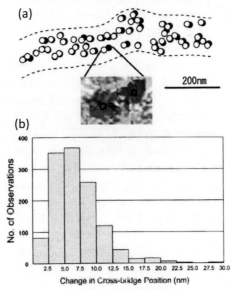

Figure 1.12   ATP-induced myosin head movement. (a) Comparison of myosin head position between two IP records, taken before and after ATP application. (Inset) An example of superimposed IP records showing change in position of the same gold particle, which are colored red (before ATP application) and blue (after ATP application). The center of mass position in each particle image is indicated by a small open circle. (b) Amplitude distribution of ATP-induced myosin head movement in the absence of actin filaments. From Sugi et al. (2008).

On application of ATP, individual myosin heads react rapidly with ATP molecule to form the complex, M·ADP·Pi, having average lifetime >10 s in the absence of actin filaments (Lymn and Tayler, 1971). On this basis, the ATP-induced myosin head movement may correspond to recovery stroke (C to D, in Fig. 1.4);

the long average lifetime of M·ADP·Pi enables us to record position changes of individual myosin heads despite the limited time resolution of our recording system (0.1 s).

1.3.3 Reversal in Direction of ATP-Induced Myosin Head Movement across Myosin Filament Bare Zone

As shown in Fig. 1.2d, the polarity of myosin heads extending from the myosin filament backbone is symmetrical across the bare zone (Fig. 1.2b), so that the direction of myosin head power and recovery strokes should be reversed across the bare zone. We succeeded in proving the reversal in direction of myosin head movement by recording the ATP-induced myosin head movement at both sides of the bare zone (Sugi et al., 2008). Since actin filament was absent in our experimental system, the myosin movement should be recovery stroke (from C to D in the diagram of Fig. 1.4). Typical results are shown in Fig. 1.13, in which open and filled circles are drawn around the center of mass positions of gold particles before and after ATP application, respectively. It can be seen that on ATP application, the particles, i.e., myosin heads, move away from the bare zone located at the center of myosin filaments. We emphasize that such a direct demonstration of myosin head recovery stroke is only possible by our EC experiments, in which ATP-induced movement of individual myosin heads can be recorded.

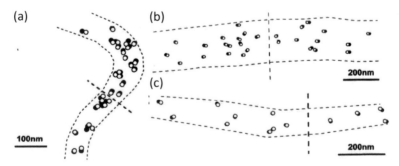

Figure 1.13 Examples of ATP-induced myosin head movement at both sides of myosin filament bare region, across which myosin head polarity is reversed. Note that individual myosin heads move away from myosin filament bare region. Approximate location of the bare region is indicated by broken line across the filament. From Sugi et al. (2008).

1.3.4 Reversibility of ATP-Induced Myosin Head Movement

To ascertain reversibility of the ATP-induced myosin head movement, we also performed experiments in which three IP records of the same myosin filament were taken in the following sequence: (1) before ATP application; (2) at 40–60 s after the onset of current pulse to ATP electrode, i.e., during ATP application; and (3) at 5–6 min after ATP application, i.e., after complete exhaustion of applied ATP. To completely remove contaminant ATP, hexokinase (50 units/ml) and D-glucose (2 mM) were added to experimental solution.

Examples of sequential changes in position of nine different pixels (2.5 × 2.5 nm each), where the center of mass positions of corresponding particles are located, are shown in Fig. 1.14. In

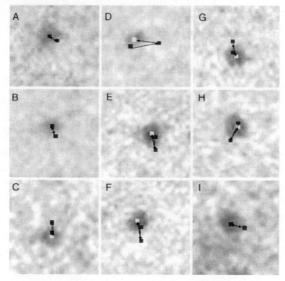

Figure 1.14   Examples showing sequential changes in position of nine different pixels (2.5 × 2.5 nm) where the center of mass positions of nine corresponding gold particles are located. In each frame, pixel positions are recorded three times, i.e., before ATP application (red), during ATP application (blue), and after exhaustion of ATP (yellow). Arrows indicated direction of myosin head movement. Note that myosin heads return towards their initial position after exhaustion of ATP. From Sugi et al. (2008).

each frame, pixel positions in the first (red), second (blue), and third (yellow) IP records indicate sequential position changes of individual myosin heads. Myosin heads, which moved in response to applied ATP, return towards their respective initial position before ATP application after exhaustion of applied ATP (Sugi et al., 2008). These results indicate reversibility of ATP-induced myosin head movement, being consistent with the diagram in Fig. 1.4, namely, the ATP-induced myosin head recovery stroke (C to D) is associated with reaction, $M + ATP \rightarrow M \cdot ATP \rightarrow M \cdot ADP \cdot Pi$, while the return of myosin heads towards their initial position is associated with reaction, $M \cdot ADP \cdot Pi \rightarrow M + ADP + Pi$ (A to B), though actin filament is absent from our experimental system. If the above explanation is correct, the return of myosin heads to their initial position corresponds to myosin head power stroke in the absence of actin filaments, implying that myosin heads can perform power and recovery strokes without being guided by actin filaments.

1.3.5   Amplitude of ATP-Induced Movement at Various Regions within a Myosin Head

We also performed experiments, in which myosin heads were position-marked with the three different antibodies attaching three different regions within a myosin head (Minoda et al., 2011). The results are summarized in Fig. 1.15, together with diagrams of myosin head power and recovery strokes. Figures 1.15a–c show histograms of amplitude distribution of ATP-induced myosin head movement position-marked with antibody 1 (anti-CAD antibody), with antibody 2 (anti-RLR antibody), and with antibody 3 (anti-LD antibody), respectively (cf. Fig. 1.8). The mean amplitude of ATP-induced myosin head movement was 6.14 ± 0.09 nm (mean ± SEM, $n = 1692$) at the distal region of myosin head CAD, and 6.14 ± 0.22 nm ($n = 1112$) at the myosin head COD, i.e., adjacent to the proximal region of myosin head CAD, and 3.55 ± 0.11 ($n = 981$) at the proximal region of myosin head LD, indicating that in the absence of actin filaments, (1) myosin head CAD remains rigid and rotates around myosin head COD, while myosin head LD rotates around the boundary between myosin head LD and myosinsubfragment-2 (S-2) (Fig. 1.15D). The reversibility of ATP-induced myosin head movement (Fig. 1.14) strongly suggests that the mode of myosin head movement is the same in the presence of actin filaments (Fig. 1.15e). As will be described later, this has recently been proved to actually be the case by our recent success in recording myosin head power stroke in the mixture of actin and myosin filaments (Sugi et al., 2015).

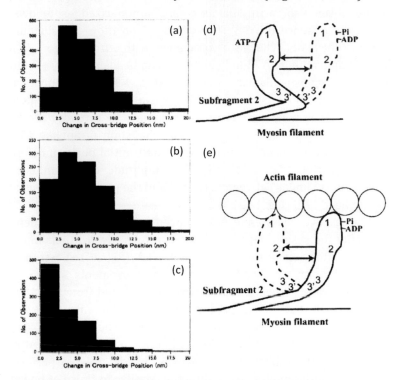

Figure 1.15 (a–c) Amplitude distribution of ATP-induced myosin head (cross-bridge) movement at three different regions within a myosin head. Individual myosin heads were position-marked by antibody 1 (a), antibody 2 (b), and antibody 3 (c). (d and e) Diagrams illustrating the mode of ATP-induced myosin head movement in the absence (d) and in the presence (e) of actin filament. Approximate points of attachment of antibodies 1, 2, and 3 are indicated by numbers 1, 2, and 3 and 3′, respectively. From Minoda et al. (2011).

1.3.6 Summary of Novel Features of ATP-Induced Myosin Head Movement Revealed by Experiments Using the EC

Here, we summarize novel features of ATP-induced myosin head movement in the absence of actin filament. First of all, we emphasize that only our experiments using the EC can record changes in position of individual myosin heads originating from hydrated myosin filaments, i.e., in the condition close to that in muscle. The novel features of ATP-induced myosin head movement in the absence of actin filaments are summarized as follows: (1) In the absence of ATP, Time-averaged position of individual myosin does not change appreciably with time (Fig. 1.11). (2) On application of ATP, individual myosin heads move by ~7 nm parallel to myosin filament long axis (Fig. 1.12). (3) Individual myosin heads move away from the myosin filament bare region, indicating that what we record corresponds to recovery stroke of myosin heads (Fig. 1.13). (4) After exhaustion of applied ATP, individual myosin heads return toward their initial position before ATP application (Fig. 1.14). (5) The amplitude of ATP-induced movement is similar between the distal and the proximal regions of myosin head CAD (Fig. 1.15).

These novel features of ATP-induced myosin head movement, especially the result that myosin heads return toward their initial position after exhaustion of applied ATP (Fig. 1.14), strongly suggest that myosin heads can perform movements similar to both power and recovery strokes (from A to B, and from C to D, in Fig. 1.4) without being guided by actin filaments. On ATP application, myosin heads binds with ATP to perform recovery stroke associated with reaction, $M \cdot ATP \rightarrow M \cdot ADP \cdot Pi$. After exhaustion of applied ATP, myosin heads might perform movement with amplitude and direction similar to those of power stroke, when Pi and ADP detach from myosin heads. This implies that myosin heads can perform power stroke without being guided by actin filaments. This point will again be discussed in the next chapter, where novel features of ATP-induced myosin head power stroke are described.

1.4 Novel Features of Myosin Head Power Stroke in the Presence of Actin Filaments

1.4.1 Preparation of Actin and Myosin Filament Mixture

To investigate myosin head power stroke in the presence of actin filaments, actin and myosin filament mixture was prepared by the method of Sugi et al. (2015). As shown in Fig. 1.16, spindle-shaped myosin filaments, with gold particles attached to individual myosin heads, are surrounded by actin filaments running approximately in parallel with myosin filament long axis.

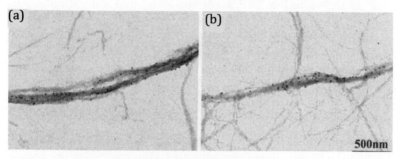

Figure 1.16 Electron micrographs showing part of actin and myosin filament mixture. Myosin heads are position-marked with gold particles via antibody 1 in a, and via antibody 2 in b. From Sugi et al. (2015).

In the absence of ATP, the time-averaged position of individual myosin heads in the filament mixture also remained unchanged with time. This indicated that the mixture was firmly fixed to the carbon sealing film, and that individual myosin heads formed tight rigor linkages with adjacent actin filaments, providing a favorable condition to record myosin head power stroke in response to applied ATP.

1.4.2 Conditions to Record ATP-Induced Myosin Head Power Stroke in the Filament Mixture

As already mentioned, the ATP concentration around myosin heads is estimated to be 5–10 µM, while the ATP concentration in muscle is 2–4 mM. Consequently, the attachment-detachment cycle between myosin heads and actin filaments proceeds rapidly in muscle, while in our EC experiments, individual myosin heads have to wait for a considerable time until next ATP comes to bind them. In this connection, our experimental condition is analogous to the optical trap experiments, in which single or a few myosin heads fixed to substratum are made to interact with an actin filament with one end attached to a bead, which is trapped in position with the laser beam, and unitary displacements or forces produced by single myosin heads can be recorded using a feedback system controlling the bead position (Finer et al., 1994; Sugiura et al., 1998). To obtain discreet unitary force productions (force spikes), experiments are performed under μ molar concentrations of ATP. As shown in Fig. 1.17, force spikes up to 1 s duration can be seen at intervals up to several seconds. These features of force spikes indicate that, under low ATP concentrations, myosin heads takes two states, i.e. force-generating state corresponds to post-power stroke configuration and non force-generating state corresponding to pre-power stroke or rigor configuration. As will be described later, we also succeeded in clearly recording positions of individual myosin heads before (non force-generating state) and after the end of ATP-induced power stroke (force-generating state) despite the limited time resolution (0.1 s) in our recording system under low ATP concentrations.

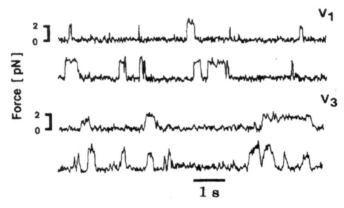

Figure 1.17   Examples of records showing force spikes up to 1 s in duration, each separated by more than 1 s. The ATP concentration was 0.5 µM. From Sugiura et al. (1998).

It should also be noted that due to low ATP concentrations around individual myosin heads, we could only activate very small proportion of myosin heads in myosin filaments, so that no gross sliding between actin and myosin filaments takes place. Instead, each ATP-activated myosin head performs power stroke by stretching adjacent elastic structures. This condition is analogous to that of myosin heads in isometrically contracting muscle fibers with two ends fixed in position, i.e., nominally isometric condition.

1.4.3 Amplitude of ATP-Induced Myosin Head Power Stroke in the Mixture of Actin and Myosin Filaments

We first examined the stability of myosin head position in the actin and myosin filament mixture in the absence of ATP, by taking two IP records of the same filament. The change in position of individual myosin heads between the two IP records was very small, being 0.6 ± 0.06 nm (mean $\pm$ SD, $n = 220$), indicating that the mixture was firmly fixed on carbon sealing film, and that myosin heads formed tight rigor linkages with adjacent actin filaments (Sugi et al., 2015).

As shown in Fig. 1.18, individual myosin heads exhibited small but distinct movement in response to ATP. Using antibodies 1 and 2 to position-mark myosin heads, it was possible to record amplitude of ATP-induced myosin head power stroke at both the distal and the proximal regions within individual myosin head CAD. Figure 1.19 shows histograms of amplitude distribution of myosin head power stroke at the distal region (Fig. 1.19a) and at the proximal region (Fig. 1.19b) of myosin head CAD. The average amplitude of myosin head power stroke was 3.3 ± 0.2 nm (mean $\pm$ SD, $n = 732$) at the distal region, and 2.5 ± 0.1 (mean $\pm$ SD, $n = 613$) at the proximal region of myosin head CAD, indicating that the power stroke amplitude is large at the distal region than at the proximal region of myosin head CAD (t test, $p < 0.01$). This result is not consistent with the lever arm mechanism of myosin head power stroke (Geeves and Holmes, 1999), in which the amplitude of power stroke is the same in both the distal and the proximal regions of myosin head CAD.

It has long been known that if the ionic strength of experimental solution is reduced by reducing KCl concentration from 125 to ≤ 20 mM, the magnitude of the maximum Ca^{2+}-activated tension in skinned skeletal muscle fibers increases approximately twofold. We have recently found that at low ionic strength, the rate of MgATPase activity of the fibers remains unchanged despite approximately twofold increase in the maximum Ca^{2+}-activated isometric force, indicating that the force generated by individual myosin head increases approximately twofold at low ionic strength (Sugi et al., 2013). In this work, skinned muscle fibers were made to contract isometrically with two ends fixed in position, so that no gross filament sliding took place. Consequently, the increased Ca^{2+}-activated force can be explained as being due to increased degree of stretch of adjacent elastic structures caused by power stroke of individual myosin heads.

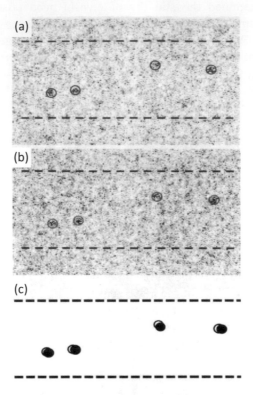

Figure 1.18 Example of records showing ATP-induced power stroke of individual myosin heads in the filament mixture. (a and b) a pair of IP records of the same myosin filament. Records a and b were taken before and after ATP application, respectively. Each image of gold particle attached to myosin head consists of small dark particles. Circles (diameter, 20 nm; red in a and blue in b) are drawn around the center of mass position of gold particles. (c) Diagram showing ATP-induced changes in position of gold particles. Open and filled circles (diameter, 20 nm) are drawn around the center of mass position of the same particles before and after ATP application, respectively. In this particular case, myosin heads were position-marked with antibody 1. From Sugi et al. (2015).

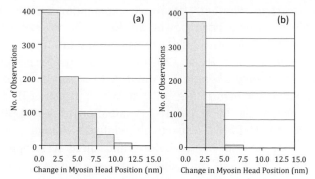

Figure 1.19   Histograms of amplitude distribution of ATP-induced myosin head power stroke at the distal region (a) and at the proximal region (b) of myosin head CAD. Myosin heads were position-marked with antibody 1 in A, and with antibody 2 in B. From Sugi et al. (2015).

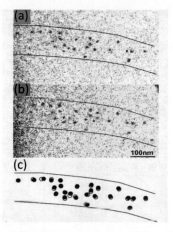

Figure 1.20   Example of records showing ATP-induced myosin head power stroke at low ionic strength. (a and b) A pair of IP records of the same myosin filament in the filament mixture. Record a (colored red) was taken before ATP application, while record b (colored blue) was taken after ATP application. Gold particles attached to individual myosin heads are seen to consist of dark small particles. In this particular case, myosin heads were position-marked with antibody 2. (c) Diagram showing ATP-induced changes in position of gold particles. Filled and open circles (diameter, 20 nm) are drawn around the center of mass position of the same particles before and after ATP application, respectively (Sugi et al., 2015).

To ascertain whether the above explanation applies to the amplitude of myosin head power stroke in the EC experiments, we reduced KCl concentration of experimental solution surrounding the filament mixture from 120 to 20 mM (corresponding to a reduction of ionic strength μ from 170 to 50 mM). An example of the results is shown in Fig. 1.20, in which a pair of IP records of the same filament taken before (a) and after (b) ATP application are colored red and blue, respectively, while the change in position of individual myosin heads before and after ATP application are shown by filled and open circles (diameter, 20 nm), respectively (c). As shown in Fig. 1.21, the amplitude of myosin head power stroke at low ionic strength showed a peak at 2.5–5 nm at both the distal and the proximal regions of myosin head CAD. The average amplitude of myosin head power stroke was 4.4 ± 0.1 nm (mean ± SD, n = 361) and 4.3 ±0.2 nm (mean ± SD, n = 305) at the distal and the proximal regions of myosin head CAD, respectively. These results indicate that the amplitude of myosin head power stroke increases markedly at both the distal and the proximal regions of myosin head CAD, being entirely consistent with our report that the maximum Ca^{2+}-activated force in skinned muscle fibers increases approximately twofold at low ionic strength.

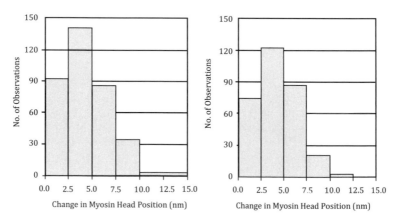

Figure 1.21 Histograms of amplitude distribution of ATP-induced myosin head power stroke at low ionic strength. Individual myosin heads were position-marked at the distal region with antibody (a), and at the proximal region with antibody (b). From Sugi et al. (2015).

1.4.4 Reversibility of ATP-Induced Myosin Head Power Stroke

The reversibility of the ATP-induced myosin head power stroke was examined by taking three IP records of the same filaments in the following sequence: (i) before ATP application, (ii) during ATP application, and (iii) after complete exhaustion of ATP. Experiments were made at standard ionic strength. Examples of sequential changes in position of six different pixels (2.5 × 2.5 nm), where the center of mass positions of six corresponding particles are located, are shown in Fig. 1.22. Pixel positions in the first, second and third IP records are colored red, blue, and yellow, respectively. Similar results were obtained irrespective of whether myosin heads were position-marked with antibody 1 or antibody 2. When the amplitude of ATP-induced power stroke was ≤ 5 nm, myosin heads returned almost exactly to the initial position before ATP application (Fig. 1.22a–d), as indicated by almost complete overlap of red and yellow pixels. With larger amplitudes of myosin head power stroke, myosin heads also tended to return to the initial position (Fig. 1.22e,f).

Considering the long interval between the time of IP recording (up to several min), individual position marked myosin heads may repeat power and recovery strokes many times. Due to the small concentrations of iontophoretically applied ATP around myosin heads (5–10 μM), individual myosin heads are expected to take either non-force-generation state or force-generating state with time course analogous to that of force spikes in the optical trap experiments (Fig. 1.7); the duration of the two myosin head states may exceed the exposure time of IP recording (0.1 s), so that the transition of individual myosin heads from non force-generating state to force-generating state (accompanied by power stroke) can be clearly recorded as the change in myosin head position between the first and the second IP records. After complete exhaustion of ATP, all myosin heads return from force-generating state to non force-generating state (accompanied by recovery stroke), and this transition in the state of myosin heads can be recorded as the return of myosin heads to their original position between the second and the third IP records. The incomplete return of myosin heads to their original position

(Fig. 1.22e,f) may arise from distortion of filament network in the filament mixture.

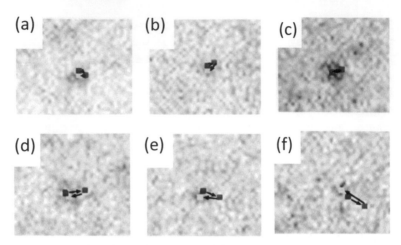

Figure 1.22   Examples of sequential change in position of different pixels (2.5 × 2.5 nm), where the center of mass positions of corresponding gold particles are located. For further explanations, see text. From Sugi et al. (2015).

1.4.5   Summary of Novel-Features of ATP-Induced Myosin Head Power Stroke Revealed by Experiments Using the EC

The novel features of ATP-induced myosin head power stroke brought about by our EC experiments can be summarized diagrammatically in Fig. 1.23. As illustrated in Fig. 1.23a, we used two different antibodies to myosin head successfully, so that we could record amplitude of power stroke at two different regions within a single myosin head. Meanwhile, a number of studies have been made to estimate the amplitude of myosin head power stroke, using single intact and skinned muscle fibers, and isolated myosin heads, attached to a glass needle or put into optical trap (for a review, see Kaya and Higuchi, 2013). In the experiments of single muscle fibers, the amplitude of myosin head power stroke is only indirectly estimated by giving mechanical perturbations, while in the experiments with isolated myosin heads their orientation in the experimental system differs too far from that in the 3D filament lattice in muscle.

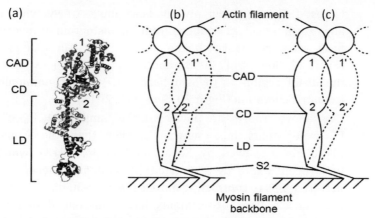

Figure 1.23   Change in the mode of myosin head power stroke depending on experimental conditions. (a) Diagram of myosin head structure consisting of catalytic (CAD), converter (CD) and lever arm (LD) domains. Approximate regions of attachment of antibodies 1 and 2 are indicated by numbers 1 and 2, respectively. (b) The mode of myosin head power stroke in the nominally isometric condition at standard ionic strength. The amplitude of power stroke is larger at the distal region (1 to 1′) than at the proximal region (2 to 2′) of myosin head CAD. (b) The mode of myosin head power stroke in the nominally isometric condition at low ionic strength. The amplitude of power stroke is similar at both the distal and the proximal regions of myosin head CAD. From Sugi et al. (2015).

As stated above, our experiments with the EC are the only means to directly record ATP-induced power stroke in individual myosin heads, extending from hydrated myosin filaments, under an electron microscope. Our direct electron microscopic recording and measurement of myosin head movement have no ambiguities arising from stiffness of myofilaments and other elastic structures within each sarcomere in single fiber experiments, and from uncertain myosin head orientations in needle and optical trap experiments.

Here we summarize the novel features of myosin head power stroke revealed by our EC experiments:

(1) The amplitude of myosin head power stroke, directly obtained by recording individual myosin head movement, changes depending on experimental conditions, being

largest (~7 nm) when myosin head move freely without any external load, as deduced from the histogram of myosin head recovery stroke in the absence of actin filaments (Fig. 1.12, Sugi et al., 2008), and being smallest (~3 nm at the distal region of myosin head CAD) under nominally isometric condition (Fig. 1.20, Sugi et al., 2015).

(2) The above range of power stroke amplitude in our EC experiments is somewhat smaller than the values (4–10 nm, reviewed by Kaya and Higuchi, 2013), which have been indirectly estimated by applying mechanical perturbations to contracting muscle fibers. It should be noted that muscle fiber experiments have inherent difficulties in evaluating contribution of elastic structures other than myosin heads.

(3) The lever arm mechanism of myosin head power stroke, constructed from crystallographic and electron microscopic work on isolated myosin heads (Geeves and Holmes, 2005) does not apply to myosin head power stroke in nominally isometric condition at standard ionic strength (Figs. 1.19 and 1.23b, Sugi et al., 2015), while it applies to myosin head power stroke in nominally isometric condition at low ionic strength (Figs. 1.21 and 1.23c, Sugi et al., 2015), in which the force generated by individual myosin heads increases approximately twofold (Sugi et al., 2013).

In conclusion, molecular mechanism of muscle contraction is still full of mysteries. We emphasize that the EC will be a most powerful tool in approaching full understanding of muscle contraction at the molecular and the submolecular levels, since it enables us to directly record structural changes of biomolecules related to their function.

Acknowledgment

We wish to thank Japan Electron Optics Laboratory Co. Ltd. for generously providing facilities to carry out experiments described in this chapter. Our thanks are also due to Drs. Kazuo Sutoh, Takeyuki Wakabayashi, and Eisaku Katayama for their contribution to our research work.

References

Buttler WP, Hale KF (1981) Dynamic experiments in the electron microscope. In: *Practical Methods in Electron Microscope* (Glauert AM, ed), North Holland, Amsterdam, vol 9, pp 1–457.

Finer JT, Simmons RM, Spudich JA (1994) Single myosin molecule mechanics: Piconewton forces and nanometer steps. *Nature,* 366: 113–119.

Fukami A, Adachi K (1965) A new method of preparation of a self-perforated microplastic grid and its applications. *J Electron Microsc,* 14: 112–118.

Fukami A, Fukushima K, Kohyama N (1991) Observation techniques for wet clay minerals using fine-sealed environmental cell equipment attached to high-resolution electron microscope. Microstructure of fine-grained sediments from mud to shale. In: *Frontiers in Sedimentary Geology,* pp. 321–331.

Fukushima K (1987) Production and application of the gas environmental chamber (in Japanese), Doctoral thesis, Nihon University, Tokyo.

Geeves MA, Holmes KC (2005) The molecular mechanism of muscle contraction. *Ann Rev Protein Chem,* 71: 161–193.

Huxley HE (1969) The mechanism of muscular contraction. *Science,* 164: 1356–1366.

Huxley HE, Hanson J. (1954) Changes in the cross-striations of muscle during contraction and stretch and their structural interpretation. *Nature,* 173: 973–976.

Kaya M, Higuchi H (2010) Nonlinear elasticity and an 8 nm working stroke of single myosin molecules in myofilaments. *Science,* 329: 686–689.

Lymn RW, Taylor EW (1971) Mechanism of adenosine triphosphate hydrolysis by actomyosin. *Biochemistry,* 10: 4617–4624.

Minoda H, Okabe T, Inayoshi Y, Miyakawa T, Miyauchi Y, Tanokura M, Katayama E, Wakabayashi T, Akimoto T, Sugi H (2011) Electron microscopic evidence for the myosin head lever arm mechanism in hydrated myosin filaments using the gas environmental chamber. *Biochim Biophys Res Commun,* 405: 651–656.

Oiwa K, Chaen S, Sugi H. (1991) Measurement of work done by ATP-induced sliding between rabbit muscle myosin and algal cell actin cables in vitro. *J Physiol,* 437: 751–763.

Oiwa K, Kawakami T, Sugi H. (1993) Unitary distance of actin-myosin sliding studied using an in vitro force-movement assay system combined with ATP iontophoresis. *J Biochem,* 114: 28–32.

Suda H, Ishikawa A, Fukami A. (1992) Evaluation of the critical electron dose on the contractile activity of hydrated muscle fibers in the film-sealed environmental cell. *J Electron Microsc,* 41: 223–229.

Sugi H (1992) Molecular mechanism of actin-myosin interaction in muscle contraction. In *Muscle Contraction and Cell Motility* (Sugi H, ed), Advances in Comparative & Environmental Physiology, Springer, Berlin and Heidelberg, vol 12, pp 132–171.

Sugi H (2012) The gas environmental chamber as a powerful tool to study structural changes of living muscle thick filaments coupled with ATP hydrolysis. In *Current Basic and Pathological Approaches to the Function of Muscle Cells and Tissues: From Molecules to Humans,* InTech, Rijeka, Croatia, pp 3–26.

Sugi H, Abe T, Kobayashi T, Chaen S, Ohnuki Y, Saeki Y, Sugiura S (2013) Enhancement of force generated by individual myosin heads in skinned rabbit psoas muscle fibers at low ionic strength. *PLoS ONE,* 8: e63658.

Sugi H, Akimoto T, Sutoh K, Chaen S, Oishi N, Suzuki S (1997) Dynamic electron microscopy of ATP-induced myosin head movement in living muscle thick filaments. *Proc Natl Acad Sci U S A,* 94: 4378–4382.

Sugi H, Chaen S, Akimoto T, Minoda H, Miyakawa T, Miyauchi Y, Tanokura M, Sugiura S (2015) Electron microscopic recording of myosin head power stroke in hydrated myosin filaments. *Sci Rep,* 5: 15700.

Sugi H, Minoda H, Inayoshi Y, Yumoto F, Miyakawa T, Miyauchi Y, Tanokura M, Akimoto T, Kobayashi T, Chaen S, Sugiura S (2008) Direct demonstration of the cross-bridge recovery stroke in muscle thick filaments in aqueous solution by using the hydration chamber. *Proc Natl Acad Sci U S A,* 105: 17396–17401.

Sugi H, Tsuchiya, T. (1998) Muscle mechanics I. Intact single fibres. In: *Current Methods in Muscle Physiology: Advantages, Problems and Limitations* (Sugi H, ed), Oxford, Oxford University Press, Oxford, pp 3–31.

Sugiura S, Kobayakawa N, Fujita H, Yamashita H, Momomura S, Chaen S, Omata M, Sugi H (1998) Comparison of unitary displacements and forces between 2 cardiac myosin isoforms by the optical trap technique: Molecular basis for cardiac adaptation. *Circ Res,* 82: 1029–1034.

Sutoh K, Tokunaga M, Wakabayashi T. (1989) Electron microscopic mapping of myosin head with site-directed antibodies. *J Mol Biol,* 206: 357–363.

Chapter 2

Studies of Muscle Contraction Using X-Ray Diffraction

John M. Squire[a] and Carlo Knupp[b]

[a]*Muscle Contraction Group, School of Physiology,*
Pharmacology & Neuroscience, Faculty of Biomedical Sciences,
University of Bristol, Bristol BS8 1TD, UK
[b]*Biophysics Group, School of Optometry and Vision Sciences,*
Cardiff University, Cardiff CF10 3NB, UK

j.m.squire@bristol.ac.uk

X-ray diffraction was one of the early methods used by muscle biologists in the middle of the 1900s to investigate the behaviour of muscle contractile proteins and a great deal of fundamental and unambiguous information was obtained. Since then ever-improving X-ray methods (stronger X-ray sources, faster X-ray detectors, increased computing power) have enabled the design and execution of ever more sophisticated diffraction experiments on contracting muscles. But there is an experimental challenge with diffraction methods in that the observed patterns need to be interpreted and this can lead to erroneous claims about what the observations mean. Here we illustrate where the power of X-ray diffraction lies and also where over-interpretation of X-ray

Muscle Contraction and Cell Motility: Fundamentals and Developments
Edited by Haruo Sugi
Copyright © 2017 Pan Stanford Publishing Pte. Ltd.
ISBN 978-981-4745-16-1 (Hardcover), 978-981-4745-17-8 (eBook)
www.panstanford.com

diffraction patterns can lead down incorrect paths. We show how proper modelling of various parts of the diffraction pattern must be limited to those occasions where the number of observations exceeds the number of independent model parameters to be fitted. We also discuss where future progress can be made.

2.1   Introduction

The application of X-ray diffraction to study muscle was pioneered by such people as Astbury (1947) and Selby and Bear (1956, and earlier references), but the first one to make real progress with the technique was Hugh Huxley in his PhD project work (1953) and in a multitude of seminal papers that followed. The idea was quickly taken up by Gerald Elliott and his collaborators (1960, 1964, 1967) (see Squire (2013) and Hitchcock-DeGregori and Irving (2014) for a summary of this early work).

One of the attractive features of studying striated muscles, unlike any other cell type, is that they are so regularly organised and so massive that they behave almost like miniature diffraction gratings and they give rise to wonderfully rich X-ray diffraction patterns that can provide a wealth of useful and unambiguous structural information about muscle components (see Huxley and Brown, 1967; Squire, 1981; Harford and Squire, 1997; Squire and Knupp, 2005).

To set the scene for this review, we first summarise in this section the basic structure of the muscle sarcomere and then in the next section [2] we outline some of the fundamental ideas behind the X-ray diffraction technique. We go on to discuss how these ideas apply to particular parts of the diffraction pattern (Sections 2.3 and 2.4) and consider what can and what cannot be deduced from analysis of these patterns. We continue (Section 2.5) by discussing what appears to be the most fruitful approach to making full use of the X-ray diffraction technique in the future.

Figure 2.1 illustrates the sarcomere, the basic contractile unit in striated muscles. The sarcomere consists of a side-by-side array of bipolar myosin filaments in the A-band together with two sets of actin filaments that overlap each end of the myosin filament array (Fig. 2.1a). The myosin filament is composed mainly of myosin molecules each of which is a long two chain α-helical rod with two globular myosin heads at one end (Fig. 2.2a).

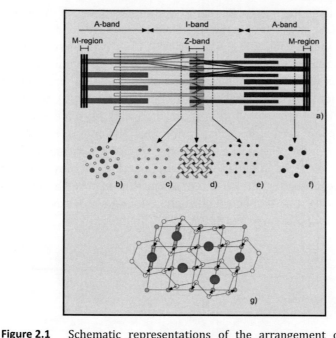

Figure 2.1   Schematic representations of the arrangement of myosin and actin filaments in the sarcomere. (a) Longitudinal view centred around the Z-band. In the Z-band, anti-parallel actin filaments (light blue and dark green) are joined together by α-actinin bridges. Also in this region, titin strands from the two halves of the sarcomere (purple, green, blue and magenta) are anchored to actin. Myosin filaments (dark and light brown) extend to the M-region where they continue their course into adjacent sarcomeres, but with changing polarity. The region occupied by myosin filaments is called the A-band, while the remaining portion of the sarcomere is called the I-band. Titin runs from the Z-band, through the I-band and along the myosin filaments to the central M-band in the middle of the M-region. (b–f) Transverse views of the actin and myosin filaments across different regions of the sarcomere. (b) Actin and myosin filaments overlapping in the A-band where they form a hexagonal lattice. (c,e) Actin filaments towards the Z-band end of the I-band form a pseudo square lattice. (d) Anti-parallel actin filaments are cross-linked by α-actinin in the Z-band often showing a so-called basketweave pattern. (f) The myosin filaments in the A-band form a hexagonal lattice. (g) There is a systematic transition of the actin filaments from a hexagonal lattice at the outer end of the A-band to a square lattice at the Z-band end of the I-band (adapted from Knupp et al., 2002).

The rods form the myosin filament backbone from which the heads project in a quasi-helical array (Fig. 2.2b). It is these heads that are enzymes which bind and hydrolyse ATP, the main energy source for contraction, and which during muscle contraction bind to the adjacent actin filaments. The actin filaments are helical arrangements of actin monomers (Holmes, et al., 1990), together with the regulatory protein tropomyosin and in some muscles (including all vertebrate striated muscles) troponin (Fig. 2.2c). Troponin binds calcium ions released into the sarcomere as a result of nervous stimulation of the muscle, thereby moving tropomyosin on the actin filament surface to expose the myosin binding sites and to allow contraction to occur (Huxley, 1972; Haselgrove; 1972; Parry and Squire, 1973).

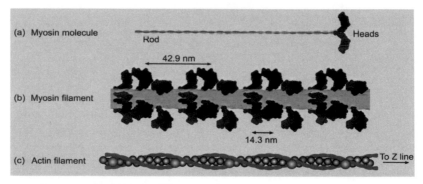

Figure 2.2   Schematic representation of the main components of the sarcomere. (a) Myosin molecule consisting of a long coiled-coil rod ending with two globular heads. (b) Parallel myosin molecules come together to form half of a myosin filament, in which the heads (red, dark red), roughly on a helical track, project out of a backbone made of the coiled-coil region of the molecules. (c) Actin filaments made of actin monomers on a helix (green spheres), decorated by tropomyosin (blue strands) and troponin (light blue spheres).

The ATPase activity of the heads is activated by actin as in the scheme in Fig. 2.3 proposed by Lymn and Taylor (1971). In the absence of ATP (state AM) the myosin heads become permanently attached to actin in what is known as the rigor conformation.

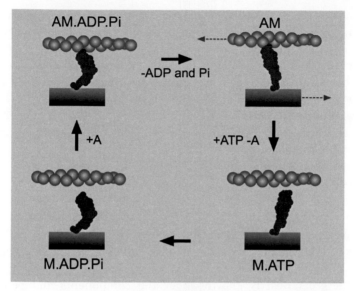

Figure 2.3   Scheme representing ATPase activity in the sarcomere as proposed by Lymn and Taylor (1971). In the top-left corner a myosin head (M, red) on a myosin filament backbone (brown) is attached to an actin filament (A, green). The head has ATP hydrolysis products ADP and inorganic phosphate Pi bound giving AM.ADP.Pi. The dissociation of ADP and the phosphate group to give AM is linked with a configurational change of the myosin head attached to actin and a relative axial shift of the actin and myosin filaments (top-right corner) if they are free to move. The binding of a new ATP molecule to the myosin head causes detachment of the head from the actin filament (M.ATP: bottom-right corner). Finally, the hydrolysis of the ATP into ADP and phosphate causes the resetting of the myosin head to be ready to attach again to actin and repeat the cycle (bottom-left corner).

In cross sections through different parts of the sarcomere, the myosin and actin filaments form varying kinds of 2D lattice. In the part of the A-band not overlapped by actin, the myosin filaments form a hexagonal lattice generated and stabilised at the M-band (middle part of the M-region) by regular cross-linking between the filaments (Fig. 2.1a). In the overlap region of the A-band the actin filaments fit into the myosin filament hexagonal lattice by occupying the so-called trigonal positions which are centrally located in a triangle formed by three adjacent myosin filaments (Fig. 2.1b). In the muscle Z-band where actin filaments of opposite polarity in successive sarcomeres are linked together by α-actinin bridges, the actin filament lattice is almost square, but not quite (Fig. 2.1c–e). Between their pseudo-square array at the Z-band and their hexagonal organisation in the A-band the actin filaments located in the I-band are thought to meander laterally, but on average to shift systematically as in Fig. 2.1g. In addition to myosin and actin filaments, the sarcomere also contains the giant molecule titin (molecular weight about 3 mDa) which binds to the myosin filaments in the A-band and then passes through the I-band to interact with the Z-band assembly (Fig. 2.1a). There may be six titin molecules on each half myosin filament (Liversage et al., 2001). These come together at the myosin filament tip to form end-filaments (Trinick, 1981) and then link to the Z-band, possibly as illustrated in Fig. 2.1a (Knupp et al., 2002).

A great deal is known about the components and general organisation of the sarcomere (Squire, 1981; 1986; 1997; Squire et al., 2005). Crystal structures are available for the myosin head (Rayment and Holden, 1993; Dominguez et al., 1998; Fig. 2.4a), for the actin monomer (Kabsch et al., 1990; Fig. 2.5a), for tropomyosin (Whitby and Phillips, 2000; Orzechowski et al., 2014; von der Ecken et al., 2015; see Fig. 2.5), for a good part of troponin (Vinogradova et al., 2005; Fig. 2.5) and for α-actinin (Ribeiro Ede et al., 2014). The myosin head shape shown in Fig. 2.4a is of particular interest in that it shows a motor domain that binds to actin and is the enzymatic part of the head and a long alpha-helical neck with two associate light chains. It is the neck which may relay to the myosin rod the structural changes in the motor domain associated with the ATPase cycle; it acts as a kind of lever and is often called the lever arm. Between the lever arm and motor domain is the converter domain which may be thought of as a kind of gear box linking the motor to the lever arm.

Detailed models are now available from electron microscopy for the myosin head distribution on resting myosin filaments (AL-Khayat et al., 2013; Fig. 2.4b), together with the likely location of titin strands and part of C-protein (MyBP-C), and for the actin filaments with troponin plus or minus bound calcium (Paul et al., 2016; Fig. 2.5a,b). However, although there are plenty of ideas about exactly what the myosin heads and the lever arms are doing to generate force and movement in muscle based on analysis of static structures (Holmes and Geeves, 2005), there is very little detailed and unambiguous information about what actually happens in active muscle. In other words, even now, we do not really know how muscular force is produced. This Review tries to answer the question: What can X-rays tell us about how muscles actually work?

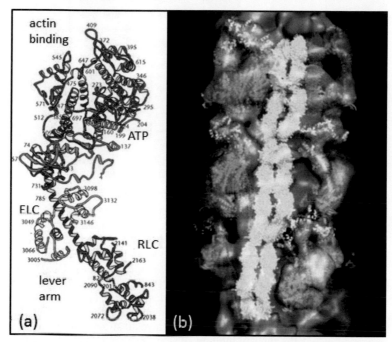

Figure 2.4   (a) Details of the myosin head structure as determine by Rayment et al. (1993) showing the motor domain with its actin binding face at the top and the long tail region or lever arm which links to the myosin rod at the bottom of the figure. RLC and ELC are the regulatory and essential light chains. (b) 3D reconstruction using electron microscopy and single particle analysis of the human cardiac muscle myosin filament. Two titin strands are shown in yellow, part of the myosin binding protein MyBP-C (myosin binding protein-C, C-protein) in mauve, and the myosin head pairs on the filament surface in a variety of colours. From AL-Khayat et al. (2013) with permission.

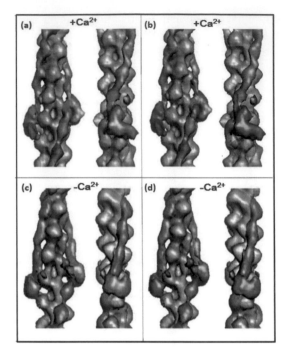

Figure 2.5   Modelling from 3D reconstructions of the actin filament in the presence (a,b) and absence (c,d) of bound Ca^{2+} ions (the right hand figure in each pair is 90° rotated from the left hand image). Two tropomyosin strands (orange) can be seen following roughly helical tracks. Troponin (blue in (b,d)) labels the tropomyosin strands every 38.5 nm along the filament. Globular actin molecules (shades of grey) aggregate to form the helical backbone of the filament. The effect of Ca^{2+}-binding to troponin is to shift the tropomyosin strands across actin so that the myosin binding sites are exposed. Reproduced from Paul et al. (2016) with permission.

2.2   Basic Concepts in Diffraction

X-ray diffraction is the result of interference between X-ray beams scattered by different parts of the diffracting object (Fig. 2.6). X-rays are electromagnetic waves characterised by a wavelength in most muscle applications of around 0.1 nm. Every atom in the diffracting object experiences the incoming X-ray beam as oscillations of the electrons in the atom. Since electrons are charged and since accelerated charged particles radiate photons,

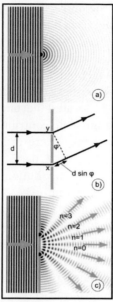

Figure 2.6   Simple two-dimensional schematic representation of wave diffraction. (a) A plane sinusoidal wave approaches a wall (orange) with a small slit in its centre. The plane wave is moving from left to right in the direction shown by the arrow. Once it reaches the wall, the plane wave is let through the small slit and propagates to the right side of the wall as a cylindrical wave. (b) Summary of the geometric variables to be taken into account in deriving the formula for wave diffraction from two slits. Here, the slits (x and y) are separated by a distance d. A plane wave, represented by black lines on the left of the wall, travels from left to right. Cylindrical waves are generated at the two apertures and they interfere with each other. Whether they interfere constructively or destructively depends on their relative phase. The relative phase depends on the extra distance travelled by the wave from the second slit ($d*\sin(\varphi)$), which in turn depends on the angle φ that the observer makes with the wall (represented here by the black lines on the right side of the wall). If the extra path length is a whole number of wavelengths ($n\lambda$) then constructive interference occurs ($d*\sin(\varphi) = n\lambda$). (c) The cylindrical waves from two slits, arising when a plane wave travelling from left to right hits the wall, interfere to give rise to well defined beams marked by the index $n = 0, 1, 2$, etc. The angles that these beams make with the incident beam are given by the formula: $\sin(\varphi) = n\lambda/d$.

each stimulated atom in the diffracting object acts as a new source of X-rays radiating in all directions with the same wavelength as the incoming beam (Fig. 2.7a,b). If the incoming X-ray beam is a narrow collimated monochromatic (single wavelength) beam directed at the diffracting object, as is usually the case in all muscle experiments, and a detector (a fluorescent screen, film or electronic detector) is located beyond the diffracting object (Fig. 2.7c), then every atom will contribute to what is recorded in every part of the detector. There are no exceptions to this; every atom contributes to every part of the X-ray diffraction pattern. In some cases the contributions from different atoms cancel, and in some cases they reinforce.

Figure 2.6b,c illustrate the second main principle in diffraction using light as an analogy. Two atoms (or scattering objects x and y, in this case slits in an opaque card) separated by a distance d will each scatter light in all directions, but in certain directions the two waves will be oscillating in step with each other and will reinforce (constructive interference), whereas in other directions they will be exactly out of step and will cancel each other out (destructive interference). From simple geometry in the 2D diagram in Fig. 2.6b it can be seen that the angle (φ) of scattering where reinforcement occurs is related to the wavelength (λ) of the incident beam and the inter-object spacing (d) by the rule: $d \sin (\varphi) = n \lambda$, where n is any integer. Putting this another way, $\sin (\varphi) = n \lambda/d$. This clearly means that the larger the value of d, the smaller the angle φ will be to get constructive interference. This is commonly referred to as the '*reciprocal nature of diffraction*'. Figure 2.6c is a simple 2D illustration of the principle of interference. Constructive interference can occur for different values of n, but the angle involved will then be different.

In the case of X-ray diffraction the scattering is from atoms, but the same ideas about interference apply in three dimensions to give what is known as Bragg's law ($n\lambda = 2d \sin (\theta)$) where in this case the angle of diffraction for constructive interference is 2θ and the scattering is from *planes* of atoms (or objects) where d is the inter-planar spacing (Fig. 2.7a).

How does all this apply to muscle? First, in muscle, the repeating distances in the actin and myosin filaments are quite large. In his early studies (summarised in the seminal paper

Huxley and Brown, 1967), Hugh Huxley found that myosin filaments showed two characteristic axial repeats: 42.9 and 14.3 nm. With a wavelength of about 0.1 nm, the angle 2θ for a 42.9 nm repeat is only about 0.134°. For this reason, diffraction associated with such large d spacings is known as low-angle or small-angle diffraction; special X-ray diffraction cameras with detectors placed several metres away from the specimen, are often needed to record good diffraction patterns from muscle (see Harford and Squire, 1997). The main periodicity in actin filaments is around 36 nm, which also gives peaks in the same low-angle region.

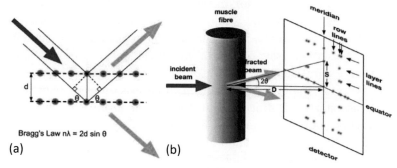

Figure 2.7   (a) Summary of the geometric variables to be taken into account in deriving the formula for diffraction of, for example, X-rays or neutrons from an array of atoms (circles) on planes a distance d apart. The direction of the incoming wave, making an angle θ with the planes of atoms, is shown in blue. Orange arrows show the directions of the scattered ('reflected') beam (also making an angle θ with the plane of atoms) and of the undeviated transmitted beam. Because of the interference from all the secondary waves from successive planes of atoms, constructive interference only occurs when $2d \sin(\theta) = n\lambda$ (known as Bragg's law), with n an integer number and λ the wavelength of the incoming radiation. (b) Schematic representation of the X-ray experimental set up during a muscle diffraction experiment. Incoming monochromatic X-radiation (blue arrow) is directed towards a muscle. Scattered radiation (yellow arrows) emerges from the muscle along defined directions dictated by the position of the atoms inside the muscle. When recorded on a detector, the diffracted beams appear as a series of well-defined spots on regularly spaced layer lines (horizontal; i.e. perpendicular to the muscle fibre axis) and row lines (vertical).

If the diffracting object is a muscle placed with its long axis vertical in a horizontal X-ray beam (Fig. 2.7b), then diffraction up and down through the centre of the diffraction pattern is said to be along the *meridian* and peaks on the meridian are meridional reflections. The portion of the pattern perpendicular to the meridian and passing through the centre of the diffraction pattern is called the *equator*.

Turning now to more specific structures, the myosin and actin filaments are both roughly helical. There is too little space here to go into details of helical diffraction theory in the case of muscle filaments (see, for example, Squire, 1981; Harford and Squire, 1997; Squire and Knupp, 2005 for this). To summarise, a so-called discontinuous helix can be characterised by a subunit axial translation (say of length p), analogous to the rise per step in a spiral staircase, and a pitch (say of length P), where in the staircase analogy one has risen to be exactly over where the staircase started. Diffraction features appear in muscle low-angle diffraction patterns from both myosin and actin filaments which relate to the values of p and P. For myosin filaments p is 14.3 nm. The pitch P is 3 × 42.9 nm, but since the myosin heads appear on three co-axial helical strands the observed axial repeat is cut down to $P/3$ = 42.9 nm. These repeats are apparent in Figs. 2.2b and 2.4b. A rule from helical diffraction theory is that only reflections corresponding to the p spacing and its orders appear on the meridian. Peaks relating to the pitch or true axial repeat of the object appear on horizontal lines perpendicular to the meridian known as layer lines (Fig. 2.7c). So, from myosin filaments in resting muscles there is a meridional reflection at an angle in Bragg's law given by the spacing 14.3 nm (see Fig. 2.8) together with meridional peaks at subsequent orders such as 14.3/2 = 7.15 nm; 14.3/3 = 4.77 nm and so on. There are also layer lines related to orders of the spacing 42.9 nm. The 14.3 nm meridional reflection is therefore on the 3rd layer line and is often referred to as the M3 reflection (M for myosin). In the case of actin, the value of p is about 2.75 nm and the pitch P is around 5.9 nm. However, in this case, there is not a whole number of actin subunits in one turn and the whole structure repeats after around 36 nm.

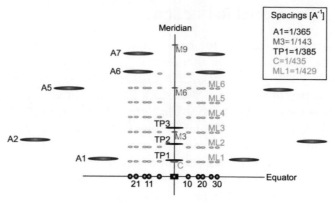

Figure 2.8   Schematic representation of an X-ray diffraction pattern from a fish muscle. Only the top half portion of the pattern is shown. The bottom half is related to the top half by mirror symmetry across the equator. The right and left portions of the pattern are also related by mirror symmetry, but across the meridian. The equatorial reflections (marked with indices 10, 11, 20, 21, and 30: see Fig. 2.9 for the meaning of the indices) are shown in black and are produced by radiation scattered from both the myosin and the actin filaments. The yellow layer-line reflections are produced only by the myosin filaments and are labelled ML1 to ML6 (ML stands for Myosin Layer-line). They relate to successive orders of a '*d*' spacing in Bragg's law of 43 nm. The layer lines in blue are produced by the actin filaments (orders of $d \sim 36$ nm) and are labelled A1 to A7 (A stands for Actin reflections). The main meridional reflections, in green, named M3, M6 and M9 (M for Myosin; orders of $d = 14.3$ nm) tell us about the one dimensional projection of the density of the myosin heads onto the muscle fibre axis (heads on myosin and in active or rigor muscle heads on actin too). Reflections on the meridian from troponin on the actin filaments (Fig. 2.2c), labelled TP1 to TP3 (orders of $d = 38.5$ nm) are shown in red.

A simple description has actin filaments with 13 actin subunits in 6 turns of the helix; 2.75×13 is 35.75 nm. In this case there is a meridional peak at 27.5 nm (on the 13th layer line from actin; referred to as the A13 peak) and there are layer lines at orders of $d = 35.75$ nm. There are particularly strong peaks from actin filaments off the meridian on layer lines 1, 2, 6 and 7 (labelled A1, A2, A6 and A7; Fig. 2.8).

2.3 Equatorial Reflections

Diffraction on the equator comes from planes of filaments that can be drawn through the A-band (and Z-band) in striated muscles when looking down the filament axis. Figure 2.9 shows a 2D lattice of points and illustrates the ways in which different lattice planes can be described in terms of what are called Miller indices *h, k, l*. In simple terms the 100 planes cut the **a** axis of the unit cell into one piece, but they are parallel to **b** (and to **c** in 3D). The 110 planes cut the **a** axis of the unit cell into one piece, and the **b** axis into one piece (and in 3D are parallel to **c**). Often the muscle reflections on the equator are just described in terms of *h* and *k*, since on the equator *l* is always zero.

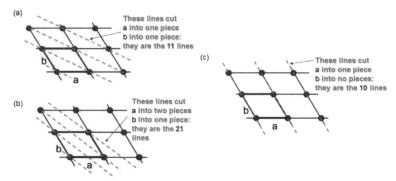

Figure 2.9 Diagrams showing the meaning of Miller indices (*h,k,l*) describing different planes through a lattice. In this 2D example the dashed lines in (a) cut each of the unit cell sides 'a' and 'b' into one piece, so these are called the 11 lines (*h* = 1, *k* = 1). In (b) the 'a' unit cell side is cut into two pieces, whereas 'b' is cut into one piece so these are the 21 lines. (c) is an example where the lines are parallel to a unit cell side 'b', so the *k* index is 0. The figure shows the 10 lines. The same system arises in planes through a 3D lattice, and then 3 indices are used. In the case of fibre patterns like those from muscle, *h* and *k* describe the equatorial reflections, *l* describes the meridional peaks and all three indices *h*, *k* and *l* are needed to identify peaks on the layer lines off the meridian.

Figure 2.10a–c illustrates typical equatorial intensity profiles, in this case from fish muscle, and Table 2.1 lists the indices and the approximate *d* spacings of these reflections. In early observations of the equator from frog muscles (e.g. Huxley, 1953; Elliott, 1960) it was soon discovered that the relative intensities of the different equatorial reflections changed when a resting muscle was activated or put into rigor (no ATP). In particular the strong 10 reflections from resting muscle became weaker and the 11 reflections which started relatively weak became stronger (Fig. 2.10). These changes were intermediate in patterns from active muscle and strongest in patterns from rigor muscle. Since it is known that all myosin heads bind strongly to actin in rigor frog muscle (Cooke and Franks, 1980; Lovell et al., 1981), it has been presumed that the smaller changes seen in patterns from active muscle suggest that some, but not all, of the heads are bound to actin in active muscle.

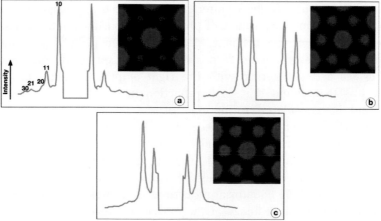

Figure 2.10 (a) Equatorial intensity profile obtained from shining X-rays through a resting muscle. The 10, 11, 20, 21 and 30 reflections (showing as intensity peaks in the diffraction pattern) are highlighted. Inset, Fourier synthesis obtained from combining the density waves associated with each of these equatorial peaks. This represents a view of the projection of the electron density of the muscle proteins down the muscle long axis. The density from a myosin filament is shown in the centre, and it is immediately surrounded by six actin filaments. (b) Equatorial profile as in (a) but this time from an active muscle, with its Fourier synthesis in the inset. (c) The same for rigor muscle, with its Fourier synthesis in the inset. Note the gradually increasing intensity of the 11 peaks and the gradual increase of mass at the actin positions on going from (a) to (b) to (c).

In a number of studies the relative intensities of the 10 and 11 peaks have been used as a measure of attachment number (the percentage of heads attached to actin). But how was this done? Can one get an image of the diffracting object from the reflections in the X-ray diffraction pattern? This would be a wonderful thing to do. In a light microscope the specimen being studied actually diffracts the light into the objective lens of the microscope. In the back focal plane of the objective lens of a light microscope this diffraction pattern can be seen. What the eyepiece then does is to recombine all these diffracted beams to recreate the image for the observer. The key thing here is that each diffracted beam not only has an intensity (or its square root the amplitude), but also a phase. We have seen (Fig. 2.6) that beams oscillating in phase reinforce each other whereas those oscillating out of phase interfere destructively. There are, of course, intermediate values of phase where partial reinforcement occurs. One of the brilliant ideas of Fourier (1822) was that periodic objects can be reconstructed by a series of sines and cosines added together with the right amplitudes and phases. It was then found that each reflection in a diffraction pattern could be considered to be one of the terms in the Fourier calculation. What the eyepiece does is to recombine all the diffracted beams in the light microscope with the correct amplitude and phase to recreate a good image of the specimen. It is doing what is called Fourier synthesis. So, if we knew the amplitudes and phases of the equatorial reflections from muscle, we could in principle rebuild an image of the muscle. However, X-rays cannot be focussed in the same way as light; the angles of refraction for X-rays are tiny. So if we want to produce an image, the role of the eyepiece needs to be mimicked by calculation in a computer. But the observed X-ray diffraction patterns show the intensities (hence amplitudes) of the reflections, but not their phases. This is the well-known phase problem in X-ray diffraction; a problem that is overcome in protein crystallography by a number of tricks which are not usually applicable to muscle diffraction (e.g. Sawyer, 1986; Helliwell, 1992). To try to image muscle, we need to find a way to determine or guess the phases of the equatorial reflections.

Looking down the filament axis there is a feature which might help. At the resolution that we are dealing with (say 20 to

40 nm), the A-band structure in projection down the muscle axis may appear centro-symmetric such that for every density at x, y there is an equivalent density at $-x, -y$. If this is true, then it is found by calculation that the phases of the reflections, instead of being anywhere between 0 and 360°, can only be 0 or 180°. For the few reflections on the equator this means that there is a limited number of possible phase combinations, some of which may give images which, based on other data such as that from electron microscopy, may be more realistic than others. Figure 2.11 illustrates how we can use the reflections on the equator to reconstruct, by Fourier synthesis, the appearance of the sarcomere projected down its main axis. Figure 2.11a shows in red and blue the projection of a plane sine wave corresponding to the 10 reflection (red are peaks, blue troughs). The amplitude of these waves is the amplitude of the 10 reflection in the diffraction pattern, but their phase must be obtained by other means (here it is set to zero—see above). Figure 2.11b,c show the sine waves corresponding to the 21 and 11 reflections, respectively. By summing together all the waves relative to all reflections on the equator, a reconstruction of the sarcomere can be obtained.

Early work of this kind (e.g. Haselgrove and Huxley, 1973) generated what are called Fourier synthesis electron density maps from the equatorial reflections, something which has been developed since (see inset maps in Fig. 2.10a,b,c). These represent possible distributions of density in the muscle when viewed down the long axis. To start with, these maps were computed just from the 10 and 11 reflections, which were assumed to have the same phase, and it was apparent in these maps that more mass appeared at the actin positions in maps from active muscle than in resting muscle, and even more in maps from rigor muscle. All of these were apparently consistent with ever more myosin heads binding to actin. Others then took such studies to higher resolutions so that reflections included were the 10, 11, 20, 21 and 30 peaks. The various phase combinations were whittled down to two preferred options, with phases 0, 0, 180, 180, 0 (Yu et al., 1985) and 0, 0, 180, 0, 0 (Harford and Squire, 1992). Once again, with such phase combinations, it was apparent that gradually more density appeared at actin when a resting muscle was activated or put into rigor (Fig. 2.10a–c).

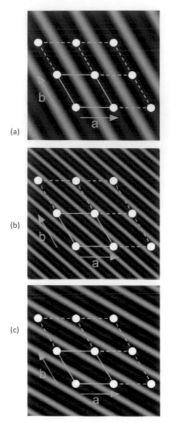

Figure 2.11 Diagrams showing how waves (sinusoidal oscillations) of density associated with different directions through a periodic lattice can combine to produce a 2D distribution of density (cf Fig. 2.10 insets) as in Fourier's theorem. Here (a), (b) and (c) show waves along the three lattice directions indicated in Fig. 2.9. (a) Projection of a two dimensional sine wave corresponding to the 10 reflection (red regions are peaks, blue troughs). The amplitude of this wave is the amplitude of the 10 reflection in the diffraction pattern (the square root of the intensity of the appropriate reflections in Fig. 2.10), but its phase must be obtained by other means (here it is set to zero). (b) As above but for the 21 reflection. (c) As above, but for the 11 reflection. The Fourier synthesis down the fibre axis is ultimately obtained by summing together all waves corresponding to all reflections on the equator. The results of combining such density waves are shown in Fig. 2.10 for different muscle states.

So what is the problem? One problem is that the relative intensities of the equatorial peaks do not just depend on the attachment number. For example, Lymn (1978) showed that for a given attachment number, simply changing the configuration of the heads on actin could have a big effect on the relative intensities of the equatorial reflections. Perhaps one could go to higher resolution because further orders have been seen from vertebrate muscle on the equator, out to the 40 reflection. However, another problem is that the assumption about centro-symmetry begins to become less solid as one goes to higher and higher resolution; it may just about be right out to the 30 reflection, is probably not quite right out to the 40 peak, but quite likely breaks down significantly beyond that. Unfortunately, we need to go beyond the 30 reflection to start to see higher resolution details of what is going on, and Fourier synthesis with these higher order reflections becomes less certain. In addition, if there are many more reflections, then deciding on the best phase combinations for the higher orders becomes much more difficult.

You might then quite reasonably say, is there not another way to do this? Can we not produce models of the myosin and actin filaments with movable myosin heads, calculate the diffraction pattern that such models would produce, and then find the best models to fit the different observed amplitudes in the equatorial diffraction patterns? Yes we can. Let us think about what needs to be done. To define such a model, one needs a number of parameters such as: the size (radius) of the myosin filament backbone and of the actin filament, the relative weights of the myosin backbone, of actin filaments and of myosin heads, the arrangement of the heads around myosin in resting muscle, the arrangement and attachment number of myosin heads in active muscle (both the attached heads around actin and the off heads around myosin) and the arrangement of myosin heads attached to actin in rigor. Add to this the fact that the A-band lattice is not perfectly ordered (Yu et al., 1985) and that the myosin and actin filament axes are not precisely located on the lattice points, but show positional fluctuations, often accounted for using a kind of temperature factor (see Harford et al., 1994), and it is easy to see that the number of parameters involved soon becomes quite large. Table 2.1 compares the number of observed equatorial reflections out to the 40 peak with the number of

Table 2.1   Parameters defining the equatorial intensities for the very simplest cylindrically-symmetric model

Equatorials				
a =	47	nm		
Observations			**Parameters**	
Indices	hh+hk+kk	Spacing		
10	1	40.7	Myosin backbone radius	1
11	3	23.5	Myosin head inner radius	1
20	5	20.4	Myosin head outer radius	1
21	7	15.4	Actin filament radius	1
30	9	13.6	Attached head inner radius	1
22	12	11.7	Attached head outer radius	1
31	13	11.3	Myosin backbone weight	known
40	16	10.2	Actin weight	known
			Head weight	known
			Number attached	1
			Lattice disorder parameter	1
			Thermal disorder of myosin backbone	1
			Thermal disorder of heads around myosin	1
			Thermal disorder of actin filament	1
			Thermal disorder of heads on actin	1
Observations		8(−1)	Parameters	12

Note: More realistic models including head shapes and configurations would require many more parameters.

parameters involved in even the simplest calculation. This simplest calculation assumes (very simplistically) that the myosin and actin filament backbones can be approximated by uniform cylinders and that myosin head density is either in a uniform cylindrical annulus around myosin or in a uniform cylindrical annulus around actin. Global searches using this minimum number of parameters show that this model fits the equator almost perfectly in all three states, resting, active and rigor. But this is not surprising since the number of free parameters involved is significantly greater than the number of observations, even out to the resolution of the 40 reflection; the problem is under-determined. It also means that such calculations give little useful information about what the myosin heads are actually doing. We know that the model is unrealistic, but to make it more sophisticated would require many more parameters than those shown in Table 2.1.

So is there a way forward with equatorial diffraction? There may be in the following way. We know that heads in the resting conformation contribute to a particular distribution of equatorial intensity. Likewise, fully active muscle and rigor muscle give their own characteristic intensity distributions. There is another well-described state too. This is the state where myosin heads with ADP and Pi bound (Fig. 2.3) attach transiently to actin in what has been termed the weak-binding state. An analogous state can be induced in some muscles such as rabbit psoas muscle by lowering the ionic strength of the salt solution bathing 'skinned' muscles, where the membrane of the muscle cell, the sarcolemma, has been rendered permeable (Brenner et al., 1984). This state is characterised by having 10 and 11 reflections of almost equal intensity. This weak binding state may be like the first rapid equilibrium attached state in the crossbridge cycle before the heads go over to strong, stereospecific states on actin.

So we have four standard states: (i) resting muscle where the heads are off actin (M.ADP.Pi), (ii) the weak binding state which may be the first attached state of heads on actin in the contractile cycle (AM.ADP.Pi), (iii) active muscle where some of the heads are strongly bound (e.g. AM.ADP) and (iv) heads which have released ADP and Pi (akin to rigor: AM). We have the intensities of some of the 10, 11, 20, 21 and 30 peaks (and some beyond that) for these four states and can take the square roots of these to get the relative amplitudes. We can also assume that the A-band structure is approximately centro-symmetric out to the resolution of the 40 peak, so that the phases associated with each of these peaks in each state will be 0 or 180°. There is also no evidence that these phases change between states. So in projection down the muscle long axis there would be only 'REST' heads in resting muscle (Fig. 2.12b). On activation these would perhaps go through a weak binding state (some heads on actin which we label WEAK and some off actin which we label OFFSET), to a strong binding state with the heads first in similar positions to the attached weak binding heads (still WEAK), and

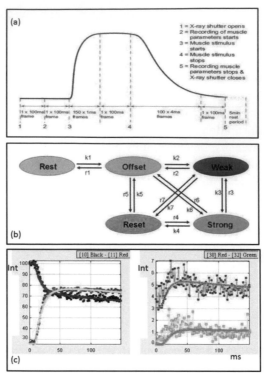

Figure 2.12 (a) Typical time-course of the tension in an isometric tetanic contraction of fish muscle, with an indication of how this time was broken up into separate time frames during which independent diffraction patterns were recorded by Eakins et al. (2016). (b) Summary of the states used in a simulation of crossbridge behaviour during a tetanic contraction starting from resting muscle. Since it takes time to restore an activated muscle back to the fully resting state, it was assumed that the muscle would start from 100% resting (Rest), would go through an activated but not yet attached state (Offset) before attaching weakly to actin (Weak). It would then go through product release and possible head movement to a strong state (Strong), would detach from actin (Reset) and go through hydrolysis to get back again to the pre-attached activated state (Offset). An algorithm was set up to allow different paths through this system to be followed. (c) Shows examples of the changing amplitudes of some of the equatorial peaks (10, 11, 30, 32 peaks) during a contraction as in (a), together with preliminary fits to the data using the model in (b). Further discussion of this approach can be found in Eakins et al. (2016).

then to the strong AM state (STRONG; akin to rigor) after release of ADP and Pi, during which the myosin head lever arm might rotate and relative movement of the myosin and actin filaments would occur if they are free to move, but we do not assume this. After this the heads would bind ATP, detach from actin and go through the hydrolysis step: M.ATP to M.ADP.Pi. We term this state RESETTING.

This idea is summarised in Fig. 2.12b where various rate constants are described for transitions between the different structural states. Note that these rate constants do not equate to rate constants between biochemical states, since, for example, the attached weak binding heads and heads in the first strong state could have identical structures (labelled WEAK), therefore the same diffraction contribution, at the resolution we are considering. It is assumed that once the actin filaments have been calcium activated the myosin heads would not return to their initial REST configuration after going through an active cycle, but would adopt a 'RESET' to OFFSET configuration during which ATP is hydrolysed to ADP and Pi after which further weak binding can occur. In this model resting muscle is therefore a starting state which leads into a self-contained four state structural cycle.

Ignoring the filament backbones for a minute, because it must be assumed that they are unchanged, the different head configurations in different states will all contribute density to the projected view down the muscle axis.

The projected density will appear like the various projected states added together with a weighting that depends on the occupancy of each state. Since we know the contribution that each type of structure makes to the equator, assuming that the backbones do not change, assuming that the phases are all 0 or 180°, and assuming that the phases don't change between states, the amplitudes of the observed equatorials at any intermediate time in a contraction ought to be a linear combination of the contributions from the various states. (Note that even if the phases are off by a few degrees, this calculation is still valid; for example, a 5° error in each phase would in the worst-case scenario change the sum of the amplitudes of a particular peak from $4 \times R$ to $4 \cos (5) \times R = 3.985\ R$, where R is the total amplitude in the model; a change of 0.375%; an average 10° phase error would in the worst case give 3.939 R instead of 4 R; an error of 1.5% in R; a 5% error in R would arise from an average worst-case phase error of 18.2°).

One kind of experiment that can be carried out with modern X-ray equipment (synchrotron X-ray sources, fast electronic 2D detectors; Harford and Squire, 1997) is to stimulate a muscle to produce, for example, a tetanic contraction (Fig. 2.12a) and to monitor the intensities of the equatorial and other reflections on a millisecond time scale (or faster). Figure 2.12c shows the changes in some of the equatorial reflections from contracting fish muscle during such a tetanus (data averaged over a number of tetani and muscles), together with the fit to the observations using the cycle in Fig. 2.12b. This fit was obtained by simulated annealing of the very rich set of observations against the model using as few free parameters as possible. The good thing about this kind of approach is that it is not necessary to know what the phases of the individual reflections actually are, as long as they do not change during the tetanus.

In this analysis, we observed the time-courses of eight equatorial reflections. As with the tension in Fig. 2.12a, each reflection showed a delay (one observation) followed by a shape that could reasonably be modelled by perhaps two exponential functions (each requiring an amplitude and a time constant), or possibly a 4 or more parameter polynomial. So each time-course would be described by a delay (the same for each one) and the equivalent of at least four other observations. This gives a minimum of (8 × 4) + 1 observations, from which we subtract one observation since the amplitudes are all relative and we can use one of them as a standard. So we have the equivalent of about 32 or more independent observations. In the state diagram of Fig. 2.12b there are up to 7 transitions each with a forward and backward rate constant (a maximum of 14 rate parameters), plus a delay and any unknown amplitude parameters. If the number of unknown amplitudes is kept less than (32 − (1 + number of rate constants)), then the problem is over-determined.

In conclusion, the equatorial reflections are consistent with a crossbridge cycle in which there are two actin-attached myosin head states (no more attached states are needed) and resetting states with the heads off actin. One can generate the relative populations of these structural states and also start to ask the question, what are the structural differences between the states? So the equatorial peaks can be very useful. The details and results from this approach are given in Eakins et al. (2016).

However, despite this progress with the equatorial intensities, much of the structural change in the heads during the contractile cycle is probably an axial swinging of the lever arms and looking down from the top, down the filament long axis, as in equatorial analysis, is not the ideal way to detect such changes. We need to get more information by studying the contractile events from the side; something that will contribute to the meridian and layer lines of the diffraction pattern. This is discussed next.

2.4   Meridional Reflections

Turning now to the meridian of the muscle diffraction pattern, peaks of particular interest are the M3 peak at 14.3 nm and the M6 peak at 7.15 nm (Fig. 2.8). It has been known for some time that these peaks are crossed by an interference function (Figs. 2.13 and 2.14), because diffraction from the myosin head arrays in the two halves of a myosin filament interferes as in Fig. 2.6b, but here with a very long spacing d. Here d is about 700 to 800 nm (Fig. 2.14). A great deal of experimental work and analysis of the M3 and M6 peaks has been carried out in recent years (see, for example, Dobbie et al., 1998; Linari et al., 2000; Irving et al., 2000; Bagni et al., 2001; Juanhuix, et al., 2001; Piazzesi et al., 2002; Reconditi et al., 2004; Ferenczi et al., 2005; Brunello et al., 2006; Griffiths et al., 2006; Huxley, et al., 2006a,b; Colombini et al., 2007; Oshima et al., 2007) and these authors have concluded that they can detect very precise movements of the lever arms of the myosin heads in different types of muscle experiment. A whole edifice of models of the contractile cycle was built up using this approach. The implication was that there was only one way to interpret the observations and therefore that this edifice must be right.

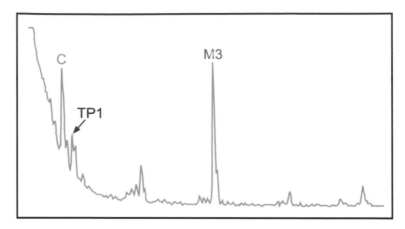

Figure 2.13 Profile of the meridian from an X-ray diffraction pattern of frog Sartorius muscle (adapted from Haselgrove, 1975). Apart from prominent peaks from myosin heads (M3), C-protein (C) and troponin (TP1), it is evident from the presence of fine sampling of all the intensity along the meridian that the meridian is modulated by interference functions arising from the two halves of the sarcomere.

Unfortunately, there was a fundamental problem in this analysis, as we showed (Knupp et al., 2009). The intensity changes in the M3 and M6 peaks that were observed in X-ray patterns from active muscle certainly include effects of interference between the two halves of the A-band, but they also include interference effects within one half A-band between myosin heads attached to actin with a roughly 14.3 nm axial periodicity and the 'OFF' myosin heads detached from actin and still showing the original 14.3 nm myosin filament repeat. Gradual axial shifting of these two populations could reproduce all of the observed changes of the M3 and M6 peaks, without involving any change in the shape of the myosin head at all (Fig. 2.14c). Therefore no reliable conclusion can be drawn on how the myosin heads change their configuration from a study of the M3 and M6 reflections alone. In our paper, we were not saying that shape changes in the myosin heads did not occur, rather that the M3 and M6 observations on their own provide little unambiguous information about any changes that might occur.

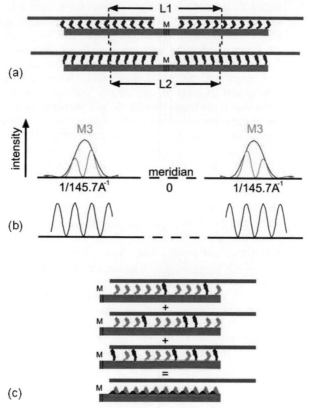

Figure 2.14 (a) Schematic representation of the two halves of a sarcomere, with the actin filament in blue, the myosin backbone in brown and the myosin heads in red. A change of configuration of the heads from a pre power stroke state to a post-power stroke state would bring the centre of mass of the heads closer together (from a distance L1 to a distance L2), as would a change in overlap of the myosin and actin filaments. (b) The M3 reflection (red) is sampled by the interference function (black), to give rise to the two sub-peaks (blue). The frequency of the sinusoidal interference function and therefore the intensity of the sampled M3 sub-peaks depends on the relative distance of the two arrays of heads on the two halves of the A-band (a). A change in this distance will cause the M3 sampling to be different, with different intensities and positions for the sub-peaks. By monitoring the changes of the M3 reflections during muscle shortening, it could be possible, in principle, to have

indirect evidence for how the myosin head lever arm swings. However, the lever arm swing would remain unproven if the M3 reflections changes could also be explained by an alternative model with no swing. (c) Alternative model for the role of the myosin heads during muscle contraction that does not require a measurable lever arm swing. In this model (Knupp et al., 2009), the electron density of several myosin heads from adjacent sarcomeres is summed together to give rise to a relatively featureless head distribution represented with Gaussians at the bottom of the panel. During muscle shortening the distribution of the heads attached to actin and that of the detached heads, move past each other, and this, along with the interference from the two halves of the sarcomere, can explain the changes of the M3 reflection remarkably well, although no specific allowance has been made for the way the lever arm moves. Since this alternative model, along with many other model variations, can explain the observations about the M3 and M6 changes, any firm conclusion about the way the myosin heads behave during muscle contraction cannot be drawn just from the M3 and M6 peaks, and further discriminatory evidence must be sought from other regions of the diffraction pattern. As shown in Table 2.2, the number of parameters needed to model the M3 and M6 intensity changes far exceeds the number of observations being fitted; the problem is under-determined.

It is easy to see why this conclusion must be true. Table 2.2 lists the various parameters which are needed in a realistic model to calculate what the changes in the M3 and M6 peaks might signify. What becomes immediately apparent is that the number of parameters involved to do a proper job is very large. But the M3 and M6 reflections, even when split into two peaks by interference effects, only provide about eight observations (the intensities and positions of each of the four component peaks). Since in the modelling the meridional intensities are all relative to each other, one can fix the intensity of one peak, so that only seven independent observations are available. Against this there are more than 13 parameters to be fitted (Table 2.2). As with modelling the equator, this meridional problem is also under-determined. As we showed in our paper (Knupp et al., 2009), with this many parameters there are very many different ways to fit the same observations and, unfortunately, simple unambiguous conclusions about the movement of the lever arm cannot be reached.

Table 2.2 Parameters defining the intensities of the M3 and M6 meridional reflections

Meridionals (M3 and M6)					
Observations			**Parameters**		
M3	I M3i	14.32	Population of head array on myosin	x	1
	I M3o		Distribution of head array on myosin (includes periodicity and degree of order)		>2
	d M3i		Shape of heads on myosin (motor domain angle, lever arm angle)		>2
	d M3o		Population of head array on actin	1-x	>2
			Distribution of head array on actin (includes periodicity and degree of order)		>2
M6	I M6i	7.16	Shape of head on actin (motor domain angle, lever arm angle)		>2
	I M6o		Relative positions of head arrays on actin and myosin		1
	d M6i		Contribution of backbone to M3 (includes weight and axial shift)		2
	d M6o		Length of the interference function		1
			Length of the bare zone		1
Observations	**8(−1)**		**Parameter set >**		**14**

2.5 The Full 2D Diffraction Pattern: Identifying Structural Mechanisms

From the kind of analysis in the last two sections, we have been led to the conclusion that to define properly the structural behaviour of the myosin heads in a contractile cycle it will be necessary to do model fitting against as many observations as possible; in other words model the whole of the low-angle 2D X-ray diffraction patterns from active muscles. This means taking the same full model of the muscle sarcomere, including all of our current knowledge of the structures of the myosin and actin filaments (Figs. 2.4 and 2.5) and the way that the heads might interact with actin, and optimising this model against all of the diffraction data that we have; the equator, the meridian and all of the layer lines. Ideas about the geometry of interaction of heads with actin, including the 3D factors involved, have already been put together into a computer program MusLabel (Squire and Knupp, 2004), and this has already shown itself to be useful in a number of muscle studies (Squire, 1992; Squire, et al., 2006; Luther et al., 2011).

In addition we have been exploring methods to model parts of the 2D low-angle diffraction data from both fish muscle and insect flight muscles (Hudson et al., 1997; AL-Khayat et al., 2003). The particular advantage of these two muscle types, unlike the muscles of frog or rabbit or any other higher vertebrate, is that they are particularly well organised in 3D (Figs. 2.15 and 2.16). This organisation arises from the fact that in bony fish muscles all of the myosin filaments are in rotational register around their long axes (Harford and Squire, 1986; Luther and Squire, 1980; Luther et al., 1981), whereas myosin filaments from higher vertebrates (frog, rabbit, chicken, humans, etc.) have myosin filaments in two different rotations around their long axes with the rotations distributed in a quasi-random manner (a statistical superlattice; Luther and Squire, 1980; 2014). Because of the randomness of the higher vertebrate lattices the layer lines are not broken up regularly into diffraction peaks, whereas the layer lines from fish and insect flight muscle are beautifully 'sampled'; they show discrete and well-defined peaks along the layer lines (Fig. 2.14; see Luther and Squire, 2014). What was done in this modelling, which in the first instance was applied solely to the arrangement of the myosin heads on the myosin filaments in resting muscle (Hudson et al., 1997; AL-Khayat and Squire, 2006), was to describe the possible myosin head shapes and positions in terms of a number of parameters (e.g. radius, tilt, slew, bend of the heads, etc.) and to calculate the diffraction pattern that such arrangements would produce. The parameters were then varied and the optimal set chosen in a simulated annealing process. The myosin head arrangement is the main contributor to the myosin layer lines in Fig. 2.8.

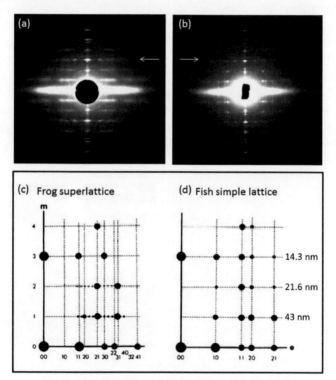

Figure 2.15 Comparison of the low-angle X-ray diffraction patterns from frog muscle (a,c) and bony fish muscle (b,d) showing the same layer lines from the myosin head arrangement (see Fig. 2.8), but the clearly different way in which these layer lines are sampled on the vertical row lines. The indexing of the peaks is shown in (c) and (d), which are equivalent to the top right hand quadrants of the diffraction patterns in (a) and (b) with the left hand vertical axis being the meridian (m). They are called reciprocal lattice rotation diagrams. (a) is from Huxley, H. E., and Irving, T. C., by permission and (b) is from data in Harford, J. J., and Squire J. M. (1986).

The program used was called MOVIE. What MOVIE does is (i) to set up the model using all the unambiguous data that is available, including the myosin head crystal structure (Rayment and Holden, 1993), (ii) to use properly stripped and corrected X-ray amplitudes using a program like FibreFix (Rajkumar et al., 2007) and then (iii) to compare the observed and calculated amplitudes in a simulated annealing procedure which gradually homes in on the best set of model parameters to explain the observations. We were able to get good fits to these layer lines from fish muscle using 18 parameters and 56 observations (i.e. the amplitudes of 56 independent X-ray layer line reflections). So, unlike modelling of the meridian and equator, this approach is well over-determined (there are many more observations than parameters to fit). Similar analysis was carried out on the low-angle diffraction patterns from insect flight muscle (Reedy, 1968) recorded by M. K. Reedy and his colleagues (Fig. 2.16; AL-Khayat et al., 2003).

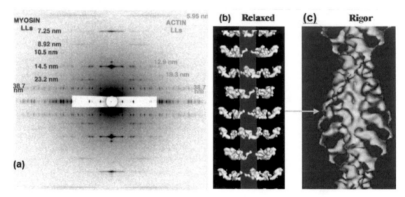

Figure 2.16 The beautifully sampled low-angle X-ray diffraction pattern from insect flight muscle (a), together with 3D modelling of the myosin head arrangement in resting muscle (b) and (c) is a reconstruction of heads on actin in rigor muscle. (b) is an example of modelling that can be done where the number of observations far exceeds the number of parameters to be fitted. Adapted from AL-Khayat et al. (2003) where there are further details.

What can be done in the future is to use the whole low-angle X-ray diffraction pattern, including not just the myosin layer lines, but the equator, the meridian and the actin layer lines as well, and model the 2D diffraction data taken through a time series like that in Fig. 2.12a, or through a complete cycle of oscillations of insect flight muscle (Perz-Edwards et al., 2011). This promises to be a very powerful approach to finding out what the myosin heads are actually doing in contracting muscle; there are many more observations than parameters to fit and the analysis can, in principle, be done right now with present experiments and computer programs. It is hard to identify any other technique which will give the same information.

2.6 Conclusion

In conclusion, X-ray diffraction methods applied to muscle are immensely powerful; they are fast, non-invasive, can be applied to living contracting muscle and patterns can be recorded with synchrotron radiation using fast detectors in milliseconds or less. But it is necessary to be wary in interpreting the observed diffraction data and to produce models of what is going on where the number of parameters required to describe a sensible model is smaller than the number of independent observations being fitted. There are tricks that can be used to study the equatorial reflections in a time-resolved manner (Fig. 2.11), and there is the potential soon to define myosin head movements in active muscle by modelling the whole low-angle X-ray diffraction pattern from contracting fish or insect flight muscles (using programs such as MOVIE), for example, during tetanic contractions or active oscillations. The future for the technique is very exciting.

Acknowledgements

We acknowledge support for our X-ray diffraction studies over the years from the MRC, the BBSRC, the EPSRC, the Wellcome Trust and the British Heart Foundation some of it through funding of the CCP13 computing project which was very fruitful. JMS is currently funded on BHF grant FS/14/18/3071. We also acknowledge the great contributions in earlier work of our colleagues Dr. Jeff Harford, Dr. Richard Denny, Dr. Hind AL-Khayat, Dr. Liam Hudson, Dr. Felicity Eakins and our collaborators, particularly Professors Michael Reedy and Tom Irving. We also acknowledge our debt to the late Professors Hugh Huxley and Gerald Elliott, who pioneered the field.

References

AL-Khayat, H. A., Hudson, L., Reedy, M. K., Irving, T. C., and Squire, J. M. (2003) Myosin head configuration in relaxed insect flight muscle: X-ray modelled resting cross-bridges in a pre-powerstroke state are poised for actin binding. *Biophys. J.,* **85**: 1063–1079.

Al-Khayat, H. A., Kensler, R. W., Squire, J. M., Marston, S. B., and Morris, E. P. (2013) Atomic model of the human cardiac muscle myosin filament. *Proc. Natl. Acad. Sci. U. S. A.,* **110**: 318–323.

AL-Khayat, H. A., and Squire, J. M. (2006) Refined structure of bony fish muscle myosin filaments from low-angle X-ray diffraction data. *J. Struct. Biol.,* **155**: 218–229.

Astbury, W. T. (1947) On the structure of biological fibres and the problem of muscle. *Proc. R. Soc. Lond. B Biol. Sci.,* **134**: 303–328.

Bagni, M. A., Colombini, B., Amenitsch, H., Bernstorff, S., Ashley, C. C., Rapp, G., and Griffiths, P. J. (2001) Frequency-dependent distortion of meridional intensity changes during sinusoidal length oscillations of activated skeletal muscle. *Biophys. J.,* **80**: 2809–2822.

Brenner, B., Yu, L. C., and Podolsky, R. J. (1984) X-ray diffraction evidence for cross-bridge formation in relaxed muscle fibers at various ionic strengths. *Biophys. J.,* **46**: 299–306.

Brunello, E., Bianco, P., Piazzesi, G., Linari, M., Reconditi, M., Panine, P., Narayanan, T., Helsby, W. I., Irving, M., and Lombardi, V. (2006) Structural changes in the myosin filament and cross-bridges during active force development in single intact frog muscle fibres: Stiffness and X-ray diffraction measurements. *J. Physiol.,* **577**: 971–984.

Colombini, B., Bagni, M. A., Cecchi, G., and Griffiths, P. J. (2007) Effects of solution tonicity on crossbridge properties and myosin lever arm disposition in intact frog muscle fibres. *J. Physiol.,* **578**: 337–346.

Cooke, R., Franks, K. (1980) All myosin heads form bonds with actin in rigor rabbit skeletal muscle. *Biochemistry,* **19**: 2265–2269.

Dobbie, I., Linari, M., Piazzesi, G., Reconditi, M., Koubassova, N., Ferenczi, M. A., Lombardi, V., and Irving, M. (1998) Elastic bending and active tilting of myosin heads during muscle contraction, *Nature,* **396**: 383–387.

Dominguez, R., Freyzon, Y., Trybus, K. M., and Cohen, C. (1998) Crystal structure of a vertebrate smooth muscle myosin motor domain and its complex with the essential light chain: Visualization of the pre-powerstroke state. *Cell,* **94**: 559–571.

Eakins, F., Pinali, C., Gleeson, A., Knupp, C., and Squire, J. M. (20154) Probing muscle myosin motor action: X-ray diffraction evidence for low force actin-attached cross-bridges in active bony fish muscle. (in preparation).

Elliott, G. F. (1960) Electron microscope and X-ray diffraction studies of invertebrate muscle fibres. PhD Thesis, University of London.

Elliott, G. F. (1964) X-ray diffraction studies on striated and smooth muscles. *Proc. R. Soc. Lond. B Biol. Sci.,* **160**: 467–472.

Elliott, G. F. (1967) Variations of the contractile apparatus in smooth and striated muscles: X-ray diffraction studies at rest and in contraction. *J. Gen. Physiol.,* **50**: 171–184.

Ferenczi, M. A., Bershitsky, S. Y., Koubassova, N., Siththanandan, V., Helsby, W. I., Panine, P., Roessle, M., Narayanan, T., and Tsaturyan, A. K. (2005) The "roll and lock" mechanism of force generation in muscle. *Structure,* **13**: 131–141.

Fourier, J. B. J. (1822), *Théorie Analytique de la Chaleur*, Paris: Chez Firmin Didot, père et fils.

Griffiths, P. J., Bagni, M. A., Colombini, B., Amenitsch, H., Bernstorff, S., Funari, S., Ashley, C. C., and Cecchi, G. (2006) Effects of the number of actin-bound S1 and axial force on X-ray patterns of intact skeletal muscle. *Biophys. J.,* **90**: 975–984.

Harford, J. J., Luther, P. K., and Squire, J. M. (1994) Equatorial A-band and I-Band X-ray diffraction from relaxed and active fish muscle: Further details of myosin crossbridge behaviour. *J. Mol. Biol.,* **239**: 500–512.

Harford, J. J., and Squire, J. M. (1986) The 'crystalline' myosin crossbridge array in relaxed bony fish muscles. *Biophys. J.,* **50**: 145–155.

Harford, J. J., and Squire, J. M. (1992) Evidence for structurally different attached states of myosin crossbridges on actin during contraction of fish muscle. *Biophys. J.,* **63**: 387–396.

Harford, J. J., and Squire, J. M. (1997) Time-resolved studies of muscle using synchrotron radiation. *Rep. Prog. Phys.,* **60**: 1723–1787.

Haselgrove, J. C. (1972) X-ray evidence for a conformational change in the actin filaments of vertebrate striated muscle. *Cold Spring Harbor Symp. Quant. Biol.,* **37**: 341–352.

Haselgrove, J. C. (1975) X-ray evidence for conformational changes in the myosin filaments of vertebrate striated muscle. *J. Mol. Biol.,* **92**: 113–143.

Haselgrove, J. C., and Huxley, H. E. (1973) X-ray evidence for radial cross-bridge movement and for the sliding filament model in actively contracting skeletal muscle. *J. Mol. Biol.,* **77**: 549–568.

Helliwell, J. R. (1992) *Macromolecular Crystallography with Synchrotron Radiation*. Cambridge University Press.

Hitchcock-DeGregori, S. E., and Irving, T. C. (2014) Hugh E. Huxley: The compleat biophysicist. *Biophys. J.*, **107**: 1493–1501.

Holmes, K. C., Popp, D., Gebhard, W., and Kabsch, W. (1990) Atomic model of the actin filament. *Nature,* **347**: 44–49.

Holmes, K. C., and Geeves, M. (2005) The molecular mechanism of muscle contraction. *Adv. Protein Chem.,* **71**: 161–193.

Hudson, L., Harford, J. J., Denny, R. J., and Squire, J. M. (1997) Myosin head configurations in relaxed fish muscle: Resting state myosin heads swing axially by 150 Å or turn upside down to reach rigor. *J. Mol. Biol.,* **273**: 440–455.

Huxley, H. E. (1953) X-ray analysis and the problem of muscle. *Proc. R. Soc. Lond. B Biol. Sci.*, **141**: 59–62.

Huxley, H. E. (1972) Structural changes in actin- and myosin-containing filaments during contraction. *Cold Spring Harbor Symp. Quant. Biol.,* **37**: 361–376.

Huxley, H. E., and Brown, W. (1967) The low-angle X-ray diagram of vertebrate striated muscle and its behaviour during contraction and rigor. *J. Mol. Biol.,* **30**: 383–434.

Huxley, H. E., Reconditi, M., Stewart, A., and Irving, T. (2006a) X-ray interference studies of crossbridge action in muscle contraction: Evidence from muscles during steady shortening. *J. Mol. Biol.,* **363**: 762–772.

Huxley, H. E., Reconditi, M., Stewart, A., and Irving, T. (2006b) X-ray interference studies of crossbridge action in muscle contraction: Evidence from quick releases. *J. Mol. Biol.,* **363**: 743–761.

Irving, M., Piazzesi, G., Lucii, L., Sun, Y. B., Harford, J. J., Dobbie, I. M., Ferenczi, M. A., Reconditi, M., and Lombardi, V. (2000) Conformation of the myosin motor during force generation in skeletal muscle. *Nat. Struct. Biol.,* **7**: 482–485.

Juanhuix, J., Bordas, J., Campmany, J., Svensson, A., Bassford, M. L., and Narayanan, T. (2001) Axial disposition of myosin heads in isometrically-contracting muscles *Biophys. J.,* **80**: 1429–1441.

Kabsch, W., Mannherz, H. G., Suck, D., Pai, E. F., Holmes, K. C. (1990) Atomic structure of the actin:DNase I complex. *Nature,* **347**: 37–44.

Knupp, C., Luther, P. K., and Squire, J. M. (2002) Titin organisation and the 3D architecture of the vertebrate striated muscle I-band. *J. Mol. Biol.,* **322**: 731–739.

Knupp, C., Offer, G., Ranatunga, K. W., and Squire, J. M. (2009) Probing muscle myosin Motor action: X-ray (m3 and m6) interference measurements report motor domain not lever arm movement. *J. Mol. Biol.*, **390**: 168–181.

Linari, M., Piazzesi, G., Dobbie, I., Koubassova, N., Reconditi, M., Narayanan, T., Diat, O., Irving, M., and Lombardi, V. (2000) Interference fine structure and sarcomere length dependence of the axial X-ray pattern from active single muscle fibers. *Proc. Natl. Acad. Sci. U. S. A.*, **97**: 7226–7231.

Liversage, A. D., Holmes, D., Knight, P. J., Tskhovrebova, L., and Trinick, J. (2001) Titin and the sarcomere symmetry paradox. *J. Mol. Biol.*, **305**: 401–409.

Lovell, S. J., Knight, P. J., and Harrington, W. F. (1981) Fraction of myosin heads bound to thin filaments in rigor fibrils from insect flight and vertebrate muscles. *Nature*, **293**: 664–666.

Luther, P. K., Munro, P. M. G., and Squire, J. M. (1981) Three-dimensional structure of the vertebrate muscle A-band III: M-region structure and myosin filament symmetry. *J. Mol. Biol.*, **151**: 703–730.

Luther, P. K., and Squire, J. M. (1980) Three-dimensional structure of the vertebrate muscle A-band II: The myosin filament superlattice. *J. Mol. Biol.*, **141**: 409–439.

Luther, P. K., and Squire, J. M. (2014) The intriguing dual lattices of the myosin filaments in vertebrate striated muscles: Evolution and advantage. *Biology*, **3**: 846–865.

Luther, P. K., Winkler, H., Taylor, K., Zoghbi, M. E., Craig, R., Padrón, R., Squire, J. M., and Liu, J. (2011) Direct visualization of myosin-binding protein C bridging myosin and actin filaments in intact muscle. *Proc. Natl. Acad. Sci. U. S. A.*, **108**: 11423–11428.

Lymn, R. W. (1978) Myosin subfragment-1 attachment to actin. Expected effect on equatorial reflections. *Biophys. J.*, **21**: 93–98.

Lymn, R. W., and Taylor, E. W. (1971) Mechanism of adenosine triphosphate hydrolysis by actomyosin, *Biochemistry*, **10**: 4617–4624.

Orzechowski, M., Li, X. E., Fischer, S., and Lehman, W. (2014) An atomic model of the tropomyosin cable on F-actin. *Biophys. J.*, **107**: 694–699.

Oshima, K., Takezawa, Y., Sugimoto, Y., Kobayashi, T., Irving, T. C. and Wakabayashi, K. (2007) Axial dispositions and conformations of myosin crossbridges along thick filaments in relaxed and contracting states of vertebrate striated muscles by X-ray fiber diffraction. *J. Mol. Biol.*, **367**: 275–301.

Parry, D. A. D., and Squire, J. M. (1973) Structural role of tropomyosin in muscle regulation: Analysis of the x-ray diffraction patterns from relaxed and contracting muscles. *J. Mol. Biol.,* **75**: 33–55.

Paul, D. M., Squire, J. M., and Morris, E. P. (2016) Relaxed and active thin filament structures reveal a new regulatory mechanism: Varying tropomyosin shifts within a regulatory unit. (submitted).

Perz-Edwards, R. J., Irving, T. C., Baumann, B. A., Gore, D., Hutchinson, D. C., Kržič, U., Porter, R. L., Ward, A. B., and Reedy, M. K. (2011) X-ray diffraction evidence for myosin-troponin connections and tropomyosin movement during stretch activation of insect flight muscle. *Proc. Natl. Acad. Sci. U. S. A.,* **108**: 120–125.

Piazzesi, G., Reconditi, M., Linari, M., Lucii, L., Sun, Y. B., Narayanan, T., Boesecke, P., Lombardi, V., and Irving, M. (2002) Mechanism of force generation by myosin heads in skeletal muscle, *Nature,* **415**: 659–662.

Rajkumar, G., AL Khayat, H. A., Eakins, F., Knupp, C., and Squire, J. M. (2007) The CCP13 *FibreFix* program suite: Semi-automated analysis of diffraction patterns from non-crystalline materials. *J. Appl. Cryst.,* **40**: 178–184.

Rayment, I., and Holden, H. M. (1993) Three-dimensional structure of myosin subfragment-1: A molecular motor. *Science,* **261**: 50–58.

Reconditi, M., Linari, M., Lucii, L., Stewart, A., Sun, Y. B., Boesecke, P., Narayanan, T., Fischetti, R. F., Irving, T., Piazzesi, G., Irving, M., and Lombardi, V. (2004) The myosin motor in muscle generates a smaller and slower working stroke at higher load. *Nature,* **428**: 578–581.

Reedy, M. K. (1968) Ultrastructure of insect flight muscle. I. Screw sense and structural grouping in the rigor cross-bridge lattice. *J. Mol. Biol.,* **31**: 155–176.

Ribeiro Ede, A. Jr, Pinotsis, N., Ghisleni, A., Salmazo, A., Konarev, P. V., Kostan, J., Sjöblom, B., Schreiner, C., Polyansky, A. A., Gkougkoulia, E. A., Holt, M. R., Aachmann, F. L., Zagrović, B., Bordignon, E., Pirker, K. F., Svergun, D. I., Gautel, M., and Djinović-Carugo, K. (2014) The structure and regulation of human muscle α-actinin. *Cell,* **159**: 1447–1460.

Sawyer, L. (1986) Synchrotrons and approaches to the phase problem. *Biochem. Soc. Trans.,* **14**: 535–538.

Selby, C. C., and Bear, R. S. (1956) The structure of actin-rich filaments of muscles according to X-ray diffraction. *J. Biophys. Biochem. Cytol.,* **2**: 71–85.

Squire, J. M. (1981) *The Structural Basis of Muscular Contraction*. Plenum Press, New York.

Squire, J. M. (1986) *Muscle: Design, Diversity and Disease*. Benjamin Cummings, Menlo Park, California.

Squire, J. M. (1992) Muscle filament lattices and stretch-activation: The match/mismatch model reassessed. *J. Muscle Res. Cell Motil.*, **13**: 183–189.

Squire, J. M. (1997) Architecture and function in the muscle sarcomere. *Curr. Opt. Struct. Biol.*, **7**: 247–257.

Squire, J. M. (2013) Obituary: Professor Gerald Elliott. *J. Muscle Res. Cell Motil.*, **34**: 429–436.

Squire, J. M., Al-Khayat, H. A., Knupp, C., and Luther, P. K. (2005) 3D molecular architecture of muscle. In: *Muscle & Molecular Motors* (Squire, J. M., and Parry, D. A. D., eds.), Advances in Protein Chemistry, 71: 17–87.

Squire, J. M., Bekyarova, T., Farman, G., Gore, D., Rajkumar, G., Knupp, C., Lucaveche, C., Reedy, M. C., Reedy, M. K., and Irving, T. C. (2006) The myosin filament superlattice in the flight muscles of flies: A-band lattice optimisation for stretch-activation? *J. Mol. Biol.*, **361**: 823–838.

Squire, J. M., and Knupp, C. (2004) MusLABEL: A program to model striated muscle a-band lattices, to explore crossbridge interaction geometries and to simulate muscle diffraction patterns. *J. Muscle Res. Cell Motil.*, **25**: 423–438.

Squire, J. M., and Knupp, C. (2005) X-ray diffraction studies of muscle. *Adv. Protein Chem.*, **71**: 195–255.

Trinick, J. A. (1981) End-filaments: A new structural element of vertebrate skeletal muscle thick filaments. *J. Mol. Biol.*, **151**: 309–314.

Vinogradova, M. V., Stone, D. B., Malanina, G. G., Karatzaferi, C., Cooke, R., Mendelson, R. A., and Fletterick, R. J. (2005) Ca^{2+}-regulated structural changes in troponin. *Proc. Natl. Acad. Sci. U. S. A.*, **102**: 5038–5043.

von der Ecken, J., Müller, M., Lehman, W., Manstein, D. J., Penczek, P. A., and Raunser, S. (2015) Structure of the F-actin-tropomyosin complex. *Nature*, **519**: 114–117.

Whitby, F. G., and Phillips, G. N. Jr. (2000) Crystal structure of tropomyosin at 7 Angstroms resolution. *Proteins*, **38**: 49–59.

Yu, L. C., Steven, A. C., Naylor, G. R., Gamble, R. C., and Podolsky, R. J. (1985) Distribution of mass in relaxed frog skeletal muscle and its redistribution upon activation. *Biophys. J.*, **47**: 311–321.

Chapter 3

Muscle Contraction Revised: Combining Contraction Models with Present Scientific Research Evidence

Else Marie Bartels

The Parker Institute, Copenhagen University Hospital Bispebjerg and Frederiksberg, 2000 Frederiksberg, Denmark

else.marie.bartels@regionh.dk

Striated muscle contraction is needed for survival in most animals, humans included. Contraction is described in part by various models, building on mechanics, biochemistry, protein chemistry and mathematics. Although there is some truth and scientific evidence in all of these, a full-scale merger of it all is still missing.

Grounded in the fact that a muscle consists of living interacting structured proteins in a predefined environment, which is perturbed at the start and completion of a contraction, I intend to look at the combined anatomical, mechanical, physical-chemical and biochemical findings. My overall aim is to attempt to create a closer to real life model which will describe the contraction of the striated muscle cell.

Muscle Contraction and Cell Motility: Fundamentals and Developments
Edited by Haruo Sugi
Copyright © 2017 Pan Stanford Publishing Pte. Ltd.
ISBN 978-981-4745-16-1 (Hardcover), 978-981-4745-17-8 (eBook)
www.panstanford.com

Based on the premise that ion movements may be the key to a more complete understanding of the contractile processes, a model has been suggested. The model is based on general principles for biological systems, following physical-chemical laws, and considering effects of smaller ions on protein charges and conformational changes in the contractile unit. This unit, consisting of collaborating partners of giant proteins and smaller proteins inter-twined in a complicated structured network, is able to produce force repeatedly in a controlled way, using energy from ATP breakdown. The interaction of the elements in the contractile unit necessary to fulfil this objective is described.

The aforementioned model is able to explain data on muscle contraction to a large extent. However, the model proposed here should not be seen as the final truth, and it is still important to keep one's mind open for future discoveries and to continue to ask questions when diverging results present themselves.

3.1   Introduction

With muscle being a major tissue in all animals, and with the importance of this tissue to keep the animal alive, muscle has always been a focus of interest, take for example the very early interpretation of muscle function by Galen (AD 129–200) with his 'balloonist theory' (Hodgson, 1990; Pearn, 2002), or Vesalius's (1543) very detailed anatomical descriptions of the human muscular-skeletal system, and Galvani (1791) showing that a muscle contracts upon electrical stimulation.

What characterizes a muscle is the ability to contract and relax and thereby create movement and force where required. Being a living tissue, this raises the problem about transformation of metabolic energy to create a mechanical force, and the problem about repeatable structural changes between the contracted and the relaxed muscle states, and repeatable and controlled changes of the environment around the involved proteins. This leads further to the three aspects which must be considered in order to understand the changes from one condition to the other: (1) the structure of the contractile system and the alterations that happens to these during the contractile process, (2) the systems producing the energy involved in the contractile process, and (3) the process by which the chemical energy is transferred into force production (Sten-Knudsen, 1953).

Since force production, and thereby mechanical energy output, is the unique feature of the muscle cell, several mechanically based models have been used to try to explain muscle contraction, where many build on A. V. Hill's model and this model's consideration of the viscoelastic elements in a muscle (Hill, 1938). Hill also wished to explain the link between observed energy consumption and recorded force and heat production, as was previously investigated by Fenn (1923, 1924).

When electron microscopy and X-ray diffraction gave a better insight into the very structured contractile system (Huxley, 1953; Huxley et al., 1965; Huxley and Niedergerke, 1954; Elliott, 1960), mechanical models based on the internal cell structures evolved (Huxley, 1957; Huxley and Hanson, 1954). The 'sliding filament' theory is still the most well-known model when considering muscle contraction and when teaching muscle physiology (Huxley, 1957), and this model, mainly a mechanical model, has been the basis for the development of some of the biochemical interpretations of the processes of ATP-hydrolysis behind muscle contraction (Taylor, 1977). Over the years this model has been further developed (i.e. Ma and Taylor, 1994) to include structural changes of the involved myosin heads (Trayer and Trayer, 1988; Levine et al., 1991), and various conformational states between the relaxed, the contracted, and the rigor state. Being a fairly simple, but in many ways very successful, description of the contractile process, the sliding filament theory does not take the complete internal and external cell environment into consideration, and some of the later discovered muscle proteins, which may be regulatory in the contraction process, were not known at the time of the development of the model, and the contribution of these to contraction is therefore not included (Huxley, 1957). The interesting thing is though, that the existence of a protein like titin (Maruyama, 1994; Trinick, 1996; Kontrogianni-Konstantopoulus et al., 2009) was predicted by Huxley (1957), yet it was not incorporated into the model.

One of the main issues in these more mechanistic models is that the proteins are not considered as the charged structures, proteins in general are in a physiological environment, and the interaction with the environment is only seen as enzymatic reactions in connection with ATP splitting to get the required energy to start a contraction. The biochemical aspects of muscle research did though try to explain a change in involved structures as being due to changes in the internal environment, especially created by ATP binding and splitting, and phosphate binding and release (Harrington and Himmelfarb, 1972; Levine et al., 1991; Stepkowski, 1995; Geeves and Holmes, 1999). Such changes in internal structures of the contractile system have been known to occur for decades as shown by X-ray diffraction (Elliott et al., 1963; Huxley, 1968, 1973; Rome, 1972; Bartels and Elliott, 1985), and on fractions of the involved proteins like the myosin head (Trayer and Trayer, 1988; Rüegg, 2002), and on proteins in the I-filament (Poole et al., 2006).

Various electrostatic models have existed for many years (Davies, 1963; Elliott, 1968; Dragomir, 1970; Elliott and Worthington, 1994) and the importance of the present ionic strength and present ions have amongst others been shown by Kawai and Brandt (1976), Bartels and Elliott (1988), Veigel et al. (1998) and Linari et al. (1998), but somehow the merger of mechanistic, metabolic, biochemical and biophysical models has been a difficult process, probably due to the difficulty of the task. The inspiration to start from the basic process, behaviour of charged molecules like proteins and small ions in a physiological environment, has in some cases like ourselves (Elliott et al., 1986) come from Szent-Györgyi's (1942) remark: "I was always led in research by my conviction that the primitive, basic functions of living matter are brought about by ions, ions being the only powerful tools which life found in the sea water where it originated. Contraction is one of the basic primitive functions".

To look into the contractile system from a more overall biophysical point of view does not leave out considering the many studies looking at a particular detail in the contractile process in a simpler, artificial, laboratory setup with only part of the system localized, and with perturbations of the system which would never take place in real life. Many in vitro studies working with either single protein molecules, or only part of the proteins, or two-dimensional protein layers, are easier to interpret from a simple sliding filament model combined with the Lymn–Taylor scheme for the energy cycle for ATP breakdown and build-up, which would follow the cross-bridge cycle and pass through a 'rigor state' (Lymn and Taylor, 1970). These systems are interesting and may reveal some of the possible functions of the involved proteins, but the proteins are taken out of their environment to a degree that may change their conformation, and thereby the model contractile system, beyond any likeness with a real life muscle cell, and it may for this very reason be worthwhile looking at other interpretations of the described observations. Furthermore, the recent studies by Sugi et al. (2014) demonstrate clearly that the more simple in vitro systems do not compare with the three-dimensional real-life contractile system. It is therefore time to start to put together the puzzle of these many new observations in an attempt to describe the complete real-life system.

My aim with this chapter is to bring together the observed behaviour of the contractile system and the proteins connected with the 'classical' contractile system (Huxley, 1957) and look at the system in the changing internal environment through the physiological states of the muscle cell. Stimulated by a recent review with the title 'to understand muscle you must take it apart' (Batters et al., 2014), I wish to follow the advice of my old PhD supervisor, Professor Ove Sten-Knudsen: 'if you take things apart and look at a minute corner of a biological process, you must always bring the animal together again to see if your findings make sense'. With many of my colleagues in the muscle world, myself included, having taken the muscle apart and looked at small details, the process of putting it all together should hopefully lead to a more true-to-life physiological model of the contraction of striated muscle, although it will still only ever be a model. Yet, such a model may finally bring us towards such an understanding that we can start to apply the collective work to physiological and clinical problems.

3.2 Findings and Facts That Must Be Part of—or Explained by—a Model for Contraction

The protein structures of a muscle cell and its intra- and extra-cellular environment are all known factors which interact with each other in predictable ways. The changes in dimensions and structure of the elements in the contractile unit during change in environment have, as mentioned above, been described in many settings. Similarly, the mechanical force production (Jewell and Wilkie, 1960; Edman, 1966; Gordon et al., 1966; Matsumoto, 1967) and the development and behaviour of the latency relaxation (Sandow, 1946; Bartels et al., 1976, 1979, 1982; Haugen and Sten-Knudsen, 1976; Haugen, 1983; Yagi, 2011) are well described. Furthermore, the metabolic energy need and production during rest and contraction (Fenn, 1924; Hill, 1938; 1953, 1964a–c; Carlson et al., 1963; Linari and Woledge, 1995; Lou et al., 1998; West et al., 2004; Barclay et al., 2010) are known in various conditions and situations. Stiffness and compliance of the system are also well-described (Sten-Knudsen, 1953; Carlsen et al., 1961; Kawai and Brandt, 1976, 1980; Hatta et al., 1984; Tsuchiya et al., 1993; Trombitas and Pollack, 1993; Bartoo et al., 1997; Linke et al., 1997; Mantovani et al., 1999; Kobayashi et al., 2004; Coomber et al., 2011; Altman et al., 2015; Rassier et al., 2015), as are electric charge changes on the involved protein structures with these other described changes (Sarkar, 1950; Collin and Edwards, 1971; Elliott and Bartels, 1982; Godt and Baumgarten, 1984; Elliott et al., 1984; Aldoroty et al., 1985; Naylor et al., 1985; Bartels and Elliott, 1985; Bartels et al., 1993).

3.2.1 Structure of the Contractile Apparatus

3.2.1.1 General structure

The regularity of skeletal muscle is unique. The contractile system may in connection with X-ray diffraction measurements be considered a liquid crystal (Huxley, 1953; Elliott, 1960; Elliott, 1973; Aldoroty et al., 1985). Since the structure is affected by change in the ionic environment, changes in distances can be seen when changing pH or external ionic environment. This is due to ion binding or release of bound ions to the proteins in the filaments (Sarkar, 1950; Bartels et al., 1993; Coomber et al., 2011). When looking at changes between the physiological states, changes in distance between filaments and change in mass distribution between filaments can also be seen (Huxley and Hanson, 1954; Elliott et al., 1963; Huxley et al., 1965; Huxley, 1973). It was this observed regularity in the X-ray diffraction recordings, together with electron microscopy images showing the A- and the I-filaments, that lead to the sliding filament theory, which clearly presents some of the events in a muscle contraction, but which does not consider the processes behind changes in protein conformation and filament distances seen when changing the environment and/or physiological state (Elliott and Bartels, 1982; Bartels and Elliott, 1983; Naylor et al., 1985; Bartels and Elliott, 1988; Sugi et al., 2014). Protein conformation changes are here considered as being changes in the secondary or tertiary structure.

The discovery of the structures and location of the sarcoplasmic reticulum and the T-tubules shed light on the pre-contraction stimulation of the contractile system, the excitation-contraction coupling (Sandow, 1952; Jöbsis and O'Connor, 1966; Winegrad, 1968; Ashley and Ridgway, 1968; von Wegner et al., 2007; Calderón et al., 2014), and the regulation of the intercellular environment concerning Ca^{2+} following the contractile processes (Caputo et al., 1994, 1999, 2004; von Wegner et al., 2007; Calderón et al., 2014), but has left the processes behind force production less well explained at the molecular level. Furthermore, the more detailed regulatory processes of released Ca^{2+} have first lately been further interpreted (Coomber et al., 2011). As part of the force development there is additionally the unexplained phenomenon of latency relaxation (Sandow, 1946; Bartels et al., 1976, 1979, 1982; Haugen and Sten-Knudsen, 1976; Haugen, 1983; Yagi, 2011), which is somewhat ignored in the sliding filament theory, although well described and recorded.

This chapter will consider the contractile unit as being the sarcomere.

A schematic diagram of the sarcomere is shown Fig. 3.1.

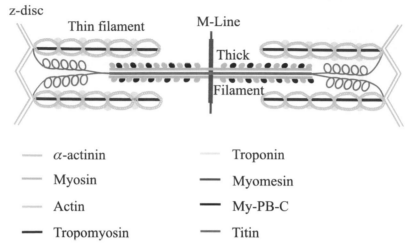

Figure 3.1   Schematic of the sarcomere.

3.2.1.2   Proteins making up the contractile unit

When looking at what may be described as the 'contractile apparatus' consisting of a series of sarcomeres, the number of involved proteins has been growing with better extraction methods and more understanding of the biochemical processes. The now very complex contractile unit, consisting of the A- and I-filaments and the sarcomeric cytoskeleton, therefore comprises myosin, actin, tropomyosin, troponin, titin, M-line protein, C-protein and α– and β–actinin, as well as nebulin, obscurin and myomesin, and some minor proteins (Huxley, 1957; Ebashi et al., 1969; Gautel, 2011).

3.2.1.2.1   The A-filament

Myosin, the protein forming the A-filaments in skeletal muscle, has a well-described structure with the helical structured rod part consisting of a double helix, light meromyosin (LMM) and the S-2 part of heavy meromyosin (HMM), and two globular heads making the S-1 part of HMM with two arms, where four light chains are attached (Dreizen et al., 1967). At each arm there is a regulatory and an essential light chain, and the regulatory function involving ATP hydrolysis and phosphate release involving the ATPase in the head section of myosin is well described (Szczesna et al., 2002; Nieznanski et al., 2003; Duggal, 2014), as is the conformational changes related to this (Trayer et al., 1987; Holmes, 1996; Malnasi-Csizmadia et al., 2007; Duggal, 2014). Parallel lined up myosin filaments can be created as a pure myosin gel of similar myosin concentration to that found in a sarcomere (Harrington and Himmelfarb, 1972; Cooke et al., 1987). An important function of myosin is as an ATPase, and ATPase sites exist both at the head end and at the rod end of the molecule, where the enzyme most active at physiological pH is the one at the head end of the molecule (Harrington and Himmelfarb, 1972). Myosin filaments are shown to bind ions at physiological conditions, and a change in ion binding is probably the main cause of the change in fixed electric charge on the filament between rigor, relaxed and contracted muscle states (Naylor et al., 1985; Bartels et al., 1985, 1993; Bartels and Elliott, 2007). An interesting finding here is that the charge change from the rigor to the relaxed state already takes place at an ATP concentration of 100 to 200 µM. Compared to the I-filament, a larger charge change takes place between relaxation and rigor on the thick filaments, and this change occurs whether or not the myosin heads are cross-linked to the thin filaments (Bartels et al., 1985, 1993). The charge changes seen in the aforementioned studies are confirmed by the studies of Scordilis (1975). The charge change seen in the A-band during contraction is like that of rigor, a more negative charge than that of the relaxed state, and the contracted state is even more negative than the rigor state (Bartels and Elliott, 2007).

Along the length of the A-filaments, myosin-binding protein C (My-BP-C) is regularly arranged, binding to the backbone of the myosin. A projecting part of the My-BP-C may interact with the actin filaments in a resting muscle fibre (Squire et al., 1982, 2003; Luther et al., 2011) since My-BP-C is shown in vitro to be able to bind to both myosin and actin (van Dijk et al., 2014). The My-BP-C areas of the A-filaments have a slightly larger periodicity than the myosin heads, and when looking at the myosin crown repeats, the My-BP-C regions are perturbed while the regions with no My-BP-C binding are regular, giving a different type of interaction between myosin heads in the perturbed regions in resting muscle (Oshima et al., 2012).

3.2.1.2.2 The I-filament

The I-filament consists of actin, tropomyosin, and troponin, with a double-stranded F-actin helix as the outer boundary, an embedded tropomyosin consisting of two α-helical sub-units, and the globular troponin, consisting of troponin C, I and T units, attached at the crossing point for the actin strands (Hanson and Lowy, 1964; Ebashi et al., 1969). The complex of troponin and tropomyosin forms the structure that steers the initiation of the contraction via calcium binding, the 'calcium switch' (Squire, 1981; Moir et al., 1983). The charge change on the I-filament between the various physiological states is not as large as that seen for the A-band, but the charge concentration in the relaxed and the contracted states follows the one recorded in the A-band (Bartels et al., 1985; Bartels and Elliott, 2007).

β-actinin is also present at the end of the filaments as a capping protein (Maruyama, 1977, 1990; Funatsu et al., 1988), as are tropomodulins (Yamashiro et al., 2012), all of which contribute to keep the filament length and structure constant.

3.2.1.2.3 The sarcoplasmic reticulum

The sarcoplasmic reticulum with its involved proteins (Porter and Palade, 1957; Franzini-Armstrong, 1970) and the function of the organelle during excitation-contraction (Ebashi, 1976; Treves et al., 2009; Rebbeck et al., 2014) has been described in detail, and the complete excitation-contraction coupling will not be part of the further discussions on models for force production. It is worth mentioning, however, that when looking at a skinned muscle preparation with an intact sarcoplasmic reticulum and a glycerinated preparation of the same muscle, both the A- and the I-band of the skinned muscle is more negatively charged than the same regions in the glycerinated preparation (Bartels and Elliott, 1982, 1985). This is believed to be due to a change in protein charge on the active sarcoplasmic reticulum, when ATP is added. In the skinned muscle, the sarcoplasmic reticulum is preserved and runs parallel to the filaments throughout the sarcomere, while it is broken down and washed out of the glycerinated muscle preparations. A confirmation of the differences in measured charges in the two muscle preparations is to be found from densely packed pellets of sarcoplasmic reticulum, which show a similar charge change when changed from a rigor solution to a relaxing solution containing 2.5 mM ATP (Bartels et al., 1987).

3.2.1.2.4 The sarcomeric cytoskeleton

Apart from the sarcomeric structures which are considered as being the classical 'contractile system' consisting of the A- and the I-filament, the importance of the sarcomeric cytoskeleton, as a part of the protein structures participating in the contractile process and maintenance of the well-organised contractile apparatus, has for many years become more and more apparent both as a mechanical component and as a signalling system (Agarkova and Perriard, 2005; Kontrogianni-Konstantopoulos et al., 2009; Gautel, 2011; Meyer and Wright, 2013). The cytoskeleton of the sarcomere is present in parallel to and around the contractile filaments, as well as being directly linked to these, and must therefore be taken into account when considering contraction and force production in the sarcomere. There are the two main zones, the Z-disc and the M-line placed vertically to the horizontal filament direction. The Z-disc is the anchoring point for the actin filaments via α-actinin which is also bound to titin via telethonin (Kontrogianni-Konstantopoulos et al., 2009; Gautel, 2011). The M-line shows a network of titin bound to either obscurin or obscurin-like 1 (OBSL1) and forms a quaternary complex together with myomesin and myosin, thereby keeping the participating structures in place (Agarkova and Perriard, 2005; Pernigo et al., 2010).

Another component of the sarcomeric cytoskeleton is nebulin which is tightly associated with the I-filaments and associated at either end with the actin filament capping proteins (Kontrogianni-Konstantopoulos et al., 2009; Meyer and Wright, 2013).

Looking at the sarcomeric cytoskeleton, the particularly interesting protein from a contraction/relaxation point of view is titin, which stretches from the Z-disc to the M-line and is therefore the only component constantly spanning over the whole sarcomere, even when stretched beyond overlap. Titin is therefore the main component of the gap filaments seen in overstretched muscle. Being the third most abundant protein in the contractile unit, with a length over 1 µm, this modular giant protein consisting of multiple immunoglobulin (Ig) domains and fibronectin type III domains (FnIII), as well as some special highly elastic regions and two protein kinase domains, has all what can be asked for in terms of possibilities for interaction with the rest of the sarcomeric proteins and signalling along the whole length of the sarcomere. The Ig domains are binding sites for proteins and appear along the entire length (Tskhovrebora and Trinick, 2003; Lange et al., 2006; Benian and Mayans, 2015). Furthermore, titin is believed to be the main player in mechano-sensing through the sarcomere, and the kinase in the M-band is believed to play a role in this (Tskhovrebora and Trinick, 2008; Gautel, 2011; Meyer and Wright, 2013). The question has recently been raised as to whether the kinase does not act as a kinase, but as a pseudo-kinase, by acting as a scaffold for assembly of signalling complexes (Bogomolovas et al., 2014). Further descriptions of structure and physiology of titin can be found in Fukuda et al. (2008) and Meyer and Wright (2013).

When looking at the gap filaments/titin at a sarcomere length with no overlap, the electric charge on titin appears to be a function of the calcium concentration, with a more negative charge at a lower calcium concentration, and showing a sharp transition at a pCa of 6.8. This indicates that titin is the component controlling the calcium dependence along the whole length of the sarcomere (Coomber et al., 2011). The Hill coefficient for this change was in this study calculated to be at least 3.4, which is in agreement with the observation that the Hill coefficient for calcium activation of force was found to be much larger than what could be explained by a two-ion calcium switch when calcium binds to troponin C (Boussouf and Geeves, 2008). The important region of titin could in this connection be the unique PEVK region (Linke et al., 1998), which has phosphorylation sites and which in heart muscle has been suggested to be the region regulating the Frank-Starling behaviour of heart muscle (Fukuda, 2010), possibly by its phosphorylation sites. If these sites are regulated by pCa, this may affect the ion binding along the molecule. The PEVK region has earlier been suggested as being the Ca-regulating site in muscle, by both Labeit et al. (2003) and Joumaa et al. (2008).

3.2.2   The Internal Environment in a Muscle Cell

The intercellular environment in a muscle cell can be considered to be the intercellular free solution containing the ions. Under physiological conditions this will be a solution with a pH of around 7.0–7.4 and concentrations of the main smaller ions will typically be around 140 mM for K^+, 9.2 mM for Na^+, and 3–4 mM for Cl^- (Boyle and Conway, 1941). The rest of the anions balancing the small-ion positive charges are organic phosphates and amino acids, which will generally be part of the proteins. Apart from this basic solution, there will be a varying concentration of Ca^{2+} in connection with initiation of contraction and following relaxation, and a small amount of Mg^{2+}, where the actual free concentration will be low. In reality this is not quite so simple. With the large proteins present in the structured sarcomere, and with these proteins forming a network of charged side chains arranged along the polypeptide chains, there is a possibility of many 'Saroff sites' ready for absorption of small ions along the muscle filaments and larger structures like titin (Loeb and Saroff, 1964). Binding of smaller ions will take place by hydrogen bonding to these sites (Elliott et al., 1986), and this is supported by the electric charge changes seen with a change of ionic strength and change in pH (Naylor et al., 1985). Any binding of calcium, ATP or phosphates will change the conformation of the protein at the binding site and further along the protein, and this may cause a change of Saroff sites in the vicinity, leading to either stripping off bound ions or opening of new sites for ion binding. Since the effect is cooperative between the sites, the effect of ion binding or release may reach far along the filament.

When setting up an experiment in vitro with systems like skinned or glycerinated fibres, or pure protein experiments, one must be aware of when the system is perturbed away from the actual physiological possibilities. With an open system like the aforementioned one, there will still be conditions where the Planck solution must continue to apply, and the whole system may be considered a Donnan regime (Sten-Knudsen, 2002), but it may not be comparable to the muscle cell in vivo. In all cases, ionic strength, osmolarity, pH and the actual concentration of the present ions must be taken into account. An effect of such non-physiological setups, choosing a Ringer solution with high chelating agents (EGTA or EDTA), has already been demonstrated in Matsubara et al.'s (1984) X-ray studies which diverged from data measured in more physiological conditions, and also in the Kawai and Brandt's studies with 'High' and 'Low' rigor (Kawai and Brandt, 1976). A more in depth explanation of the findings in the aforementioned studies can be found in Bartels and Elliott (1988) and Coomber et al. (2011), where the change in electric charges on the filaments separate from the possible changes under physiological conditions can justify both Matsubara's X-ray data and the two rigor conditions.

At rest, the internal environment in the sarcomere will be at a steady state, but any perturbation of the system like temperature, change in pH, ionic strength, or small ions may affect conformation and electric charge of the proteins and thereby behaviour of the contractile system (Ranatunga, 2015; Naylor et al., 1985). With calcium release during stimulation, charge changes on the A- and I-filaments are demonstrated (Bartels and Elliott, 2007), and the system is therefore perturbed to enable contraction.

In physiological rigor many changes take place. The defined in vitro rigor solutions are clearly described, as are the different muscle preparations used to illustrate rigor, but in vitro rigor must always be considered an interesting non-physiological condition which will help to explain the behaviour of the many components in the contractile unit. It is an important way of getting into the understanding of a complicated system, it gives a window into the mechanisms, but like all in vitro systems, it may also lead us astray.

Since the chapter is concerned with the contractile unit, this is not the place to extend to the effects of changes in a muscle's external environment.

3.2.3   Energy Consumption During Contraction

When looking at energy demands during contraction, and the consideration of a muscle as a machine, there are two basic effects which seem to be fulfilled in all types of muscle, and even in actomyosin gels. A muscle follows the force velocity equation set out by A. V. Hill (1938), and it shows the Fenn effect, where extra energy is consumed when a muscle shortens against a load, and this extra energy is proportional to the extra work carried out by the muscle. The phenomenon is shown as extra heat release (Fenn, 1923, 1924). A recent study has used magnetic resonance spectroscopy to confirm the Fenn effect in muscle (Ortega et al., 2015).

When considering energy consumption in a muscle cell in connection with molecular models describing the underlying process for contraction, it is valid only to consider ATP as the source of energy.

3.2.3.1 ATP consumption and ATPase rates during contraction

The scheme for actomyosin interaction and ATP binding and consumption during the contractile process, originally suggested by Lymn and Taylor (1970), has been discussed in many contexts. Even when not mentioned, it is always assumed that the binding site in question is the binding site on the globular domain of the myosin head which contains the actin binding site and the ATP binding site with the ATPase (Taylor, 1977; Levine et al., 1991). The binding and release of phosphate will involve small conformational changes at the site, which will give rise to larger displacements of more distant segments (Levine et al., 1991). Such a movement of the myosin head was demonstrated by Sugi et al. with the very elegant Hydration Chamber technique, where the addition of ATP showed movement of myosin heads away from the bare region of the A-filament and a different polarity on the two sides of the bare region, while ADP had no effect (Sugi et al., 2008). It is worth noticing that the equilibrium constant for the initial ATP binding step is dependent on ionic strength, which according to Sugi et al. is entirely expected due to the fact that we are dealing with a charged substrate. Furthermore, the rate constants for the ATP binding and dissociation are large, and under physiological conditions it seems that the rate-limiting step is the product release (Ma and Taylor, 1994).

This scheme is very much interpreted in connection with the original sliding filament theory where all cross-bridges work together in unison. Other studies described below have shown that the whole process of involvement of cross-bridges and other protein structures in the sarcomere in connection with contraction and relaxation must be reconsidered.

In unloaded myofibrils, following a quick initial burst of ATP consumption, the rate of ATP usage has been found to be uniform during the actual shortening process, implicating that the ATPase rate does not change during shortening, while the rate does decrease to a lower, but steady value, when shortening has ended. The first energy burst appears as an inorganic phosphate burst from the release of protein-bound phosphate. When looking at the total ATP used during shortening, it was not dependent on the extent of shortening—or the sarcomere length at which the contraction started. The result is that there are two uniform rates of ATPase, first the higher rate during shortening then a lower rate during recovery from shortening, right after shortening has ceased (Lionne et al., 1996; Barman et al., 1998). Yanagida et al. also found a constant ATPase rate during shortening (Yanagida et al., 1985). The uniform energy use during shortening demonstrated in the aforementioned studies has later been confirmed in studies on heart during shortening and lengthening, here measured as heat production (Barclay et al., 2003).

When looking at ATP consumption during isometric contraction, the situation is different since ATP consumption drops with a lesser degree of overlap between filaments (He et al., 1997).

A series of studies have looked at effects of quick release on ATPase rate/ATP hydrolysis. When force was released to zero, extra ATP was hydrolysed, but the amount was largely independent of the amplitude of the release (Potma and Stienen, 1996). This was confirmed by Siththanandan et al. (2006), who reported that the release caused a substantial increase in the ATPase rate. During the recovery phase where the force was restored, the ATPase rate went down to a lower rate than prior to the release. The phosphate transient, indicating raised ATP use, takes place after the release and during the recovery of tension to the pre-release value, indicating that there may be extra interactions between actin and myosin at this stage. An interesting finding is that the phosphate release takes place after the force has started to develop (Smith and Sleep, 2004).

Another finding which needs to be explained is the observation that calcium-activated isometric force in skinned fibres at an ionic strength of 50 mM is more than twice that observed at 170 mM. Yet, despite this, the ATPase activity under the two conditions appears to be the same. This does not agree with simple kinetics (Sugi et al., 2013).

Like quick release, temperature jumps also affect the ATPase rate, which increases with temperature, giving rise to an increase in force (Mutunga and Ranatunga, 1998; Ranatunga, 2016).

In myofibrils, ATP in concentrations exceeding the Mg^{2+} concentration showed substrate inhibition of the ATPase in the relaxed condition, while this was not the case with calcium present in concentrations below, but close to, the pCa, which would stimulate contraction of the muscle. Pyrophosphate, or ATP and pyrophosphate together, showed the same inhibitory effect as ATP alone (Perry and Grey, 1956). This is in accordance with the charge effects seen when adding ATP or pyrophosphate to a glycerinated rigor muscle. A drop in A-band charge is seen both with ATP and with pyrophosphate, while neither ADP nor AMP-PNP has that effect (Bartels and Elliott, 1983).

When looking at the importance of actin binding to the myosin head during contraction, Sugi et al. (2014) have recently presented some interesting data. When the actin-binding site of the catalytic domain of the myosin head was blocked with an antibody, it has no effect on in vitro actin-myosin sliding or contraction in calcium-activated muscle. This raises the question as to whether the rigor-type 'mechanical' binding is present during contraction of a live muscle. This may be the solution to the question about a third conformational state to explain X-ray findings concerning actin binding to myosin (Geeves and Holmes, 1999), a state with no direct docking of actin during contraction. Blocking the domain between the catalytic domain and the lever-arm domain of the head with another antibody affected the actin–myosin preparation by inhibiting sliding, while muscle contraction happened as prior to antibody-binding, and the ATPase activity was unchanged in both preparations. Upon finally blocking the lever arm domain with a third antibody, which attached itself in the area of the regulatory light chains, force development was reduced in the muscle fibre, while there was no effect on the actin–myosin preparation, and the ATPase was not affected in either preparation (Sugi et al., 2014). This is accordance with earlier studies on glycerinated muscle fibres, where the S2 region of myosin was blocked with antibodies, leading to force reduction and stiffness reductions, but no change in ATPase activity (Sugi et al., 1992). Since pH has not changed in the mentioned studies, and since the ATPase site should not be affected by the antibody blockade or by the earlier mentioned changes in ionic strength, the findings make good sense from a biochemical point of view, especially due to the fact that the ATPase in question is very active at physiological pH (Harrington and Himmelfarb, 1972).

Phosphorylation of the regulatory light chain of the myosin head has been suggested to play a role in regulation of relaxation of a contracted muscle (Cooke, 2007). This was rejected by Duggal (2014), who could not find any effects of phosphorylation of the regulatory light chain on the cross-bridge cycle. Phosphorylation of My-BP-C may be the link between this discrepancy, since My-BP-C can bind to the S2 part of myosin, and phosphorylation of My-BP-C is found to modulate contractility in skinned fibres (Kunst et al., 2000).

3.2.3.2   Electric charge changes initiated by ATP

In a muscle cell there exist structured charged proteins in an ionic environment. The environment and any alterations in the environment will decide how conformational changes can and will happen in a particular protein, and thereby how the protein can bind/detach and/or create forces in all directions—in principle projected in three dimensions—and dependent on time, since the charge changes often happen with time due to temporary changes in free calcium, osmolality, free small ions or changes created by metabolic products.

Present electrostatic forces, which will affect the distances between the charged proteins, will also determine possible reactions like binding of ATP or actin binding to myosin—or the distances kept between neighbouring proteins. Since binding of any kind between filaments must be dependent on charges, since charges determine both conformation (availability of binding sites) and electrostatic binding/forces, then charge changes must be seen as a main player in those effects observed when the contractile unit is perturbed in any way. Movement of ions through the sarcomere will also be affected by the presence of charged interfaces everywhere in the sarcomere (Chan and Halle, 1984).

When relaxing solution containing ATP is added to a glycerinated or skinned muscle, the electric charge concentration in the A- and the I-band changes from the A-band region having a much more negative charge concentration than the I-band region to the two regions showing the same charge concentration, closer to, but higher than the rigor I-band charge concentration (Bartels and Elliott, 1985). The charge concentration is in both bands, and in both rigor and the relaxed state, dependent on ionic strength in a similar way. Repeating the measurements in myosin, myosin rod and LMM gels showed the same charge effect of ATP as seen for whole muscle (Bartels et al., 1993). Furthermore, pyrophosphate showed a similar effect to ATP, while ADP did not create this effect on the electric charge concentration. It was also shown that the change was completed with only 100–200 μM ATP, indicating that this is a ligand interaction with one or more sites on the myosin molecule. This interaction most likely causes a conformational change along the myosin molecule leading to a decrease of absorption of anion. In all studies pH was kept constant and within the physiological range.

During contraction, the charge concentration was, like in the relaxed muscle, found to be the same in the A- and the I-band (Bartels and Elliott, 2007). First the charge concentration increased to a concentration higher than that seen in the A-band during rigor. Following this, the charge concentration decreased to a value below the one measured in the relaxed state. Clearly a conformational change must take place following the Ca^{2+} binding to troponin, which causes the actin–myosin interaction. These conformational changes will accommodate an increased absorption of anions in both filaments. The changes happen very quickly, and as ATP is consumed during the process, but still is sufficiently present for further turnover of the ATPase, phosphate will be accumulated and may bind to both A- and I-filaments to give rise to a negative charge concentration higher than that seen in the A-band during rigor. The later drop in charge is probably due to pyrophosphate binding, since pyrophosphate will accumulate and give the same effect as ATP-binding seen in the relaxed state. Despite a good buffering capacity in muscle, acidosis is likely to occur as the ATP continues to break down, which will cause a decrease in negative charges (Naylor et al., 1985). An increase in Mg^{2+} has also been suggested (Westerblad and Allen, 1996), which would also cause a decrease in negative charge. Such findings are therefore explainable by ion binding and release, and it has, furthermore, proved possible to reverse the muscle back to the relaxed state by changing back to a relaxing solution, indicating that such a perturbation does not upset the contractile units studied (Bartels and Elliott, 2007).

To support the suggestion that anion binding and release is the cause of charge changes when changing the environment, a temperature study was set up, where charge measurements were carried out in both the rigor and relaxed state at a temperature range from 10–35°C (Regini and Elliott, 2001). The charge concentrations in the A- and the I-band in rigor were found to have different concentrations as earlier reported (Bartels and Elliott, 1985), and stayed constant over the range 10–27.5°C. Subsequently, a drop in charge concentration occurred, again giving a stable charge concentration in the range 30–35°C. The same result was found for relaxed muscle, although the shift with the drop in charge concentration with increased temperature was not so clear cut, but was in the range of 20–25°C. The findings illustrate that the most likely cause for the charge changes is hydrogen-bonded anions located along the length of the filaments at sites like Saroff sites formed by the electrically effective surfaces of the binding proteins. As the thermal energy of the bound ions goes up, it will be possible for the ions to escape from their binding site. The process is a cooperative one, indicating linked binding sites (McLaclan and Karn, 1982), and will restore more order to the filament lattice, perhaps due to reduced stress along the proteins in the filaments, or perhaps due to diminished electrostatic charge interaction.

3.2.4 Active Force Development

Going back to basics, the length-tension curve for isometric tension, where the length is set prior to both stimulation and subsequent contraction, describes the dependency of produced force on sarcomere length, i.e. on the amount of overlap between the A- and I-filaments (Gordon et al., 1966). Although this curve has been the basis for the acceptance of the sliding filament theory, there has always been a question concerning the two end points of the curve, since they are difficult to explain using the simple cross-bridge theory.

Another phenomenon often ignored when looking at contraction is the latency relaxation (tension relaxation), the small drop in tension preceding contraction (Sandow, 1946). During contraction the time from stimulation to start of the latency relaxation is constant and independent of sarcomere length, while the amplitude is dependent on sarcomere length and follows the force with a size of around 1/1000 of the isometric force, and the maximum amplitude was seen at a sarcomere length of around 3.0–3.1 µm in both frogs and mammals. With no filamentous overlap, there is no measurable latency relaxation (Bartels et al., 1979). Latency relaxation must be part of any model which aims to explain the contractile process. Latency relaxation has been related to change in osmotic pressure created by calcium release (Peachey and Schild, 1968), a reduction in electrostatic repulsive forces between the filaments by shielding these with released calcium (Huxley and Brown, 1967), breaking of bound cross-bridges (Hill, 1968), or release of tension in the sarcoplasmic reticulum (Sandow et al., 1975).

Looking at tension development during quick stretch/release experiments in rigor and during contraction (Linari et al., 1998), the two situations appear very different in terms of recovery, where the recovery time to get back to the starting tension is several tens of milliseconds in rigor, while the contracting muscle returns to the start tension prior to the stretch/release in a few milliseconds (Linari et al., 1998).

Recently, Sugi et al. (2012) looked at the effect of ionic strength on calcium activated isometric contraction. The decrease in ionic strength was created by decreasing the KCL concentration from 140 to 0 mM, creating a change in ionic strength from 170 to 50 mM. The maximum isometric force was seen to increase linearly with decreasing ionic strength to a final value of more than twice the force at physiological ionic strength. Interestingly, when registering a force–velocity curve under the same ionic strength conditions, the maximum unloaded shortening velocity remained unchanged. The question remains then as to whether a change in electric charges on the involved filaments creates a more optimal condition for force production in the low ionic strength, especially since the unchanged maximum shortening velocity suggests that the elements creating the force are the same in the two situations.

3.2.5 Stiffness and General Elastic Properties of the Contractile Unit

The muscle cell looked upon as an elastic body is both viscous-elastic and anisotropic. With the complicated structure of the contractile unit, this is not surprising, but it does mean that one has to define clearly which direction is considered when talking about measured compliance/stiffness, and which protein structures are the main ones influencing the compliance/stiffness in question, especially when simple linear elasticity cannot be considered in most situations.

In a recent review by Kaya and Higuchi (2013), the problem concerning stiffness of the sarcomere and the part each participating component plays in this is thoroughly discussed, pointing out that a substantial part (60–70%) of the half-sarcomere compliance is situated in the filaments, and that stiffness cannot be taken as an expression of present force-generating myosin heads. X-ray experiments have confirmed this, showing that more than 50% of muscle compliance derives from the A- and I filaments themselves and not from the cross-bridges (Huxley et al., 1998). Furthermore, the elasticity of myosin is shown to be non-linear (Kaya and Higuchi, 2010). Another important component is titin, which clearly plays a major role in the passive force of the sarcomere (Higuchi, 1992; Linke et al., 1994). It is also more than likely that other structures from the sarcomeric cytoskeleton are components that add to the compliance of the contractile unit.

When studying quick stretch of skinned fibres by applying length-change oscillations, stiffness was in calcium-activated fibres seen to temporarily decrease at oscillations above 1 Hz, while stiffness increased temporarily in the same fibres at oscillation below 1 Hz. When actin–myosin interaction was inhibited, no decrease in stiffness was seen at frequencies above 1 Hz, but the temporary increase was now seen at frequencies up to 20 Hz, indicating that the decrease in stiffness was cross-bridge dependent, while the increase, probably hidden in the 1–20 Hz range by the decrease when actin–myosin interaction was possible, was caused by a non-cross-bridge structure (Altman et al., 2015).

3.3   The Dynamic Contractile Unit

Having discussed the important elements necessary to understand the molecular events of a contraction, we are now set to try to piece everything together. From a biophysical point of view, the contractile unit may be considered a polyelectrolyte gel consisting of insoluble aggregates of protein molecules. These aggregates form structural fibrils and a skeleton which secures the precise position of the filaments by setting a frame which defines the possible general movements of the filaments. The fibrils exist in an aqueous solution, such that the fibril structures with their chemically bound side chains have a net electric charge, and around these structures are diffusible ions like Na^+, K^+, Ca^{2+}, Cl^-, Pi^{3-} and PPi^{2-}. The fixed charge density is in this system decided by both the charges on the proteins themselves and the soluble ions bound to the filaments at the Saroff sites mentioned earlier in this chapter. As an example of one of the main proteins, myosin at physiological pH has about 110 negative groups and 100 positive groups per 10^5 Da, giving a negative net charge of 10 per 10^5 Da. There are two governing factors represented by the osmotic potential and the chemical potential which will decide the stability of the gel. There are other forces helping to stabilize the gel, such as cross-linking and van der Waals forces. In an intact muscle there will also be ion pumps driven by metabolism, which will help stabilize the ionic environment.

3.3.1   What Happens during a Contraction?

In a live muscle, the contractile unit will be in the relaxed state prior to stimulation. This will in physiological conditions mean that the A- and the I-band charge concentration is the same (Bartels and Elliott, 1985) and that the titin filaments will be in the more negative state (Coomber et al., 2011).

Other proteins in the contractile unit may add to the charge concentration, but the major participants in keeping this charge concentration must be the three major proteins, forming the filaments and creating the surfaces to which the smaller ions are absorbed along the filament lengths. Since ATP in the relaxed state must be expected to bind to the ATPase site at the head end of the myosin molecules, this binding will lead to a conformational change affecting the whole A-filament to a conformation with less sites for anion binding than the rigor state (Bartels and Elliott, 1985). This change is independent of the presence of actin (Bartels et al., 1993). The conformational change of the head region is in accordance with Sugi et al.'s finding that the myosin head will move away from the bare region of the filament when ATP is introduced to myosin filaments in rigor, taking up another position 'ready for contraction' (Sugi et al., 2008).

When the relaxed muscle is stimulated, the sudden rise in Ca^{2+} will bind to troponin and thereby via tropomyosin change the conformation along the I-filament, lining the filament up for the start of contraction (Squire, 1981; Moir et al., 1983). Parallel to this, Ca^{2+} binding to the titin filament, will cause a charge decrease along the filament stretching throughout the sarcomere (Coomber et al., 2011). This change in charge concentration will affect the electrical potential which exists between the filaments (Chu, 1967; Elliott and Hodson, 1998). The activity of the ATPase may be assumed to be sensitive to the potential profile, which will imply that the ATPase situated on the heads projecting towards the minima in the potential profiles may not be activated. Thus the effect, starting at the A-I junction, will run along the train of myosin heads around a given I-filament, such that each event will initiate the following one, causing sequential operation, and not a simultaneous one as earlier believed (Elliott and Worthington, 2012). Since this is a general regulatory system set by the potential distributions inside the sarcomere, it does not go against the findings of Sugi et al. (2014) that the ATPase activity of the muscle is unaffected by blocking various regions of the myosin head, or the finding that the ATPase activity is not affected by ionic strength (Sugi, 2013).

The sharp change in charge concentration when calcium binds to titin may also explain the occurrence of the latency relaxation (Bartels et al., 1976, 1979). Titin can be considered as an integral part of the A-filament, binding to myosin and MyBP-C (Labeit et al., 1992), and as the calcium binding occurs along the titin filament, the sudden—and short-lived—drop in charge concentration at the binding point will cause a drop in electrostatic forces, which will result in a drop in force prior to the following rise in charge concentration during the contraction. This is all in accordance with the time course and length dependency seen for the latency relaxation (Bartels et al., 1976, 1979). It also explains the fact that phosphate release first takes place after the force has started to develop, since the very start, the latency relaxation, is not ATP dependent (Smith and Sleep, 2004).

At the start of shortening, more myosin heads will probably be activated until the sequential process starts, and force/shortening is produced. This will be in accordance with the first burst of ATP hydrolysis at the start of shortening (Lionne et al., 1996). Moreover, the results from the quick release studies also make sense. Here extra ATP was hydrolysed at a higher rate than the one seen during the preceding phase, following release during contraction, but the ATP used was independent of the size of the release (Potma and Stienen, 1996; Siththanandan et al., 2006). The quick release will simply reset the filaments to their starting position, with a higher number of actin–myosin interactions, until the contraction is restored to the force prior to the release, and the myosin heads subsequently return to sequential operation (Elliott and Worthington, 2012). During the recovery phase where force is restored, the ATPase rate decreases to a lower rate than that prior to the release. This could be due to a higher phosphate concentration following the extra energy burst, leading to phosphate binding to the filaments, a higher negative charge binding, and a change in ATPase activity as a result.

Effects of temperature jumps with increased ATPase rate and increase in force is possibly a similar effect to that described for quick release (Mutungi and Ranatunga, 1998; Ranatunga, 2015).

As the contraction starts, an increase in charge concentration is seen. This must be due to conformational changes in both the A- and the I-filament following the Ca^{2+} binding to troponin and to titin, changing the ability to bind anions in both filaments. As phosphate is released through the quick ATP turnover, phosphate will accumulate and may bind to both A- and I-filaments to give rise to a negative charge concentration higher than that seen in the A-band during rigor. Following this, a drop in charge occurs, most probably due to pyrophosphate binding, since pyrophosphate will accumulate and give rise to the same effect as ATP-binding seen during the relaxed state (Bartels and Elliott, 2007). If acidosis follows, despite a good buffering capacity afforded by muscle proteins, binding of hydrogen ions will add to the drop in the charge concentration (Naylor et al., 1985). Another ion which could add to a drop in charge concentration by binding to the filaments is Mg^{2+}. Mg^{2+} has been suggested as being a fatigue-causing ion upon accumulation in muscle cells during periods of strenuous contraction (Westerblad and Allen, 1996). As calcium is removed, the muscle goes back to the charge conditions seen in a relaxed muscle (Bartels and Elliott, 2007).

With this muscle model, the change in force of more than two occurring between ionic strengths of 170 and 50 mM, as described by Sugi et al. (2013), can easily be explained by a difference in charge concentration in the two situations, resulting in a potential difference inside the sarcomere, which in turn gives rise to a difference in ion binding to the filaments. In Sugi et al.'s experiment, there will probably be some ATP^{2-} binding and some Mg^{2+} binding, albeit at very low concentrations. The protein charge concentration will be substantially lower at the low ionic strength than at the high ionic strength. Since the ATPase activity is the same in the two situations, an explanation must surely be found in the possible force that can be produced between actin and myosin. Elliott and Worthington (1996) described the forces capable of comprising the net force obtained between the filaments. This latter force consists of the impulsive force between the actin and myosin and the viscoelastic forces trying to stop the motion created by the impulsive force. With a different conformation of the filaments due to a different charge concentration, the impulsive force is likely to be larger than that at physiological ionic strength, while there is no reason to believe that the viscoelastic forces will change between these two situations. This would give rise to the higher force seen in Sugi et al.'s study. Following this through, Elliott and Worthington (1996) demonstrated cleverly that this was in accordance with the Hill equation, and that some form of frictional force between the actin and myosin exists. The obtained Hill's equation also lead to a precise value for the time within which the impulsive contractile force is delivered, again supporting the description of the model for muscle contraction presented above.

During isometric force generation, ATP consumption drops as the filament overlap is reduced (He et al., 1997). This indicates that less actin–myosin interactions are occurring, something that is far from surprising with less possibility of interaction. The model described is therefore applicable to this finding and to the length-tension curve. As the myosin filaments reach the area close to the Z-disc, the conformation of the A-filaments and the titin filaments must change, and actin–myosin interaction may not be possible at all. It is likewise probable that the interaction between filaments and sarcomeric cytoskeleton prevents any possible shortening that would result in the destruction of the well-organized sarcomere. Indeed, yet another reason for the finding could be that the calcium binding sites on titin are distorted to such a degree that calcium binding is not possible.

Results obtained from contracting muscles do not speak for an actin–myosin interaction that is the same as that observed in the rigor muscles. It is clear from Sugi et al.'s (2014) data that an actin binding to the site on the myosin head is not needed to create a contraction. This is in accordance with H. E. Huxley's (1973) X-ray data, where contraction data did not resemble rigor data. It is more likely that the interaction during contraction is of an electrostatic nature. The results from Bartels and Elliott (2007) support this proposal, since these particular data are much better explained by ion binding and release from the filaments, involving Ca^{2+} binding and binding of ATP and hydrolysis products from ATP.

As ATP is hydrolysed during contraction, phosphate will be released and bind to the filaments. In heart muscle, phosphorylation of My-PB-C promotes detachment of the actin–myosin interaction (Rosas et al., 2015) probably due to a conformational change affecting the A-filaments. Such a modulating effect is very likely also present in skeletal muscle and may, as well as bringing down the calcium concentration in the sarcomere, lead to the end of contraction and the return to the relaxed state. Another similar way of modulating contraction and inducing relaxation could be through phosphorylation of the PEVK regions of titin (Fukuda, 2008). This causes a reduction in tension, which may be caused by a conformational change of the protein which could release calcium from titin and thereby change the potential profiles in the sarcomere towards the situation in the relaxed muscle.

There are probably more undiscovered regulatory mechanisms hidden within the complicated sarcomeric cytoskeleton, which clearly performs a far greater function than mere non-cross-bridge stiffness.

3.3.2 Importance of Considering Ion Movements as the Base for Contraction

Muscle diseases are, apart from the genetic ones, common both as heart diseases and age-related conditions leading to sarcopenia. Knowledge of ion mechanisms, like the effect caused by impaired phosphorylation of My-PB-C (Rosas et al., 2015) or defects leading to a change in the ionic environment of the cells, may in the future lead to a better understanding of what has gone wrong in the muscle at the molecular level, and what can be done to restore healthy muscle function. During long-term unloading of a muscle, sarcopenia sets in, and titin levels are found to be reduced in the muscle cells (Fitts, 2000). Titin is found to interact with MURF1 (Witt et al., 2005), which is involved in degradation of muscle proteins (Bodine and Baehr, 2014). Without understanding the function of titin in the muscle in connection with both the structural order of the sarcomere and the regulation of the contraction, it may be difficult to comprehend the importance of titin and think in new directions, which could lead to the prevention- or even halt the progress of sarcopenia. It is thus essential that we consider muscle in a less mechanistic way in the future if treatment of muscle is to make leaps and bounds, especially in a society with a growing population of elderly people predisposed to muscle dysfunction problems.

3.4 Conclusions

Muscle research has over the years revealed so much exciting knowledge concerning the contractile unit. From a very simple system and very simple mechanical analogue models, a complicated unit with collaborating partners of giant proteins and smaller proteins inter-twined in a structured network has appeared. This has demanded joint efforts in order to be able to understand the basics of how this cleverly designed unit can repeatedly produce force in a controlled way, using metabolic energy from ATP breakdown. The model proposed here is not the final truth, and it is important to keep one's mind open for future discoveries and continue to ask questions.

The model is based on general principles for biological systems, following physical-chemical principles, and considering effects of small ions (calcium, magnesium, sodium, potassium, ATP, phosphate etc.) on protein charges, protein structures and conformational changes. The repeatability of the contractile process demands that all proteins, which in the contractile unit are highly organized and completely inter-twined, must function in unison and involve simple changes which can be easily reversed.

Acknowledgements

This chapter is dedicated to my two mentors and very good friends who throughout life kept an open mind and a keen interest in muscle biophysics, Professor Ove Sten-Knudsen, Copenhagen, and Professor Gerald Elliott, Oxford. They are both sadly missed.

The Parker Institute is supported by a grant from the OAK Foundation.

References

Agarkova I, Perriard JC (2005). The M-band: An elastic web that crosslinks thick filaments in the center of the sarcomere. *Trends Cell Biol,* **15**: 477–485.

Aldoroty RA, Garty NB, April EW (1985). Donnan potentials from striated muscle liquid crystals. Sarcomere length dependence. *Biophys J,* **47**: 89–95.

Altman D, Minozzo FC, Rassier DE (2015). Thixotropy and rheopexy of muscle fibers probed using sinusoidal oscillations. *PLos One,* **10**: e0121726.

Ashley CC, Ridgway EB (1968). Simultaneous recording of membrane potential, calcium transient and tension in single muscle fibers. *Nature,* **219**: 1168–1169.

Barclay CJ, Widen C, Mellors LJ (2003). Initial mechanical efficiency of isolated cardiac muscle. *J Exp Biol,* **206**: 2725–2732.

Barclay CJ, Woledge RC, Curtin NA (2010). Is the efficiency of mammalian (mouse) skeletal muscle temperature dependent? *J Physiol,* **588**: 3819–3831.

Barman T, Brune M, Lionne C, Piroddi N, Poggesi C, Stehle R, Tesi C, Travers F, Webb MR (1998). ATPase and shortening rates in frog fast skeletal myofibrils by time-resolved measurements of protein-bound and free Pi. *Biophys J,* **74**: 3120–3130.

Bartels EM, Cooke PH, Elliott GF, Hughes RA (1993). The myosin molecule–charge response to nucleotide binding. *Biochim Biophys Acta,* **1157**: 63–73.

Bartels EM, Elliott GF (1982). Donnan potentials in rat muscle: Differences between skinning and glycerination. *J Physiol,* **327**: 72–73.

Bartels EM, Elliott GF (1983). Donnan potentials in glycerinated rabbit skeletal muscle: The effects of nucleotides and of pyrophosphate. *J Physiol,* **343**: 32P–33P.

Bartels EM, Elliott GF (1985). Donnan potentials from the A- and I-bands of glycerinated and chemically skinned muscles, relaxed and in rigor. *Biophys J,* **48**: 61–76.

Bartels EM, Elliott GF (1988). High and low rigor states in skeletal muscle show different A-band electric charges. *Biophys J,* **53**: 63a.

Bartels EM, Elliott GF (2007). Electric charge changes in cross-striated barnacle and rat muscle during contraction. *JPCCR,* **1**: 142–145.

Bartels EM, Elliott GF, Wall RS (1987). Donnan potential measurements from the sarcoplasmic-reticulum of rabbit muscle under rigor and relaxed conditions. *J Physiol,* **388**: 34.

Bartels EM, Jensen P (1982). Latency relaxation in frog skeletal muscle under hypertonic conditions. *Acta Physiol Scand,* **115**: 165–172.

Bartels EM, Jensen P, Sten-Knudsen OS (1976). The dependence of tension relaxation in skeletal muscle on the number of sarcomeres in series. *Acta Physiol Scand,* **97**: 476–485.

Bartels EM, Skydsgaard JM, Sten-Knudsen O (1979). The time course of the latency relaxation as a function of the sarcomere length in frog and mammalian muscle. *Acta Physiol Scand,* **106**: 129–137.

Bartoo ML, Linke WA, Pollack GH (1997). Basis of passive tension and stiffness in isolated rabbit myofibrils. *Am J.Physiol,* **273**: C266–C276.

Batters C, Veigel C, Homsher E, Sellers JR (2014). To understand muscle you must take it apart. *Front Physiol,* **5**: 90.

Benian GM, Mayans O (2015). Titin and obscurin: Giants holding hands and discovery of a new Ig domain subset. *J Mol Biol,* **427**: 707–714.

Bodine SC, Baehr LM (2014). Skeletal muscle atrophy and the E3 ubiquitin ligases MuRF1 and MAFbx/atrogin-1. *Am J Physiol Endocrinol Metab,* **307**: E469–E484.

Bogomolovas J, Gasch A, Simkovic F, Rigden DJ, Labeit S, Mayans O (2014). Titin kinase is an inactive pseudokinase scaffold that supports MuRF1 recruitment to the sarcomeric M-line. *Open Biol,* **4**: 140041.

Boussouf SE, Geeves MA (2007). Tropomyosin and troponin cooperativity on the thin filament. *Adv Exp Med Biol,* **592**: 99–109.

Boyle PJ, Conway EJ (1941). Potassium accumulation in muscle and associated changes. *J Physiol,* **100**: 1–63.

Calderón JC, Bolanõs P, Caputo C (2014). The excitiation-contraction coupling mechanism in skeletal muscle. *Biophys Rev,* **6**: 133–160.

Caputo C, Bolanos P, Escobar AL (1999). Fast calcium removal during single twitches in amphibian skeletal muscle fibres. *J Muscle Res Cell Motil,* **20**: 555–567.

Caputo C, Bolanos P, Gonzalez A (2004). Inactivation of Ca^{2+} transients in amphibian and mammalian muscle fibres. *J Muscle Res Cell Motil,* **25**: 315–328.

Caputo C, Edman KA, Lou F, Sun YB (1994). Variation in myoplasmic Ca^{2+} concentration during contraction and relaxation studied by the indicator fluo-3 in frog muscle fibres. *J Physiol,* **478**: 137–148.

Carlson FD, Hardy DJ, Wilkie DR (1963). Total energy production and phosphocreatine hydrolysis in the isotonic twitch. *J Gen Physiol,* **46**: 851–882.

Carlsen F, Knappeis GG, Buchthal F (1961). Ultrastructure of the resting and contracted striated muscle fiber at different degrees of stretch. *J Biophys Biochem Cytol,* **11**: 95–117.

Chan DY, Halle B (1984). The Smoluchowski-Poisson-Boltzmann description of ion diffusion at charged interfaces. *Biophys J,* **46**: 387–407.

Chu B (1967). *Molecular Forces.* Chapter 4. John Wiley & Sons, New York.

Collins EW, Jr., Edwards C (1971). Role of Donnan equilibrium in the resting potentials in glycerol-extracted muscle. *Am J Physiol,* **221**: 1130–1133.

Cooke R (2007). Modulation of the actomyosin interaction during fatigue of skeletal muscle. *Muscle Nerve,* **36**: 756–777.

Cooke PH, Bartels EM, Elliott GF, Hughes RA (1987). A structural study of gels, in the form of threads, of myosin and myosin rod. *Biophys J,* **51**: 947–957.

Coomber SJ, Bartels EM, Elliott GF (2011). Calcium-dependence of Donnan potentials in glycerinated rabbit psoas muscle in rigor, at and beyond filament overlap; a role for titin in the contractile process. *Cell Calcium,* **50**: 91–97.

Davies RE (1963). A molecular theory of muscle contraction. *Nature,* **199**: 1068–1074.

Dragomir CT (1970). On the nature of the forces acting between myofilaments in resting state and under contraction. *J Theor Biol,* **27**: 343–356.

Dreizen P, Gershman LC, Trotta PP, Stracher A (1967). Subunits and their interactions. *J Gen Physiol,* **50** Suppl: 118.

Duggal D, Nagwekar J, Rich R, Midde K, Fudala R, Gryczynski I, Borejdo J (2014). Phosphorylation of myosin regulatory light chain has minimal effect on kinetics and distribution of orientations of cross bridges of rabbit skeletal muscle. *Am J Physiol Regul Integr Comp Physiol,* **306**: R222–R233.

Ebashi S (1976). Excitation-contraction coupling. *Ann Rev Physiol,* **38**: 293–313.

Ebashi S, Endo M, Otsuki I (1969). Control of muscle contraction. *Q Rev Biophys,* **2**: 351–384.

Edman KA (1966). The relation between sarcomere length and active tension in isolated semitendinosus fibres of the frog. *J Physiol,* **183**: 407–417.

Elliott GF (1960). Electron microscopy and X-ray diffraction studies of invertebrate muscle fibres. PhD-thesis, University of London, London.

Elliott GF (1973). The muscle fiber: Liquid-crystalline and hydraulic aspects. *Ann NY Acad Sci,* **204**: 564–574.

Elliott GF, Bartels EM (1982a). Spacing measurements of the myosin meridional reflections using synchroton radiation at the Daresbury source. *J Muscle Res Cell Motil,* **3**: 460.

Elliott GF, Bartels EM (1982b). Donnan potential measurements in extended hexagonal polyelectrolyte gels such as muscle. *Biophys J,* **38**: 195–199.

Elliott GF, Bartels EM, Cooke PH, Jennison K (1984). A reply to Godt and Baumgarten's potential and K⁺ activity in skinned muscle fibers: Evidence for a simple Donnan equilibrium under physiological conditions. *Biophys J,* **45**: 487–488.

Elliott GF, Bartels EM, Hughes RA (1986). The myosin filament: Charge amplification and charge condensation. In *Electric Double Layers in Biology* (Blank M, ed), Plenum Publication Corporation, New York, pp. 277–285.

Elliott GF, Hodson SA (1998). Cornea, and the swelling of polyelectrolyte gels of biological importance. *Rep Prog Phys,* **61**: 1325–1365.

Elliott GF, Lowy J, Worthington CR (1963). An X-ray diffraction and light diffraction study of the filament lattice of striated muscle in the living state and in rigor. *J Mol Biol,* **6**: 295–305.

Elliott GF, Worthington CR (1994). How muscle may contract. *Biochim Biophys Acta,* **1200**: 109–116.

Elliott GF, Worthington CR (2012). Along the road not taken: How many myosin heads act on a single actin filament at any instant in working muscle? *Prog Biophys Mol Biol,* **108**: 82–92.

Fenn WO (1923). A quantitative comparison between the energy liberated and the work performed by the isolated sartorius muscle of the frog. *J Physiol,* **58**: 175–203.

Fenn WO (1924). Relation between worked performed and energy liberated in muscular contraction. *J Physiol,* **58**: 373–395.

Fitts RH, Riley DR, Widrick JJ (2000). Physiology of a microgravity environment invited review: Microgravity and skeletal muscle. *J Appl Physiol,* **89**: 823–839.

Franzini-Armstrong C (1970). Studies of the triad I: Structure of the junction in frog twitch fibers. *J Cell Biol,* **47**: 488–499.

Fukuda N, Granzier HL, Ishiwata S, Kurihara S (2008). Physiological functions of the giant elastic protein titin in mammalian striated muscle. *J Physiol Sci,* **58**: 151–159.

Fukuda N, Terui T, Ishiwata S, Kurihara S (2010). Titin-based regulations of diastolic and systolic functions of mammalian cardiac muscle. *J Mol Cell Cardiol,* **48**: 876–881.

Funatsu T, Asami Y, Ishiwata S (1988). Beta-actinin: A capping protein at the pointed end of thin filaments in skeletal muscle. *J Biochem,* **103**: 61–71.

Galvani L (1791). *De Viribus electricitatis in motu musculari commentarius.*

Gautel M (2011). The sarcomeric cytoskeleton: Who picks up the strain? *Curr Opin Cell Biol,* **23**: 39–46.

Geeves MA, Holmes KC (1999). Structural mechanism of muscle contraction. *Ann Rev Biochem,* **68**: 687–728.

Godt RE, Baumgarten CM (1984). Potential and K^+ activity in skinned muscle fibers. Evidence against a simple Donnan equilibrium. *Biophys J,* **45**: 375–382.

Gordon AM, Huxley AF, Julian FJ (1966). The variation in isometric tension with sarcomere length in vertebrate muscle fibres. *J Physiol,* **184**: 170–192.

Hanson J, Lowy J (1964). The structure of actin filaments and the origin of the axial periodicity in the I-substance of vertebrate striated muscle. *Proc R Soc Lond B Biol Sci,* **160**: 449–460.

Harrington WF, Himmelfarb S (1972). Effect of adenosine di- and triphosphates on the stability of synthetic myosin filaments. *Biochemistry,* **11**: 2945–2952.

Hatta I, Tamura Y, Matsuda T, Sugi H, Tsuchiya T (1984). Muscle stiffness changes during isometric contraction in frog skeletal muscle as studied by the use of ultrasonic waves. *Adv Exp Med Biol,* **170**: 673–686.

Haugen P, Sten-Knudsen O (1976). Sarcomere lengthening and tension drop in the latent period of isolated frog skeletal muscle fibers. *J Gen Physiol,* **68**: 247–265.

Haugen P (1983). Latency relaxation and short-range elasticity in single muscle fibres of the frog. *Acta Physiol Scand Suppl,* **519**: 1–48.

He ZH, Chillingworth RK, Brune M, Corrie JE, Trentham DR, Webb MR, Ferenczi MA (1997). ATPase kinetics on activation of rabbit and frog permeabilized isometric muscle fibres: A real time phosphate assay. *J Physiol,* **501**: 125–148.

Higuchi H (1992). Changes in contractile properties with selective digestion of connectin (titin) in skinned fibers of frog skeletal muscle. *J Biochem,* **111**: 291–295.

Hill AV (1938). The heat of shortening and the dynamic constants of muscle. *Proc R Soc B,* **126**: 136–195.

Hill AV (1953). A reinvestigation of two critical points in the energetics of muscular contraction. *Proc R Soc Lond B Biol Sci,* **141**: 503–510.

Hill AV (1964a). The variation of total heat production in a twitch with velocity and shortening. *Proc R Soc Lond B Biol Sci,* **159**: 596–605.

Hill AV (1964b). The efficiency of mechanical power development during muscular shortening and its relation to load. *Proc R Soc Lond B Biol Sci,* **159**: 319–324.

Hill AV (1964c). The effect of load on the heat of shortening. *Proc R Soc Lond B Biol Sci,* **159**: 297–318.

Hill DK (1968). Tension due to interaction between the sliding filaments in resting striated muscle. The effect of stimulation. *J Physiol,* **199**: 637–684.

Hodgson ES (1990). Long-range perspectives on neurobiology and behavior. *Am Zool,* **30**: 403–505.

Holmes KC (1996). Muscle proteins–their actions and interactions. *Curr Opin Struct Biol,* **6**: 781–789.

Huxley HE (1953). X-ray analysis and the problem of muscle. *Proc R Soc B,* **141**: 59–62.

Huxley AF (1957). Muscle structure and theories of contraction. *Prog Biophys,* **7**: 255–318.

Huxley HE (1968). Structural difference between resting and rigor muscle; evidence from intensity changes in the low angle equatorial x-ray diagram. *J Mol Biol,* **37**: 507–520.

Huxley HE (1973). Structural changes in the actin- and myosin-containing filaments during contraction. *Cold Spring Harbor Symp Quant Biol,* **37**: 361–376.

Huxley HE, Brown W (1967). The low-angle x-ray diagram of vertebrate striated muscle and its behaviour during contraction and rigor. *J Mol Biol,* **30**: 383–434.

Huxley HE, Brown W, Holmes KC (1965). Constancy of axial spacings in frog Sartorius muscle during contraction. *Nature,* **206**: 1358.

Huxley HE, Hanson J (1954). Changes in the cross-striations of muscle during contraction and stretch and their structural interpretation. *Nature,* **173**: 973–976.

Huxley AF, Niedergerke R (1954). Structural changes in muscle during contraction; interference microscopy of living muscle fibres. *Nature,* **173**: 971–973.

Huxley HE, Stewart A, Sosa H, Irving T (1994). X-ray diffraction measurements of the extensibility of actin and myosin filaments in contracting muscle. *Biophys J,* **67**: 2411–2421.

Jewell BR, Wilkie DR (1960). The mechanical properties of relaxing muscle. *J Physiol,* **152**: 30–47.

Jobsis FF, O'Connor MJ (1966). Calcium release and reabsorption in the sartorius muscle of the toad. *Biochem Biophys Res Commun,* **25**: 246–252.

Joumaa V, Rassier DE, Leonard TR, Herzog W (2008). The origin of passive force enhancement in skeletal muscle. *Am J Physiol Cell Physiol,* **294**: C74–C78.

Kawai M, Brandt PW (1976). Two rigor states in skinned crayfish single muscle fibers. *J Gen Physiol,* **68**: 267–280.

Kawai M, Brandt PW (1980). Sinusoidal analysis: A high resolution method for correlating biochemical reactions with physiological processes in activated skeletal muscles of rabbit, frog and crayfish. *J Muscle Res Cell Motil,* **1**: 279–303.

Kaya M, Higuchi H (2010). Nonlinear elasticity and an 8-nm working stroke of single myosin molecules in myofilaments. *Science,* **329**: 686–689.

Kaya M, Higuchi H (2013). Stiffness, working stroke, and force of single-myosin molecules in skeletal muscle: Elucidation of these mechanical properties via nonlinear elasticity evaluation. *Cell Mol Life Sci,* **70**: 4275–4292.

Kobayashi T, Saeki Y, Chaen S, Shirakawa I, Sugi H (2004). Effect of deuterium oxide on contraction characteristics and ATPase activity in glycerinated single rabbit skeletal muscle fibers. *Biochim Biophys Acta,* **1659**: 46–51.

Kontrogianni-Konstantopoulos A, Ackermann MA, Bowman AL, Yap SV, Bloch RJ (2009). Muscle giants: Molecular scaffolds in sarcomerogenesis. *Physiol Rev,* **89**: 1217–1267.

Kunst G, Kress KR, Gruen M, Uttenweiler D, Gautel M, Fink RH (2000). Myosin binding protein C, a phosphorylation-dependent force regulator in muscle that controls the attachment of myosin heads by its interaction with myosin S2. *Circ Res,* **86**: 51–58.

Labeit D, Watanabe K, Witt C, Fujita H, Wu Y, Lahmers S, Funck T, Labeit S, Granzier H (2003). Calcium-dependent molecular spring elements in the giant protein titin. *Proc Nat Acad Sci U S A,* **100**: 13716–13721.

Lange S, Ehler E, Gautel M (2006). From A to Z and back? Multicompartment proteins in the sarcomere. *Trends Cell Biol,* **16**: 11–18.

Levine BA, Moir AJG, Goodearl AJ, Trayer IP (1991). Mechanism of nucleotide binding and hydrolysis. *Biochem Soc Trans,* **19**: 423–425.

Linari M, Dobbie I, Reconditi M, Koubassova N, Irving M, Piazzesi G, Lombardi V (1998). The stiffness of skeletal muscle in isometric contraction and rigor: The fraction of myosin heads bound to actin. *Biophys J,* **74**, 2459–2473.

Linari M, Woledge RC (1995). Comparison of energy output during ramp and staircase shortening in frog muscle fibres. *J Physiol,* **487**: 699–710.

Linke WA, Ivemeyer M, Labeit S, Hinssen H, Ruegg JC, Gautel M (1997). Actin-titin interaction in cardiac myofibrils: Probing a physiological role. *Biophys J,* **73**: 905–919.

Linke WA, Ivemeyer M, Mundel P, Stockmeier MR, Kolmerer B (1998). Nature of PEVK-titin elasticity in skeletal muscle. *Proc Nat Acad Sci U S A,* **95**: 8052–8057.

Linke WA, Popov VI, Pollack GH (1994). Passive and active tension in single cardiac myofibrils. *Biophys J,* **67**: 782–792.

Lionne C, Travers F, Barman T (1996). Mechanochemical coupling in muscle: Attempts to measure simultaneously shortening and ATPase rates in myofibrils. *Biophys J,* **70**: 887–895.

Loeb GI, Saroff HA (1964). Chloride and hydrogen-ion binding to ribonuclease. *Biochem,* 3: 1819–1826.

Lou F, Curtin NA, Woledge RC (1998). Contraction with shortening during stimulation or during relaxation: How do the energetic costs compare? *J Muscle Res Cell Motil,* 19: 797–802.

Luther PK, Winkler H, Taylor K, Zoghbi ME, Craig R, Padron R, Squire JM, Liu J (2011). Direct visualization of myosin-binding protein C bridging myosin and actin filaments in intact muscle. *Proc Nat Acad Sci U S A,* 108: 11423–11428.

Lymn RW, Taylor EW (1970). Transient state phosphate production in the hydrolysis of nucleoside triphosphates by myosin. *Biochemistry,* 9: 2975–2983.

Ma YZ, Taylor EW (1994). Kinetic mechanism of myofibril ATPase. *Biophys J,* 66: 1542–1553.

Malnasi-Csizmadia A, Toth J, Pearson DS, Hetenyi C, Nyitray L, Geeves MA, Bagshaw CR, Kovacs M (2007). Selective perturbation of the myosin recovery stroke by point mutations at the base of the lever arm affects ATP hydrolysis and phosphate release. *J Biol Chem,* 282: 17658–17664.

Mantovani M, Cavagna GA, Heglund NC (1999). Effect of stretching on undamped elasticity in muscle fibres from Rana temporaria. *J Muscle Res Cell Motil,* 20: 33–43.

Maruyama K (1994). Connectin, an elastic protein of striated muscle. *Biophys Chem,* 50: 73–85.

Maruyama K, Kimura S, Ishi T, Kuroda M, Ohashi K (1977). beta-actinin, a regulatory protein of muscle. Purification, characterization and function. *J Biochem,* 81: 215 232.

Maruyama K, Kurokawa H, Oosawa M, Shimaoka S, Yamamoto H, Ito M, Maruyama K (1990). Beta-actinin is equivalent to Cap Z protein. *J Biol Chem,* 265: 8712–8715.

Matsubara I, Goldman YE, Simmons RM (1984). Changes in the lateral filament spacing of skinned muscle fibres when cross-bridges attach. *J Mol Biol,* 173: 15–33.

Matsumoto Y (1967). Validity of the force-velocity relation for muscle contraction in the length region, l less than or equal to l-o. *J Gen Physiol,* 50: 1125–1137.

McLachlan AD, Karn J (1982). Periodic charge distributions in the myosin rod amino acid sequence match cross-bridge spacings in muscle. *Nature,* 299: 226–231.

Meyer LC, Wright NT (2013). Structure of giant muscle proteins. *Front Physiol,* **4**: 368.

Moir AJG, Ordidge M, Grand RJA, Trayer IP, Perry SV (1983). Studies of the interaction of troponin-I with proteins of the I-filament and calmodulin. *Biochem J,* **209**: 417–426.

Naylor GR, Bartels EM, Bridgman TD, Elliott GF (1985). Donnan potentials in rabbit psoas muscle in rigor. *Biophys J,* **48**: 47–59.

Nieznanski K, Nieznanska H, Skowronek K, Kasprzak AA, Stepkowski D (2003). Ca^{2+} binding to myosin regulatory light chain affects the conformation of the N-terminus of essential light chain and its binding to actin. *Arch Biochem Biophys,* **417**: 153–158.

Ortega JO, Lindstedt SL, Nelson FE, Jubrias SA, Kushmerick MJ, Conley KE (2015). Muscle force, work and cost: A novel technique to revisit the Fenn effect. *J Exp Biol*, **218**: 2075–2082.

Oshima K, Sugimoto Y, Irving TC, Wakabayashi K (2012). Head-head interactions of resting myosin crossbridges in intact frog skeletal muscles, revealed by synchrotron x-ray fiber diffraction. *PLoS. One,* **7**, e52421.

Peachey LD, Schild RF (1968). The distribution of the T-system along the sarcomeres of frog and toad sartorius muscles. *J Physiol,* **194**: 249–258.

Pearn J (2002). A curious experiment: The paradigm switch from observation and speculation to experimentation, in the understanding of neuromuscular function and disease. *Neuromuscul Dis,* **12**: 600–607.

Pernigo S, Fukuzawa A, Bertz M, Holt M, Rief M, Steiner RA, Gautel M (2010). Structural insight into M-band assembly and mechanics from the titin-obscurin-like-1 complex. *Proc Nat Acad Sci U S A,* **107**: 2908–2913.

Perry SV, Grey TC (1956). A study of the effects of substrate concentration and certain relaxing factors on the magnesium-activated myofibrillar adenosine triphosphate. *Biochemistry,* **64**: 184–192.

Poole KJ, Lorenz M, Evans G, Rosenbaum G, Pirani A, Craig R, Tobacman LS, Lehman W, Holmes KC (2006). A comparison of muscle thin filament models obtained from electron microscopy reconstructions and low-angle X-ray fibre diagrams from non-overlap muscle. *J Struct Biol,* **155**: 273–284.

Porter KR, Palade GE (1957). Studies on the endoplasmic reticulum. III. Its form and distribution in striated muscle cells. *J Biophys Biochem Cytol,* **3**: 269–300.

Ranatunga KW (2016). Characteristics and mechanic(s) of force generation by increase of temperature in active muscle. In *Muscle Contraction and Cell Motility: Fundamentals and Developments* (Sugi H, ed), Pan Stanford Publishing, Singapore.

Rassier DE, Leite FS, Nocella M, Cornachione AS, Colombini B, Bagni MA (2015). Non-crossbridge forces in activated striated muscles: A titin dependent mechanism of regulation? *J Muscle Res Cell Motil,* **36**: 37–45.

Rebbeck RT, Karunasekara Y, Board PG, Beard NA, Casarotto MG, Dulhunty AF (2014). Skeletal muscle excitation-contraction coupling: Who are the dancing partners? *Int J Biochem Cell Biol,* **48**: 28–38.

Regini JW, Elliott GF (2001). The effect of temperature on the Donnan potentials in biological polyelectrolyte gels: Cornea and striated muscle. *Int J Biol Macromol,* **28**: 245–254.

Rome E (1972). Relaxation of glycerinated muscle: Low-angle x-ray diffraction studies. *J Mol Biol,* **65**: 331–345.

Rosas PC, Liu Y, Abdalla MI, Thomas CM, Kidwell DT, Dusio GF, Mukhopadhyay D, Kumar R, Baker KM, Mitchell BM, Powers PA, Fitzsimons DP, Patel BG, Warren CM, Solaro RJ, Moss RL, Tong CW (2015). Phosphorylation of cardiac myosin binding protein-C is a critical mediator of diastolic function. *Circ Heart Fail,* **8**: 582–594.

Sandow A (1946). The causation of the latency relaxation. *Fed Proc,* **5**: 91.

Sandow A (1952). Excitation-contraction coupling in muscular response. *Yale J Biol Med,* **25**: 176–201.

Sandow A, Krishna M, Pagala D, Sphicas EC (1975). Excitation-contraction coupling: Effects of "zero"-Ca^{2+} medium. *Biochim Biophys Acta,* **404**: 157–163.

Sarkar NK (1950). The effect of ions and ATP on myosin and actomyosin. *Enzymologia,* **14**: 237–245.

Siththanandan VB, Donnelly JL, Ferenczi MA (2006). Effect of strain on actomyosin kinetics in isometric muscle fibers. *Biophys J,* **90**: 3653–3665.

Smith DA, Sleep J (2004). Mechanokinetics of rapid tension recovery in muscle: The myosin working stroke is followed by a slower release of phosphate. *Biophys J,* **87**: 442–456.

Squire J (1981). Muscle regulation: A decade of the steric blocking model. *Nature,* **291**: 614–615.

Squire JM, Harford JJ, Edman AC, Sjostrom M (1982). Fine structure of the A-band in cryo-sections. III. Crossbridge distribution and the axial structure of the human C-zone. *J Mol Biol,* **155**: 467–494.

Squire JM, Luther PK, Knupp C (2003). Structural evidence for the interaction of C-protein (MyBP-C) with actin and sequence identification of a possible actin-binding domain. *J Mol Biol,* **331**: 713–724.

Sten-Knudsen O (1953). Torsional elasticity of the isolated cross striated muscle fibre. *Acta Physiol Scand,* **28**: 3–240.

Sten-Knudsen O (2002). *Biological Membranes, Theory of Transport, Potentials and Electrical Impulses.* Chapter 3, Cambridge University Press, Cambridge, pp. 3.3–3.6.

Stepkowski D (1995). The role of the skeletal muscle myosin light chains N-terminal fragments. *FEBS Lett,* **374**: 6–11.

Sugi H, Abe T, Kobayashi T, Chaen S, Ohnuki Y, Saeki Y, Sugiura S (2013). Enhancement of force generated by individual myosin heads in skinned rabbit psoas muscle fibers at low ionic strength. *PLoS One,* **8**: e63658.

Sugi H, Chaen S, Kobayashi T, Abe T, Kimura K, Saeki Y, Ohnuki Y, Miyakawa T, Tanokura M, Sugiura S (2014). Definite differences between in vitro actin-myosin sliding and muscle contraction as revealed using antibodies to myosin head. *PLoS One,* **9**: e93272.

Sugi H, Minoda H, Inayoshi Y, Yumoto F, Miyakawa T, Miyauchi Y, Tanokura M, Akimoto T, Kobayashi T, Chaen S, Sugiura S (2008). Direct demonstration of the cross-bridge recovery stroke in muscle thick filaments in aqueous solution by using the hydration chamber. *Proc Nat Acad Sci U S A,* **105**: 17396–17401.

Szczesna D, Zhao J, Jones M, Zhi G, Stull J, Potter JD (2002). Phosphorylation of the regulatory light chains of myosin affects Ca^{2+} sensitivity of skeletal muscle contraction. *J Appl Physiol,* **92**: 1661–1670.

Szent-Györgyi A (1942). Myosin and muscular contraction: Discussion. *Stud Inst Med Chem Univ Szeged,* **1**: 69–72.

Taylor EW (1977). Transient phase of adenosine triphosphate hydrolysis by myosin, heavy meromyosin, and subfragment 1. *Biochemistry,* **16**: 732–739.

Trayer HR, Trayer IP (1988). Fluorescence resonance energy transfer within the complex formed by actin and myosin subfragment 1. Comparison between weakly and strongly attached states. *Biochemistry,* **27**: 5718–5727.

Trayer IP, Trayer HR, Levine BA (1987). Evidence that the N-terminal region of A1-light chain of myosin interacts directly with the C-terminal region of actin. A proton magnetic resonance study. *Eur J Biochem,* **164**: 259–266.

Treves S, Vukcevic M, Maj M, Thurnheer R, Mosca B, Zorzato F (2009). Minor sarcoplasmic reticulum membrane components that modulate excitation-contraction coupling in striated muscles. *J Physiol,* **587**: 3071–3079.

Trinick J (1996). Titin as a scaffold and spring. Cytoskeleton. *Curr Biol,* **6**: 258–260.

Trombitas K, Pollack GH (1993). Elastic properties of connecting filaments along the sarcomere. *Adv Exp Med Biol,* **332**: 71–79.

Tskhovrebova L, Trinick J (2003). Titin: Properties and family relationships. *Nat Rev Mol Cell Biol,* **4**: 679–689.

Tskhovrebova L, Trinick J (2008). Giant proteins: Sensing tension with titin kinase. *Curr Biol,* **18**: R1141–R1142.

Tsuchiya T, Iwamoto H, Tamura Y, Sugi H (1993). Measurement of transverse stiffness during contraction in frog skeletal muscle using scanning laser acoustic microscope. *Jpn J Physiol,* **43**: 649–657.

van Dijk SJ, Bezold KL, Harris SP (2014). Earning stripes: Myosin binding protein-C interactions with actin. *Pflugers Arch,* **466**: 445–450.

Vesalius A (1543) *De humani corporis fabrica.*

Von Wegner F, Both M, Fink RH, Friedrich O (2007). Fast XYT imaging of elementary calcium release events in muscle with multifocal multiphoton microscopy and wavelet denoising and detection. *IEEE Trans Med Imaging,* **26**: 925–934.

West TG, Curtin NA, Ferenczi MA, He ZH, Sun YB, Irving M, Woledge RC (2004). Actomyosin energy turnover declines while force remains constant during isometric muscle contraction. *J Physiol,* **555**: 27–43.

Westerblad H, Allen DG (1996). Mechanisms underlying changes of tetanic $[Ca^{2+}]i$ and force in skeletal muscle. *Acta Physiol Scand,* **156**: 407–416.

Winegrad S (1968). Intracellular calcium movements of frog skeletal muscle during recovery from tetanus. *J Gen Physiol,* **51**: 65–83.

Witt SH, Granzier H, Witt CC, Labeit S (2005). MURF-1 and MURF-2 target a specific subset of myofibrillar proteins redundantly: Towards understanding MURF-dependent muscle ubiquitination. *J Mol Biol,* **350**: 713–722.

Worthington CR, Elliott GF (1996). Muscle contraction: The step-size distance and the impulse-time per ATP. *Int J Biol Macromol,* **18**: 123–131.

Yagi N (2011). Mechanism of latency relaxation in frog skeletal muscle. *Prog Biophys Mol Biol,* **105**: 180–186.

Yamashiro S, Gokhin DS, Kimura S, Nowak RB, Fowler VM (2012). Tropomodulins: Pointed-end capping proteins that regulate actin filament architecture in diverse cell types. *Cytoskeleton,* **69**: 337–370.

Yanagida T, Arata T, Oosawa F (1985). Sliding distance of actin filament induced by a myosin crossbridge during one ATP hydrolysis cycle. *Nature,* **316**: 366–369.

Zoghbi ME, Woodhead JL, Moss RL, Craig R (2008). Three-dimensional structure of vertebrate cardiac muscle myosin filaments. *Proc Nat Acad Sci U S A,* **105**: 2386–2390.

Chapter 4

Limitations of in vitro Motility Assay Systems in Studying Molecular Mechanism of Muscle Contraction as Revealed by the Effect of Antibodies to Myosin Head

Haruo Sugi,[a] Shigeru Chaen,[b] Takuya Miyakawa,[c] Masaru Tanokura,[c] and Takakazu Kobayashi[d]

[a]Department of Physiology, Teikyo University School of Medicine, Tokyo, Japan
[b]Department of Integrated Sciences in Physics and Biology,
College of Humanities and Sciences, Nihon University, Tokyo, Japan
[c]Graduate School of Agricultural and Life Sciences,
University of Tokyo, Tokyo, Japan
[d]Department of Electronic Engineering,
Shibaura Institute of Technology, Tokyo, Japan

sugi@kyf.biglobe.ne.jp

Muscle contraction results from ATP-dependent attachment detachment cycle between myosin heads extending from myosin filaments and corresponding sites in actin filaments. Despite extensive studies on extracted muscle protein samples, the molecular mechanism of muscle contraction still remains to be

Muscle Contraction and Cell Motility: Fundamentals and Developments
Edited by Haruo Sugi
Copyright © 2017 Pan Stanford Publishing Pte. Ltd.
ISBN 978-981-4745-16-1 (Hardcover), 978-981-4745-17-8 (eBook)
www.panstanford.com

a matter of debate and speculation. A myosin head consists of catalytic (CAD), converter (COD) and lever arm (LD) domains. To give information about the role of these myosin head domains in muscle contraction, we compared the effect of three site-directed monoclonal antibodies between ATP-dependent in vitro actin–myosin sliding and Ca^{2+}-activated muscle fiber contraction. Antibody 1, attaching to junctional peptide between 50k and 20k myosin heavy chain segments in the CAD, had no appreciable effect on both in vitro actin–myosin sliding and muscle contraction. As these actin-binding sites in the CAD are necessary for rigor actin–myosin linkage formation, it is suggested that myosin heads do not form rigor linkages with actin during muscle contraction. Antibody 2, attaching to reactive lysin residue in the COD, inhibited in vitro actin–myosin sliding without changing actin-activated myosin head ATPase activity, but did not affect muscle contraction. Antibody 3, attaching to two peptides of regulatory light chain in the LD, had no significant effect on in vitro actin–myosin sliding, but reduced isometric tension of muscle fibers without changing their MgATPase activity. The above definite differences in the effect of antibodies 2 and 3 between in vitro actin–myosin sliding and muscle fiber contraction can be accounted for by the random orientation of myosin heads in the former and the regular arrangement of myosin heads in the 3D myofilament lattice in the latter. These results clearly indicate limitations of in vitro motility assay systems in giving insights into molecular mechanism of muscle contraction.

4.1   Introduction

Although more than 50 years have passed since the monumental discovery of Huxley and Hanson (1954) that muscle contraction results from relative sliding between actin and myosin filaments, the mechanism of myofilament sliding still remains to be a matter for debate and speculation. From the standpoint of muscle physiology, intact single muscle fibers are the best material in studying mechanisms of muscle contraction, and vast experimental results are accumulated on contraction characteristics of muscle fibers (for a review, see Sugi, 1998). Since it is difficult to change environment around myofilaments within intact single fibers, single skinned muscle fibers, from which surface membrane is removed to enable chemical substances to diffuse into muscle fibers, are frequently used to bridge muscle physiology and muscle biochemistry (for a review, see Brenner, 1998).

The results obtained from intact and skinned muscle fibers are, however, insufficient to elucidate mechanisms of muscle contraction at the molecular level, because of complex muscle fiber structures as well as huge numbers of myofilaments involved in muscle contraction. In order to study muscle contraction mechanisms without complications, it is desirable to construct simplified experimental systems, consisting of a small number of actin and myosin filaments or molecules. The in vitro motility assay systems have been developed to comply with this desire. A number of novel facts and knowledge have been obtained by these systems (for a review, see Sugi, 1998).

Although the results obtained are interesting, it is not entirely proved whether the in vitro assay systems give information effective in elucidating muscle contraction mechanisms; muscles consist of hexagonal lattice composed of actin and myosin filaments, and it is therefore necessary to ascertain whether actin–myosin interaction taking place in muscle is the same as that taking place in motility assay systems. In this chapter, we describe definite differences in the effect of antibodies to myosin head, indicating the definite limitations of using in vitro motility assay systems to study muscle contraction mechanism.

4.2 Historical Background

Muscle consists mainly of two contractile proteins, myosin and actin. As illustrated in Fig. 4.1, a myosin molecule is divided into two parts; heavy meromyosin (HMM) with two pear-shaped heads (subfragement-1 or S-1) and a rod of 34 nm long, and light meromyosin (LMM) forming a rod of 113 nm long (a). In myosin filament, LMM aggregates to form filament backbone, while HMM extends laterally from the filament backbone with axial interval of 14.3 nm (b). Polarity of myosin filament is reversed across central bare zone, where no HMM is present. On the other hand, actin filament is composed of double helix of globular actin molecules (pitch, 35.5 nm long), and also contains two regulatory proteins, tropomyosin and troponin (c). In muscle, myosin and actin filaments are regularly arranged to form a sarcomere, i.e., structural and functional unit of striated muscle (d). Myosin filaments are located at the center of sarcomere, while actin filaments originate from structure called Z-line, and extend in between myosin filaments. During muscle contraction, actin filaments are further pulled in between myosin filaments, in other words, relative sliding between actin and myosin filaments takes place.

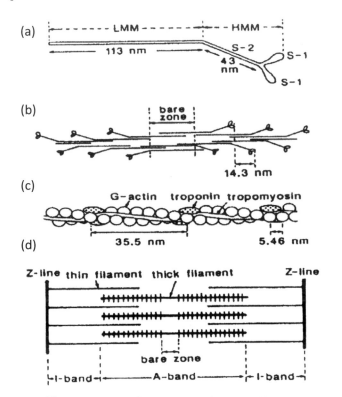

Figure 4.1 Ultrastructure of myosin and actin filament and their arrangement within a single sarcomere. For explanation, see text (Sugi, 1992).

The question arises, what makes the filaments slide past each other? The answer has been given by Huxley (1969), who ingeniously put forward a hypothesis shown in Fig. 4.2. A myosin head extending from myosin filament first attaches to a site on actin filament (upper diagram), changes its configuration to cause sliding of actin filament (middle diagram), and then detach from actin filament (lower diagram). Since axial periodicity is different between actin and myosin filaments, the attachment-detachment cycle takes place asynchronously. As muscle is regarded as an engine utilizing ATP as its fuel, the above attachment-detachment cycle is coupled with ATP hydrolysis. In fact, both actin-binding site and ATPase site are located in myosin head, indicating that myosin heads play a key role in producing muscle contraction.

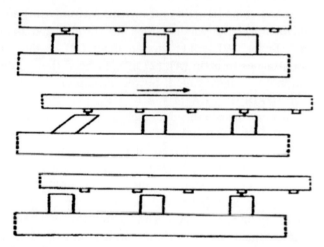

Figure 4.2   Diagrams showing hypothetical attachment-detachment cycle between myosin heads extending from myosin filaments and corresponding myosin-binding sites on actin filaments (Huxley, 1969).

Since the proposal of the cyclic attachment-detachment cycle coupled with ATP hydrolysis, a number of studies have hitherto been made to prove the performance of myosin heads producing sliding between actin and myosin filaments. Despite the extensive studies, the performance of myosin heads during muscle contraction still remains to be a matter for debate and speculation.

4.3 Development of in vitro Motility Assay Systems

Complex structures of muscle fibers prevent us to obtain clear-cut information about the performance of myosin heads producing muscle contraction. Meanwhile, mechanism of muscle contraction can also be studied by using actin and myosin extracted from muscle. In solutions containing actin and myosin, the three-dimensional (3D) filament lattice-structure is broken, so that it is impossible to measure mechanical responses of filament-lattice system. Instead, it is to some extent possible to study cyclic actin–myosin interaction coupled with ATP hydrolysis, by investigating ATPase reaction steps in actin and myosin (actomyosin) solution. The actomyosin ATPase reaction steps are summarized as follows (Lymn and Taylor, 1971): (1) myosin head (M) in the form of $M \cdot ADP \cdot Pi$ attaches to actin (A); (2) myosin head then undergoes conformational changes called power stroke, associated with reaction, $A \cdot M \cdot ADP \cdot Pi \rightarrow A \cdot M + ADP + Pi$; (3) after completion of power stroke, myosin head detaches from actin by binding with next ATP, $A \cdot M + ATP \rightarrow A + M \cdot ATP$; (4) myosin head performs recovery stroke to restore its initial position while hydrolyzing ATP, $M \cdot ATP \rightarrow M \cdot ADP \cdot Pi$, and again attaches to actin. The reaction steps explained above (generally called the Lymn–Taylor scheme) has generally been believed to reflect the attachment-detachment cycle between myosin heads and actin filament shown in Fig. 4.2, since the reaction includes attachment of M to A and detachment of M from A.

In addition to the complete loss of 3D filament lattice structures, however, the biochemical experiments have to be made in conditions that differ too far from those in muscle. The effective myosin head concentration within muscle fibers is estimated to be ~200 µM. In contrast, the myosin molecule (HMM or S-1) concentration in solutions for biochemical experiments cannot be made more than a few µM, due to high viscosity of myosin molecule. Moreover, in the experiments of Lymn and Taylor (1971), the actin concentration was much smaller than that to be required for the maximum steady-state ATPase activity. Eisenberg and coworkers (Eisenberg et al., 1972; Eisenberg and Keilly, 1973; Stein et al., 1981) studied reaction steps during the maximum steady-state ATPase activity of actomyosin solution at low ionic strength, and found an additional ATP hydrolysis pathway in which M remains attached to A. These biochemical studies stimulated muscle investigators to link biochemical studies on actomyosin ATPase and physiological studies on muscle mechanics.

In 1986, Kron and Spudich developed an in vitro motility assay system, in which fluorescently labeled actin filaments were made to slide on myosin heads fixed on a glass surface in the presence of ATP (Fig. 4.3a). This assay system had the following advantages: (1) the ATP-dependent sliding between actin filament and myosin heads can be directly observed on a monitor screen with a light microscope, and can be recorded on videotape for analysis; (2) this system is composed of extracted contractile proteins, so that they can be easily replaced by their proteolytic fragments of recombinant molecules to study molecular mechanism of in vitro actin–myosin sliding. Using this assay system, it has been shown that myosin heads can generate force by connecting a glass microneedle to a sliding actin filament (Ishijima et al., 1991). In this system, however, myosin heads are fixed in random orientations, so that actin filaments do not move in a straight line, but with winding paths. In addition, it is not clear how the leading end of moving actin filaments select myosin heads to interact with them (Huxley, 1990), though some theoretical models have been presented.

Homsher et al. (1992) carefully compared the actin filament sliding velocity on myosin heads (V_f) with the maximum unloaded shortening velocity of muscle fibers (V_u). V_f and V_u were qualitatively similar with respect to their dependence on substrate and product concentrations at temperatures above 20°C, but different with respect to some other factors, indicating that V_f is not always a good analog of V_u. Nevertheless, this assay system has been widely used to determine factors affecting ATP-dependent actin–myosin sliding up to the present time.

A more sophisticated assay system was developed by Finer et al. (1994). They held an actin filament taut between two optical traps by laser light, and brought it into contact with a single myosin head (HMM) fixed on a bead surface (Fig. 4.3b). In the presence of ATP at extremely low concentrations, intermittent displacement or force responses of fairly uniform amplitude are recorded depending on the trap force, and these phenomena only constitute evidence that mechanical responses of a single HMM molecule are actually recorded. In addition to the ambiguity whether the records actually reflect single molecule events, the optical trap method has been widely used up to the present time by many investigators to study ATP-induced mechanical response of single HMM or S-1 molecules with interesting results (Molloy et al., 1995; Debold et al., 2013; Brizendine et al., 2015; Sung et al., 2015).

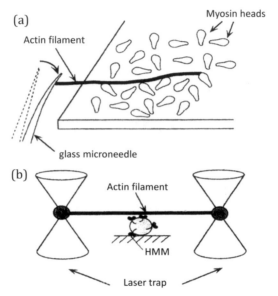

Figure 4.3   In vitro motility assay systems to study ATP-dependent actin–myosin sliding and force generation. (a) Measurement of force of actin–myosin sliding by a microneedle to which an actin filament is attached. Note random orientation of myosin heads fixed on a glass surface. (b) Measurement of force and displacement generated by a single HMM by the optical trap method (Sugi, 1998).

4.4   In vitro Force-Movement Assay Systems

All in vitro motility assay systems described above have serious shortcomings or uncertainties to correlate the results obtained to what is actually taking place in muscle. In the case of actin filament sliding on myosin (Fig. 4.3a), it is not clear how the front end of actin filaments choose myosin molecules fixed on a glass surface (Huxley, 1990). In the case of optical trap systems (Fig. 4.3b), it is not clear how a single myosin head is fixed in position; it seems likely that the mode of fixation of single myosin heads on a bead surface differs too far from the mode of their extension from myosin filaments in 3D filament lattice in muscle, so that it is uncertain whether the results obtained can be useful in giving insights into the phenomena actually taking place in contracting muscle. Although a muscle is a machine to convert chemical energy of ATP hydrolysis into force and motion, none of the above in vitro assay systems can record force and motion generated by myosin simultaneously.

In order to record force and motion generated by myosin heads simultaneously, we developed an in vitro motility assay system, in which myosin-coated glass microneedle was made to slide on well-organized actin filament arrays (actin cables) (Fig. 4.4a) (Chaen et al., 1989; Oiwa et al., 1991). With this assay system, it was possible to measure ATP-dependent displacements and forces simultaneously, so that we could obtain experimental results with respect to force–velocity (under continuously increasing load) and force-work output relations in the in vitro ATP-dependent actin–myosin sliding. In general, the results obtained were similar to those obtained from contracting muscle, indicating that a small number of myosin heads involved in in vitro actin–myosin sliding exhibit characteristics similar to those in contracting muscle (Iwamoto et al., 1990; Wolege, Curtin and Homsher, 1985).

We further developed another force-movement assay system, in which myosin-coated latex beads were made to slide along actin cables mounted on the rotor of a centrifuge microscope. As illustrated in Fig. 4.5a, myosin-coated beads were observed to move along actin cables, mounted on the rotating rotor of a centrifuge microscope under various constant load, i.e., centrifugal forces, imposed on myosin molecules on the bead surface (Fig. 4.5a) (Oiwa et al., 1990). Since the polarity of actin cables was reversed across the central indifferent zone, it was possible to record velocity of bead movement not only under positive loads against bead movement, but also under negative loads pushing the bead. Under a constant load, the bead moved with a constant velocity over many seconds. The steady-state force–velocity relation thus obtained (Fig. 4.5b) was double-hyperbolic in shape, resembling closely that obtained with intact single muscle fibers (Edman, 1988), indicating that a small number of myosin heads retains characteristics similar to single muscle fibers. On the other hand, the force–velocity relation under negative load is at present difficult to explain clearly.

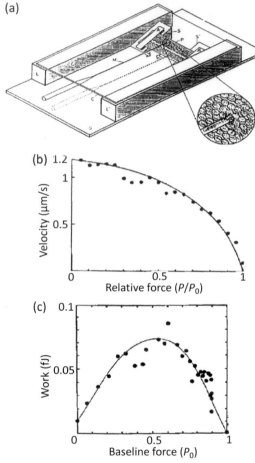

Figure 4.4 In vitro motility assay system to study characteristics of ATP-dependent sliding between myosin-coated glass microneedle and algal cell actin filament arrays (actin cables). (a) Arrangement of the assay system. (b) Dependence of actin–myosin sliding velocity on force (Chaen et al., 1989). (c) Dependence of work done by actin–myosin sliding on force. Oiwa et al. (1991).

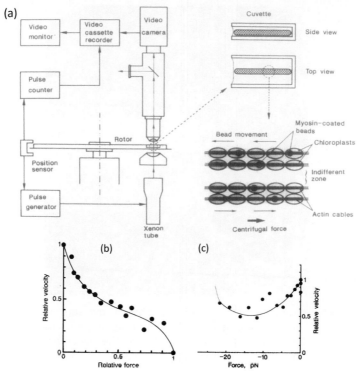

Figure 4.5 A force-movement assay system using the centrifuge microscope. (a) Diagram of experimental arrangement for video recording of ATP-dependent bead movement along actin cables under various constant centrifugal forces. (b) Steady-state force–velocity relation of a myosin-coated bead under various positive loads against bead movement. (c) Steady-state force–velocity relation of a myosin-coated bead under various negative loads (Oiwa et al., 1990).

We examined the mode of attachment of myosin molecules on the latex bead using a high-power scanning microscope (Takahashi et al., 1993). As shown in Fig. 4.6, myosin molecules formed myosin filaments, which run in random directions along the bead surface with myosin heads extending from them. The close similarities in characteristics of the myosin-coated needle or myosin-coated bead versus actin cable system to those of contracting muscle fibers suggest that actin–myosin sliding is caused mainly by myosin heads extending from myosin filaments, which are nearly parallel to actin cables.

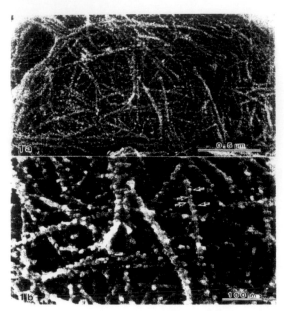

Figure 4.6   Scanning electron micrographs of myosin-coated surface of a polystyrene bead. Numerous myosin filaments run in random directions. Arrows in the lower panel indicate myosin heads extending from myosin filaments (Takahashi et al., 1993).

In this connection, it seems strange that although investigators using the optical trap system intend to explain mechanism of muscle contraction based on their results, they never attempt to examine the mode of attachment of HMM or S-1 to the bead surface electron microscopically; if HMM or S-1 is fixed to the bead surface in the mode entirely different from the mode of extension of myosin heads from myosin filaments, then it becomes uncertain whether their results with optical trap experiments actually contribute to elucidation of muscle contraction mechanism. In fact, using synthetic myosin-rod cofilaments, Tanaka et al. (1998) showed that unitary displacements and forces of myosin heads differ markedly depending on the angle between myosin and actin filaments, indicating that myosin heads exhibit different kinetics according to conditions in which they are allowed to interact with actin filaments. We emphasize that myosin head behavior should be studied in the condition analogous or similar to that in the 3D myofilament lattice structure in muscle.

4.5   Properties of Three Antibodies Used to Position-Mark Myosin Heads at Different Regions within a Myosin Head

In a series of our experimental work to record ATP-induced myosin head movement in hydrated (living) myosin heads, we always used three different monoclonal antibodies to myosin head; antibody 1, binding with junctional peptide between 50k and 20k segments of myosin heavy chain in the myosin head catalytic domain (CAD) (Sutoh et al., 1989); antibody 2 binding with peptides around reactive lysine residue in the myosin head converter domain (COD) (Sutoh et al., 1989); and antibody 3 binding with peptides in two regulatory light chains in the myosin head lever arm domain (LD) (Minoda et al., 2011). Figure 4.7 shows structure of myosin head, in which approximate binding sites of three antibodies are indicated. Figure 4.8 is a gallery of electron micrographs of rotary shadowed myosin molecules with antibodies attached. These antibodies can effectively position-mark myosin heads in hydrated myosin filaments (Sugi et al., 1997, 2008; Minoda et al., 2011).

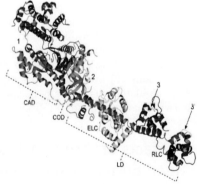

Figure 4.7   Myosin head structure showing approximate sites of binding of antibodies 1, 2, and 3, indicated by numbers 1, 2, and 3 and 3′. Catalytic domain (CAD) consists of 25k (green), 50k (red) and part of 20k (dark blue) fragments of myosin heavy chain, while lever arm domain (LD) consists of the rest of 20k fragment and essential (ELC, light blue) and regulatory (RLC, magenta) light chains. CAD and LD are connected via converter domain (COD). Location of peptides, to which antibodies bind, is colored yellow (Minoda et al., 2011).

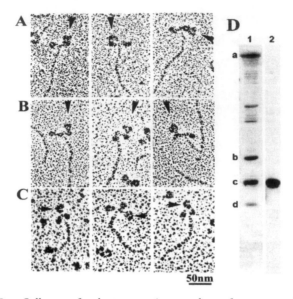

Figure 4.8   Gallery of electron micrographs of rotary shadowed antibody 1, 2, or 3 (IgG)-myosin complexes. IgG is affinity-purified antibody 1 (A), antibody 2 (B) and antibody 3 (C). IgG molecules are indicated by arrowheads. Panels A and B are from Sutoh et al. (1989), while panel C is from Minoda et al. (2011). (D) Binding specificity of antibody 3 for myosin regulatory light chains. Proteins in myosin sample were separated on SDS-PAGE, and transferred onto PVDF membrane. In lane 1, all myosin subunits, i.e., heavy chain (a), essential light chain (b), regulatory light chain (c), and other light chains (d) are visualized by Coomassie staining. In lane 2, regulatory light chain is specifically visualized by Western blot using antibody 3 (Minoda et al., 2011).

4.6 Different Effects of three Antibodies to Myosin Head between in vitro Actin– Myosin Sliding and Muscle Contraction

4.6.1 Antibody 1 (Anti-CAD Antibody) Has No Effect on Both in vitro Actin–Myosin Sliding and Muscle Contraction

As shown in Figs. 4.9 and 4.10, antibody 1, attaching to distal region of myosin head CAD (Sutoh et al., 1989), showed no appreciable effect on both the velocity of actin filament sliding on myosin and the contraction characteristics of skinned muscle fibers, even in concentrations up to 2 mg/ml (Sugi et al., 2014). This may be taken to indicate that though the myosin head CAD contains actin-binding and ATP-binding sites, the CAD is not involved in conformational change of myosin heads producing muscle contraction, in agreement with general view that the CAD remains rigid during attachment-detachment cycle between myosin heads and actin filaments (Geeves and Holmes, 1999).

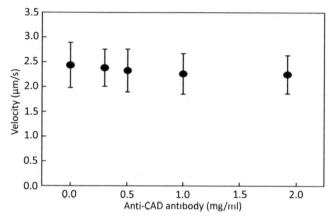

Figure 4.9   No appreciable effect of antibody 1 (anti-CAD antibody) on the velocity of actin filament sliding on myosin. In this figure and Figs. 4.11 and 4.13, vertical bars indicate SD (*n* = 80–120) (Sugi et al., 2014).

4.6.2   Antibody 2 (Anti-RLR Antibody) Inhibits in vitro Actin–Myosin Sliding, but Has No Appreciable Effect on Muscle Contraction

In contrast with no appreciable effect of ant-CAD antibody, attaching to peptides in the myosin head CAD, on both in vitro actin–myosin sliding and muscle contraction, antibody 2 (anti-RLR antibody), attaching to peptides around reactive lysine residue (Lys 83) in the myosin head COD (Sutoh et al., 1989), was found to inhibit in vitro actin–myosin sliding without affecting muscle contraction (Sugi et al., 2014). As shown in Fig. 4.10, anti-RLR antibody completely inhibited sliding movement of actin filaments on myosin even in low concentrations >0.1 mg/ml. Despite its marked inhibitory effect on in vitro actin–myosin sliding, ant-RLR antibody showed no appreciable effect on the actin-activate ATPase activity of myosin heads (S-1). On the other hand, anti-RLR antibody showed no appreciable effect on muscle contraction (Fig. 4.12).

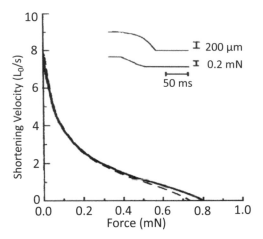

Figure 4.10 No appreciable effect of antibody 1 (anti-CAD antibody) on the force–velocity relation of Ca^{2+}-activated skinned muscle fibers. In this figure and Fig. 4.11, solid and broken lines indicate force–velocity curves before and after application of anti-CAD antibody (2 mg/ml), respectively. Inset shows the method of obtaining force–velocity curves by applying ramp decreases in force to isometrically contracting fibers (Sugi et al., 2014).

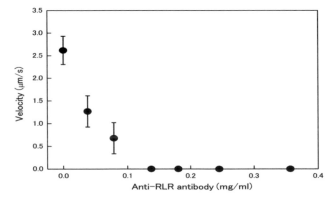

Figure 4.11 Marked inhibitory effect of anti-RLR antibody on in vitro actin filament sliding velocity on myosin (Sugi et al., 2014).

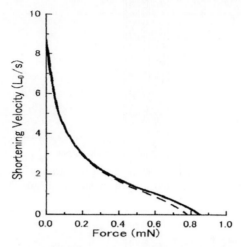

Figure 4.12 No appreciable effect of anti-RLR antibody on the force–velocity relation of Ca^{2+}-activated skinned muscle fibers (Sugi et al., 2014).

4.6.3 Antibody 3 (Anti-LD Antibody) Shows No Marked Inhibitory Effect on in vitro Actin–Myosin Sliding, but Has Inhibitory Effect on Ca^{2+}-Activated Muscle Contraction

As shown in Fig. 4.13, the velocity of actin filament on myosin did not change significantly by anti-LD antibody, attaching to two light chains in the myosin head LD in concentrations up to 2 mg/ml (Sugi et al., 2014), although the velocity of actin–myosin sliding exhibited a tendency to decrease with increasing antibody concentration. In contrast, however, the isometric tension development in Ca^{2+}-activated muscle fibers was found to decrease in the presence of anti-LD antibody in a concentration dependent manner (Fig. 4.14). On the other hand, the maximum unloaded velocity of shortening in Ca^{2+}-activated fibers remained unchanged despite the decrease of isometric tension (Fig. 4.15a). If the isometric tension was normalized with respect to the maximum value, the force–velocity curves, obtained at different antibody concentrations, were found to be identical (Fig. 4.15b), indicating that the kinetic properties of muscle fibers did not change with decreasing isometric tension development (Sugi et al., 2014).

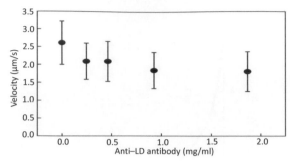

Figure 4.13 No significant effect of anti-LD antibody on the velocity of in vitro actin–myosin sliding (Sugi et al., 2014).

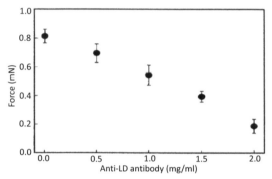

Figure 4.14 Inhibitory effect of anti-LD antibody on the maximum Ca^{2+}-activated isometric tension development of skinned muscle fibers (Sugi et al., 2014). Vertical bars, S.E.M. ($n = 4\sim6$).

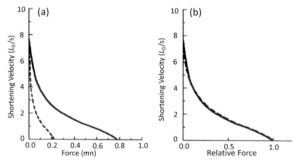

Figure 4.15 Effect of anti-LD antibody on the force–velocity relation of Ca^{2+}-activated skinned muscle fibers. (a) Force–velocity curves before (solid line) and after (broken line) application of anti-LD antibody (1.5 mg/ml). (b) The same force–velocity curves with tension expressed relative to the maximum value (Sugi et al., 2014).

Figure 4.16 shows simultaneous recordings of MgATPase activity (upper traces) and Ca^{2+}-activated isometric tension (lower traces), obtained before (a) and after (b) application of anti-LD antibody (2 mg/ml). The MgATPase activity of muscle fibers (slope of upper traces) did not change appreciably despite marked decrease in isometric tension (Sugi et al., 2014).

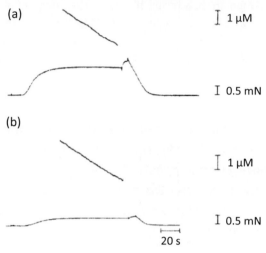

Figure 4.16 Simultaneous recordings of MgATPase activity, measured by decrease of NADH (upper traces) (Sugi et al., 1992, 2013), and Ca^{2+}-activated isometric tension development (lower traces). Note the slope of the ATPase records does not change appreciably (Sugi et al., 2014).

4.7 Definite Differences in the Mechanism between in vitro Actin–Myosin Sliding and Muscle Contraction as Revealed by the Effect of Antibodies to Myosin Head

4.7.1 Evidence That Myosin Heads Do Not Pass through Rigor Configuration during Their Cyclic Attachment-Detachment with Actin Filaments

The two main actin-binding sites in myosin head are located at both sides of junctional peptide between 50k and 20k segments of myosin heavy chain (e.g., Bagshaw, 1993). If bulky anti-CAD antibody (IgG) binds to this junctional peptide, the actin-binding sites are completely covered by the antibody molecule, so that myosin heads can no longer form rigor linkages with actin filaments. Nevertheless, both in vitro actin–myosin sliding and muscle contraction are not affected by anti-CAD antibody (Figs. 4.9 and 4.10). This implies that during cyclic actin–myosin interaction, myosin heads do not take the form of A•M in the Lymn–Taylor biochemical scheme (1971), but take another kind of linkages with actin filaments, possibly by an electrostatic mechanism. We reserve further discussions in this chapter, although experiments are going on to investigate this issue in more detail.

4.7.2 The Finding That Anti-RLR Antibody Inhibits in vitro Actin–Myosin Sliding but Not Muscle Contraction Suggests That Myosin Head Flexibility at the Converter Domain Is Necessary for in vitro Actin–Myosin Sliding but Not for Muscle Contraction

Mainly based on crystallographic studies on nucleotide-dependent structural changes of myosin head crystal detached from myosin filaments (e.g., Geeves and Holmes, 1999), it is generally believed that during cyclic interaction with actin filaments, the myosin head CAD remains rigid, while the active rotation of the LD around the COD produces sliding between actin and myosin filaments. This mechanism is called the lever arm mechanism. In fact, chemical modification of the reactive lysine residue, located central in the COD, inhibits both in vitro actin–myosin sliding and actin activated myosin head (S-1) ATPase activity (Muhlrad et al., 2003). They explained their results as being due to steric hindrance causing decrease of flexibility of the COD, resulting from the chemical modification of reactive lysine residue.

In contrast, however, we have found that binding of bulky anti-RLR antibody to the reactive lysine residue markedly inhibits in vitro actin–myosin sliding, but has no appreciable effect on Ca^{2+}-activated muscle fiber contraction (Figs. 4.11 and 4.12). This may be taken to indicate that the COD flexibility is essential for in vitro actin–myosin sliding, but not for muscle contraction. In the in vitro assay system, myosin (HMM or S-1) molecules are fixed on a glass surface in random directions with variable mode of attachment to the glass surface. As a result, the COD flexibility is essential for myosin heads to start interacting with actin filaments, approaching to myosin heads in variable directions. In muscle fibers, on the other hand, each myosin filament is surrounded by six actin filaments in the hexagonal myofilament-lattice structure, and myosin heads are always in close proximity of actin filaments. In such a condition, COD flexibility is not necessary for myosin heads to interact with actin filaments. In the work of Muhlrad et al. (2003), chemical modification of reactive lysine residue may cause structural changes not only at the COD, but also at the CAD to result in the loss of actin-activated ATPase activity of myosin heads. Meanwhile, attachment of anti-RLR antibody to reactive lysine residue is reversible (Sugi et al., 2014) and may not produce any structural changes in both the COD and the CAD. Consequently, anti-COD antibody does not inhibit actin-activated ATPase activity of the CAD in myosin (HMM or S-1) molecules in the motility assay system (Sugi et al., 2014). If the above explanations are correct, the mode of interaction of myosin heads with actin filaments in in vitro assay systems definitely differs from that in muscle fibers.

4.7.3 The Finding That Anti-LD Antibody Inhibits Muscle Contraction but Not in vitro Actin–Myosin Sliding Suggests that Movement of the LD Is Necessary for Muscle Contraction but Not for in vitro Actin–Myosin Sliding

Anti-LD antibody, attaching to the myosin head LD, had no significant effect on in vitro actin–myosin sliding (Fig. 4.13), but inhibited Ca^{2+}-activated muscle fiber tension development in a concentration-dependent manner (Fig. 4.14) without affecting the MgATPase activity (Fig. 4.15), indicating the LD movement is necessary for muscle contraction but not for in vitro actin–myosin sliding (Sugi et al., 2014).

Figure 4.17 shows the myosin head structure (a) and diagrams of hypothetical myosin head power and recovery strokes (b). Since myosin heads are connected to myosin filament backbone via myosin subfragement-2 (S-2), it is clear that myosin head CAD rotates not only around the COD, but also around the LD-S-2 junction. In the lever arm mechanism, constructed from nucleotide-dependent structural changes of myosin head crystal as well as acto-S1 complex (e.g., Geeves and Holmes, 1999), rotation of myosin heads around the LD-S-2 junction seems to be regarded as passive rotation caused by active rotation of the CAD around the COD.

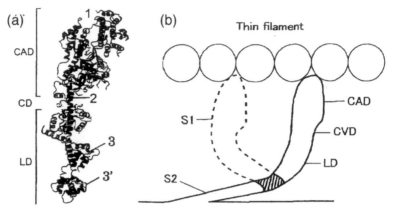

Figure 4.17 (a) Structure of myosin head with approximate attachment regions of antibodies 1, 2, and 3. (b) Diagram showing hypothetical myosin head configurations before (solid line) and after (broken line) after power stroke. Shaded area indicates the myosin head LD-S-2 junction (Sugi et al., 2014).

Evidence for the essential role of myosin head LD has been, however, presented by Sugi and Harrington. They report that polyclonal antibody to myosin head LD inhibits tension development of Ca^{2+}-activated muscle fibers without affecting the MgATPase activity (Sugi et al., 1992). The inhibitory effect of polyclonal antibody was dependent on both time and antibody concentration, being similar to the inhibitory effect of anti-LD antibody. It may be that if anti-LD antibody attaches to myosin head LD, it inhibits rotation of myosin heads around the LD-S-1 junction to result in inhibition of tension development of muscle fibers. In the presence of anti-LD antibody, the isometric tension development decreased gradually with increasing antibody concentration, while the maximum unloaded shortening velocity remained unchanged (Figs. 4.14 and 4.15). These results indicate a gradual decrease in the number of myosin heads responsible for muscle fiber contraction, reflecting gradual antibody attachment to individual myosin heads, since myosin heads, to which the antibody is not yet attached, are believed to retain their normal contraction characteristics and kinetic properties (Sugi et al., 2014).

The ineffectiveness of anti-LD antibody on in vitro actin–myosin sliding (Fig. 4.13) may result from that in the majority of myosin heads fixed in position in the assay system, their LD is fixed to the glass surface, so that the fixed LD cannot contribute to actin filament sliding on myosin. In this connection, the tendency of actin filament sliding velocity to decrease with increasing antibody concentration (Fig. 4.13) seem to suggest that in a small proportion of myosin heads, their LD is not completely fixed to the glass surface.

4.8 Conclusion

In this chapter, we first described historical background of research work to clarify mechanisms of muscle contraction at the molecular level, and then explained the development of in vitro motility assay systems to bridge biochemical experiments with physiological experiments. By using three different antibodies, which effectively position-mark three different regions within a myosin head in our research work with the gas environmental chamber (Sugi et al., 1997, 2008; Minoda et al., 2011), we studied difference in the effect of the three antibodies on in vitro actin–myosin sliding and muscle fiber contraction. The results obtained indicate definite differences in the mechanism of cyclic actin–myosin interaction between in vitro actin–myosin sliding and muscle fiber contraction, and explained the results in terms of myosin head state in the motility assay system and in muscle. It should also be noted that the same uncertainties exist with respect to the mode of fixation of single myosin heads (HMM or S-1) to a bead surface in the optical trap experiments.

Finally we emphasize that although the results obtained from in vitro motility assay systems are interesting, they are not necessarily useful in elucidating what actually taking place in contracting muscle. It is therefore absolutely necessary to develop new motility assay systems, in which three-dimensional filament-lattice structure in muscle can be preserved.

References

Bagshaw, C. R. (1993) *Muscle Contraction*, Chapman and Hall, London, pp. 155.

Brenner, B. (1998) Muscle mechanics II: Skinned muscle fibers. In *Current Methods in Muscle Physiology* (Sugi, H., ed.), Oxford University Press, Oxford, pp. 33–69.

Brizendine, R. K., Alcala, D. B., Carter, M. C., Heldeman, B. D., Facemyer, K. C., and Baker, J. E. (2015) Velocities of unloaded muscle filaments are not limited by drag forces imposed by myosin cross-bridges. *Proceedings of the National Academy of Sciences of the USA*, 112: 11235–11240.

Chaen, S., Oiwa, K., Shinnmen, T., Iwamoto, H., and Sugi, H. (1989) Simultaneous recordings of force and sliding movement between a myosin-coated glass microneedle and actin cables in vitro. *Proceedings of the National Academy of Sciences of the USA*, 86: 1510–1514.

Debold, E. P., Turner, M. A., Stout, J. C., and Walcott, S. (2011) Phosphate enhances myosin-powered actin filament velocity under acidic conditions in a motility assay. *American Journal of Physiology, Regular, Integrated and Comparative Physiology*, 300: R1401–R1408.

Eisenberg, E., Dobkin L., Kielley WW. (1992). Heavy meromyosin: evidence for a refractory state unable to bind to actin in the presence of ATP. *Proceedings of the National Academy of Science of the USA*, **69**, 667–671.

Eisenberg, E., and Kielly, W. W. (1973) Evidence for a refractory state of heavy meromyosin and subfragment-1 unable to bind to actin in the presence of ATP. *Cold Spring Harbor Symposium for Quantitative Biology*, 37: 145–152.

Finer, J. T., Simmons, R. M., and Spudich, J. M. (1994) Single myosin molecule mechanics: Piconewton forces and nanometer steps. *Nature*, 368: 113–119.

Geeves, M. A., and Holmes, K. C. (1999) Structural mechanism of muscle contraction. *Annual Review of Biochemistry*, 68: 687–728.

Homsher, E., Wang, F., and Sellers, J. R. (1992) Factors affecting movement of F-actin filaments propelled by skeletal muscle heavy meromyosin. *American Journal of Physiology*, 31: C714–C723.

Huxley, H. E. (1969) The mechanism of muscular contraction. *Science*, 164: 1356–1366.

Huxley, H. E. (1990) Sliding filaments and molecular motile systems. *Journal of Biological Chemistry*, 265: 8347–8350.

Huxley, H. E., and Hanson, J. (1954) Changes in the cross-striation of muscle during contraction and stretch and their structural interpretation. *Nature*, 173: 973–976.

Ishijima, A., Doi, T., Sakurada, K., and Yanagida, T. (1991) Sub-piconewton force fluctuations of actomyosin in vitro. *Nature*, 352: 301–306.

Kron, S. J., and Spudich, J. A. (1986) Fluorescent actin filaments move on myosin fixed to a glass surface. *Proceedings of the National Academy of Sciences of the USA*, 83: 6272–6276.

Lymn, R. W., and Tayloe, E. W. (1971) Mechanism of adenosine triphosphate hydrolysis by actomyosin. *Biochemistry*, 10: 4617–4624.

Minoda, H., Okabe, T., Inayoshi, Y., Miyakawa, T., Miyauchi, Y., Tanokura, M., Katayama, E., Wakabayashi, T., Akimoto, T., and Sugi, H. (2011) Electron microscopic evidence for the myosin head lever arm mechanism in hydrated myosin filaments using the gas environmental chamber. *Biochemical and Biophysical Research Communications*, 405: 651–656.

Molloy, J. E., Burns, J. E., Kendrick-Jones, J., Tregear, R. T., and White, D. C. S. (1995) Movement and force produced by a single myosin head. *Nature* 378: 209–212.

Muhlrad, A., Peyser, Y. M., Nili, M., Ajtai, K., Reisler, E., and Burghardt, T. P. (2003) Chemical decoupling of ATPase activation and force production from the contractile cycle in myosin by steric hindrance of lever-arm movement. *Biophysical Journal*, 84: 1047–1056.

Oiwa, K., Chaen, S., Kamitsubo, E., Shinnmen, T., and Sugi, H. (1990) Steady-state force-velocity relation in the ATP-dependent sliding movement of myosin-coated beads on actin cables in vitro studied with a centrifuge microscope. *Proceedings of the National Academy of Sciences of the USA*, 87: 7893–7897.

Oiwa, K., Chaen, S., and Sugi, H. (1991) Measurement of work done by ATP-induced sliding between rabbit muscle myosin and algal cell actin cables in vitro. *Journal of Physiology (London)*, 437: 751–763.

Stein, L. A., Chock, P. B., and Eisenberg, E. (1981) Mechanism of actomyosin ATPase: effect of actin on the ATP hydrolysis step. *Proceedings of the National Academy of Sciences of the USA*, 78: 1346–1350.

Sugi, H. (1992) Molecular mechanism of actin-myosin interaction in muscle contraction. In *Muscle Contraction and Cell Motility:*

Advances in Comparative & Environmental Physiology (Sugi, H., ed.), vol. 12, Springer, Berlin & Heidelberg.

Sugi, H., Akimoto, T., Sutoh, K., Chaen, S., Oishi, N., and Suzuki, S. (1997) Dynamic electron microscopy of ATP-induced myosin head movement in living muscle thick filaments. *Proceedings of the National Academy of Sciences of the USA*, 94: 4378–4382.

Sugi, H., Chaen, S., Kobayashi, T., Abe, T., Kimura, K., Saeki, Y., Ohnuki, Y., Miyakawa, T., Tanokura, M., and Sugiura, S. (2014). Definite differences between in vitro actin myosin sliding and musle contraction as revealed using antibodies to myosin head. *PLOS ONE*, 9: 93272.

Sugi, H., Kobayashi, T., Gross, T., Noguchi, K., and Karr, T. (1992) Contraction characteristics and ATPase activity of skeletal muscle fibers in the presence of antibody to myosin head subfragment-2. *Proceedings of the National Academy of Sciences of the USA*, 89: 6134–6137.

Sugi, H., Minoda, H., Inayoshi, Y., Yumoto, F., Miyakawa, T., Miyauchi, Y., Tanokura, M., Akimoto, T., Kobayashi, T., Chaen, S., and Sugiura, S. (2008) Direct demonstration of the cross-bridge recovery stroke in muscle thick filaments in aqueous solution by using the hydration chamber. *Proceedings of the National Academy of Sciences of the USA*, 105: 17396–17401.

Sugi, H., and Tsuchiya, T. (1998) Muscle mechanics I: Intact single muscle fibres. In *Current Methods in Muscle Physiology: Advantages, Problems and Limitations* (Sugi, H., ed.), Oxford University Press, Oxford, pp. 3–31.

Sung, J., Ng, S., Mortensen, K. I., Vestergaad, C. L., Sutton, S., Ruppel, K., Flyvbjerg, H., and Spudich, J. A. (2015) Harmonic force spectroscopy measures load-dependent kinetics of individual human β-cardiac myosin molecules. *Nature Communications*, 6: 7931.

Sutoh, K., Tokunaga, M., and Wakabayashi, T. (1989) Electron microscopic mapping of myosin head with site-directed antibodies. *Journal of Molecular Biology*, 206: 357–363.

Takahashi, I., Oiwa, K., Kawakami, T., Tanaka, H., and Sugi, H. (1993) Scanning electron microscopy of the myosin-coated surface of polystyrene beads in a force-movement assay system for ATP-dependent actin-myosin sliding. *Journal of Electron Microscopy*, 42: 334–337.

Tanaka, H., Ishijima, A., Honda, M., Saito, K., and Yanagida, T. (1998) Orientation dependence of displacements by a single one-headed myosin relative to the actin filament. *Biophysical Journal*, 75: 1886–1894.

Chapter 5

Characteristics and Mechanism(s) of Force Generation by Increase of Temperature in Active Muscle

K. W. Ranatunga

School of Physiology & Pharmacology,
University of Bristol, Medical Sciences Building,
Bristol BS8 1TD, England, UK

k.w.ranatunga@Bristol.ac.uk

This chapter reviews the basic observations made in temperature-perturbation experiments on active mammalian skeletal muscle. In the isometric muscle, a small rapid temperature-jump (T-jump) induces a characteristic biphasic tension rise, where the initial fast phase represents force generation in attached crossbridges; thus, crossbridge force rises when heat is absorbed, endothermic. Evidence suggests that a T-jump enhances a pre-phosphate release step in the acto-myosin (crossbridge) ATPase cycle. During steady shortening, the T-jump force generation is enhanced, whereas during steady lengthening, the T-jump force generation is depressed. The sigmoidal temperature dependence of steady

Muscle Contraction and Cell Motility: Fundamentals and Developments
Edited by Haruo Sugi
Copyright © 2017 Pan Stanford Publishing Pte. Ltd.
ISBN 978-981-4745-16-1 (Hardcover), 978-981-4745-17-8 (eBook)
www.panstanford.com

active force may be largely due to the endothermic nature of force generation. Compared to the isometric curve, the force versus temperature curve is sharper and shifted to higher temperatures with steady shortening and to lower temperatures with steady lengthening in a velocity dependent manner. Overall, the results would be compatible with the notion that during steady lengthening the force generation is depressed and ATPase cycle is slowed, whereas during shortening this step and the ATPase cycle are accelerated. What structural mechanism underlies this end othermic crossbridge force generation still remains unclear.

5.1   Introduction

From experiments on isolated amphibian and mammalian skeletal muscles and on in situ human muscle, it was established more than half-century ago (Hadju, 1951; Clarke et al., 1958) that the process of muscle contraction is sensitive to temperature. Subsequent studies on active mammalian muscle showed that the maximal force increases by ∼2-fold, shortening velocity increases by ∼6-fold and power increases by >10-fold in warming from 10°C to physiological (>30°C) temperatures (see review, Ranatunga, 2010). Thus, determination of temperature-sensitivity of force and force generation induced by temperature-jump (T-jump) are pertinent techniques in examining the underlying mechanisms of muscle force generation.

This review lists some of the main findings from steady-state and rapid-perturbation temperature studies on mammalian muscle. Where appropriate, they will be examined in relation to the molecular process that is thought to drive muscle contraction, namely, the ATP-driven, cyclic inter-action of myosin heads (crossbridges) of the thick (M = myosin) with thin (A = actin) filaments in a sarcomere (Huxley, 1957; Huxley, 1969; Lymn and Taylor, 1971; White and Taylor, 1976).

Our analyses show that in active muscle, the force generation in attached crossbridges is endothermic and it occurs before the release of inorganic phosphate in the acto-myosin cycle; it is strain-sensitive being enhanced during shortening and depressed during lengthening. The exact structural mechanism that underlies this endothermic crossbridge force generation, however, remains unresolved. It may involve a transition of non-stereo-specifically attached to stereo-specifically attached crossbridge states (Zhao and Kawai, 1994) and/or a "protein unfolding/folding-process" within a crossbridge (myosin head) (Davis and Epstein, 2007). Either or a combination of such molecular mechanisms is possible. These temperature-dependence studies on muscle seem to indicate some shortcomings in our understanding of the process of muscle force generation.

5.2   Methods and Materials

The experiments reviewed here are findings from electrically stimulated intact fibre bundles and chemically activated skinned fibres. Some of the essential features of methodologies and considerations given are listed below, since comprehensive details have been published elsewhere (see references in Ranatunga, 2010).

5.2.1   Experimental Techniques and Procedures

The rapid T-jump used was one in which a laser pulse heated a volume of aqueous solution (>500 × fibre volume) bathing the muscle fibres. The trough system, mounted on an optical microscope stage, consisted of three ~50 µl troughs milled in a titanium block and a front experimental trough with glass windows in the front and bottom (Ranatunga, 1996). The temperature of the whole trough system was kept <5°C but, by Peltier modules assembled on its back wall, the front trough temperature could be independently clamped to the experimental temperature. The temperature sensitivity of the force (=tension) recording was reduced by using two AE 801 elements (Akers, Norway): one cut-beam element was connected to the muscle fibre (natural resonant frequency ~14 kHz) and the other set close-by acted as a dummy, forming a full bridge (Ranatunga, 1999). A moving-coil motor built by the author was used to produce ramp and constant velocity length changes in muscle fibres.

A T-jump was induced in the front trough by a near infra-red (λ = 1.32 µm) pulse of 200 µs duration (maximum power = 2 J per pulse) from a Nd-YAG laser (Schwartz Electro-Optics Inc., Florida, U.S.A.). The energy absorption by water at this wavelength (~50%) was such that the laser pulse that entered the trough through the front window was reflected back by the aluminium foil in the back wall and raised the temperature of the 50 μl aqueous medium in the trough, and the fibre immersed in it, by 3–5° in 200 μs. The raised solution temperature remained constant for ~500 ms, but the duration could be increased by using the Peltier T-clamping (see Ranatunga, 1996).

5.2.2 Muscle Preparations

Intact fibre experiments referred to here were done on bundles of fibres isolated from fast foot muscle (flexor hallucis brevis) of adult male rats (Coupland and Ranatunga, 2003); animals were killed with an intra-peritoneal injection of an overdose (~150 mg kg^{-1} body weight) of Sodium Pentobarbitone (Euthatal, Rhône Mérieux). Small bundles of ~5–10 intact excitable fibres were dissected from the mid-belly of the whole muscle and aluminium foil T-clips were fixed on the tendons within 0.2 mm of the fibre-ends; at sarcomere length ~2.5 μm, the resting fibre length was ~2 mm and the bundle width was 100–200 μm. A preparation was super-fused with physiological saline solution containing (mM) NaCl, 109; KCl, 5; $MgCl_2$, 1; $CaCl_2$, 4; $NaHCO_3$, 24; $NaH_2 PO_4$, 1; sodium pyruvate, 10 and 200 mg l^{-1} of bovine foetal serum; the solution was bubbled with a mixture of 95% O_2 and 5% CO_2. A preparation was directly stimulated with supra-maximal voltage pulses (<0.5 ms duration) applied to two platinum plate electrodes placed on either side of a bundle (Coupland and Ranatunga, 2003).

Skinned fibre experiments were done on rabbit Psoas fibres. Bundles of fibres from psoas muscle of adult male rabbits (killed by an intravenous injection of an overdose of sodium pentobarbitone) were chemically skinned using 0.5% Brij 58. A single fibre segment (2–4 mm long) was fixed, using nitro-cellulose glue, between the force transducer hook and the motor hook. The sarcomere length in a 0.5 mm region monitored using He-Ne laser diffraction (see Ranatunga et al., 2002), near the tension transducer was set at ~2.5 μm. The buffer solutions contained 10 mM glycerol-2-phosphate as a low temperature-sensitive pH buffer and also 4% Dextran (mol. wt. ~500 kDa) to compress the filament lattice spacing to normal dimensions; major anion was acetate, ionic strength = 200 mM, pH = 7.1 (see Coupland et al. (2001, 2005a) for further details).

5.2.3 Abbreviations, Nomenclature and Data Analyses

Following abbreviations are used: L = muscle fibre length and L_0 = optimal fibre length, P = force (=tension), P_0 = maximal isometric force, P_2 = force during steady shortening/lengthening, P_{max} = P_0 at 30–35°C. T = temperature, $T_{0.5}$ = temperature at which force in a sigmoidal plot is half-maximal.

Tension data in some figures are fitted with a sigmoidal curve (see Coupland et al., 2001), $P = P_{max} - P_{max}/[1 + \exp\{(-\Delta H/R)(1/T - 1/T_{0.5})\}]$, where P_{max} is maximal force, R is 8.314 J mol^{-1} K^{-1}, T is absolute temperature (in K), ΔH is enthalpy change and $T_{0.5}$ is temperature corresponding to $0.5 P_{max}$, i.e. where ΔG (free energy change) is minimal. The sigmoidal curve is derived from the vant Hoff equation. Indeed, when the force data were examined as a vant Hoff plot, $\text{Log}(P/(P_{max} - P))$ versus $(1/T)$ (see Brown, 1957), the force data distribution was approximately linear (see Roots and Ranatunga, 2008). Taking $\text{Log}(P/(P_{max} - P)) = (-\Delta H/RT) + (\Delta S/R)$ and ΔS is entropy change, this indicated that, in the temperature range of 5 to 30–35°C, $\Delta H°$ is temperature-independent. Also, $T_{0.5}$ represents the temperature at which $\text{Log}(P/(P_{max} - P)) = 0$ in the vant Hoff plot, i.e. $\Delta G° = 0$ and $T_{0.5} = \Delta H°/\Delta S°$; for maximally active mammalian muscle in the isometric state and in different studies, this temperature is ~283 K (~10°C) and $\Delta H°$ is ~120 kJmol^{-1}.

The nomenclature used in labelling the homologous components of tension responses induced by different perturbations is also the same as in our previous papers (see Ranatunga, 2010). From bi-exponential curve fitting, the tension rise above the pre-perturbation level, induced by a T-jump (and a pressure-release, P-jump) on isometric muscle consists of two phases, phase 2b and phase 3. Phase 3 is the slower component and is insensitive to Pi and MgADP. In shortening muscle T-jump tension rise is monophasic and phase 3 is not seen (Ranatunga, 2010). A T-jump (and P-jump) can induce a concomitant drop in force due to expansion in some series elasticity (see Goldman et al., 1987; Ranatunga et al., 1990) and it compares with phase 1 in length-release experiments (Huxley and Simmons, 1971). With a length-release, phase 1 is followed by partial quick tension recovery (phase 2) and this tension recovery is biphasic (Davis and Harrington, 1993)—with phase 2a (fast) and phase 2b

(Ranatunga et al., 2002). In summary, particular emphasis is phase 2b that is identified as endothermic force generation.

5.3   Temperature Dependence of Steady Force

Three functional and mechanics states may be identified in skeletal muscle, namely the resting (or relaxed), the rigor and the active states. In resting state, the crossbridges are detached and resistance to stretch arises from stretch of non-crossbridge structures in sarcomeres, such as titin filaments. In rigor, with no ATP, all crossbridges remain attached to actin but not cycling, whereas in active state, crossbridges are cycling, i.e. they attach to actin, develop force and detach. Interestingly these three states also differ with respect to temperature-effect on their tension (Ranatunga, 1994). The resting muscle tension is largely temperature-insensitive. The rigor muscle tension decreases linearly with increase of temperature so that it drops synchronously with a T-jump, indicating thermal expansion of its elasticity by heat. An active muscle tension rises with temperature-rise and a T-jump leads to a characteristic tension generation. In sub-maximally active contractions (e.g. twitch), tension responses and behaviour to T-jump are complicated due to excitation-contraction processes (see Coupland et al., 2005b). It was shown experimentally that the tetanic tension versus (sarcomere) length relation in rat muscle remains similar at low and high temperatures (15 and 30°C) except that the tension at the full range of length was higher at the higher temperature (Elmubarak and Ranatunga, 1984); thus, ignoring small thermal expansion effect, the sarcomeric structure, filament lengths, etc. remain virtually the same at different temperatures. This review deals with temperature-effects in maximally active mammalian muscle (fibres) at optimal sarcomere length range and, in different experiments, over the temperature range of ~5 to 35–40°C.

5.3.1   Isometric Force and Force during Shortening/Lengthening

The isometric tetanic contractions in Fig. 5.1a show that the maximal steady force in an intact muscle fibre bundle is temperature-sensitive, it increases with temperature. Figure 5.1b shows pooled tension (P_0) data from several such experiments and they show that the relation between force and (reciprocal) temperature is non-linear and sigmoidal with a half-maximal tension at $\sim$10°C $(T_{0.5})$; the slope at this corresponded in different experiments to $\Delta H°$ of $\sim$120 kJmol^{-1}.

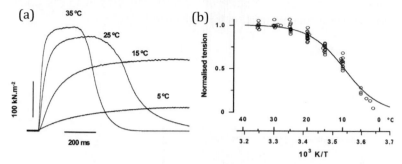

Figure 5.1 (a) Isometric tetanic contractions from an intact fibre bundle. (b) Tetanic tension data from eight bundles plotted (as a ratio of that at 35°C) against reciprocal absolute temperature (adapted from Coupland and Ranatunga, 2003).

Experimental studies show that stiffness in active isometric muscle remains unchanged (see ref in Roots and Ranatunga, 2008) or tension/stiffness ratio is increased (Galler and Hilber, 1998) with increase of temperature. Single-molecule experiments of Kawai et al. (2006) showed that the force that a crossbridge generates is independent of temperature and a number of structural and other studies have shown that the average force (or strain) per crossbridge in active muscle is higher at a higher temperature (Griffiths et al., 2002; Linari et al., 2005; Colombini et al., 2008). Hence, the basis for steady active force in isometric muscle may be simplified to a two state system—pre-force generating (low-force) and force generating (high-force) states in equilibrium; while the total number of attached crossbridges remains the same, higher temperature favours (endothermic) force generation. In principle, such a simplistic scheme can account for the sigmoidal temperature dependence as being due to the shift in the equilibrium from low- to high-force states (Davis, 1998; Roots and Ranatunga, 2008) on heating.

Figure 5.2 shows tension records from a fibre bundle at 10°C (a) and 35°C (b) in which ramp lengthening and shortening at two velocities were applied on isometric tension plateau. The resultant four traces superimposed on an isometric contraction, show that the tension declines during shortening and rises during lengthening. The tension response during shortening/lengthening decreases in slope (referred to as the P_2 transition, Pinniger et al., 2006; Roots et al., 2007). The P_2 tension is taken as steady crossbridge tension for a given velocity.

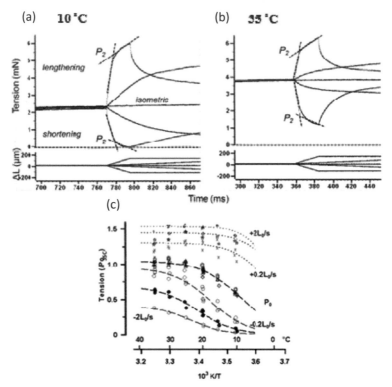

Figure 5.2 (a, b) A bundle was stimulated tetanically and ramp shortening (–) or lengthening (+) of ~6% L_0 at two velocities applied. The resultant four traces are superimposed on an isometric contraction. The P_2 tension was measured (see Roots et al., 2007, 2012) at intersection between two lines fitted to the tension record before and after the P_2 transition. The dashed line is zero active force level. (c) Normalised (to P_0 at 35°C) tension data from three fibres. P_0 (△, ◇, □) and P_2 for shortening (●, ○) at a given velocity increase with warming. For lengthening (+, ×), the tension was not correlated with temperature ($P > 0.05$).

Figure 5.2c shows P_2 and P_0 tension data plotted against $1/T$. It is seen that with warming isometric tension and the shortening tension increase sharply and each set of data can be fitted with a sigmoidal curve; the curve shifts to higher temperatures with increase of shortening velocity (as in Roots and Ranatunga, 2008). From fitted curves, $\Delta H°$ is 118 kJmol^{-1} and $T_{0.5}$ is ~9°C for isometric; they increase with shortening velocity and for $-2L_0/s$, $\Delta H°$ is 145 kJmol^{-1} and $T_{0.5}$ 23°C. In contrast, tension during lengthening is relatively temperature-insensitive and not correlated with temperature. Closer examination indicates that lengthening tension versus temperature also may be sigmoidal (curves fitted); data for temperatures below 10°C would be necessary, but due to loss of electrical excitability at such low temperatures, such data would be difficult to obtain experimentally from intact mammalian fibres. Nevertheless, we speculated that force versus $1/T$ relation may be sigmoidal and is shifted to lower temperatures with lengthening and this is perhaps the underlying cause of temperature insensitivity at higher temperatures. The sigmoidal curves can be fitted to the data as shown and $T_{0.5}$ is $<<5°C$.

5.3.2   Effects of Pi and ADP (Products of ATP Hydrolysis)

Experiments on skinned fibres enable one to examine the effects of changed chemical composition of the intracellular medium. Under control conditions, a sigmoidal temperature dependence of maximal active force, similar to intact fibres, is obtained in maximally Ca-activated skinned fibres (Ranatunga, 1994). Figure 5.3 shows that the position of the sigmoidal relation of force, with respect to temperature, is sensitive to inorganic phosphate (Pi) and MgADP, two products released during crossbridge (acto-myosin ATPase) cycling; Pi is released earlier and ADP later in the crossbridge cycle (see Goldman et al., 1984; Hibberd et al., 1985; He et al., 1997, 1999).

Figure 5.3a shows that active force is depressed by Pi so that the sigmoidal curve is shifted to higher temperatures, whereas Fig. 5.3b shows that MgADP potentiates force and the curve is shifted to lower temperatures. It is also seen that, because of the endothermic nature of force generation, the relative effects on tension at a given level of Pi or ADP is less at higher physiological temperatures.

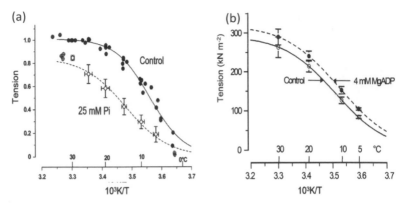

Figure 5.3   (a) Pooled data (5 fibres); a fibre was activated at ~5°C, and temperature raised by laser T-jumps and/or Peltier. Tensions normalised to control at 30°C, filled circles from control and open symbols, the mean (± SD) tensions with 25 mM Pi (from Coupland et al., 2001). (b) Pooled data from 18 fibres where tension was measured in control and with 4 mM added MgADP at one or more temperatures. Mean (± s.e.m.) specific tension are shown; open symbols—control, filled symbols—with MgADP (Coupland et al., 2005a).

5.4   Tension Response to Temperature-Jump

The above findings show that the tension in active muscle is temperature-sensitive but with some specific features associated with it. How these may be related to the crossbridge cycle may be examined by using small (3–5°C), rapid (0.2 ms) temperature-jump (T-jump), since such experiments can reveal the underlying mechanisms.

It has been shown by different research groups (Goldman et al., 1987; Davis and Harrington, 1987; Bershitsky and Tsaturyan, 1992) and using different T-jump techniques that a T-jump on isometric muscle leads to a biphasic tension rise to a new steady state. The tension responses to a T-jump on an intact muscle (Fig. 5.4a,b) confirm such findings and also complement the steady-state tension data. A T-jump induces a biphasic tension rise, both at 10 and 20°C. Data analyses show that the amplitude of tension rise to a T-jump decreases exponentially with increase temperature (see Fig. 5.4c), as expected from tetanic tension versus temperature data. However, the initial tension rise after T-jump (phase 2b) is faster than tetanic tension rise at all temperatures (see Fig. 5.4d). Phase 2b represents force generation in attached crossbridges when heat is absorbed (endothermic).

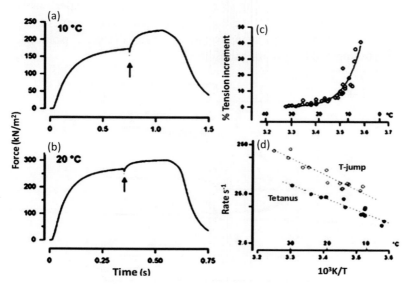

Figure 5.4   (a, b) Tension response to a ~4°C T-jump, on the isometric tetanus (different time scales); (c) amplitude of T-jump tension-rise decreases with increased temperature. (d) T-jump tension rise (open symbols) is faster than initial tetanic tension rise (filled symbols) (from Coupland and Ranatunga, 2003).

5.4.1   During Muscle Shortening and Lengthening

Figure 5.5a shows a family of superimposed tension responses (middle panel) to a T-jump from a maximally Ca-activated skinned fibre. They illustrate the T-jump tension responses in the different mechanical states. During ramp lengthening (top tension traces), the tension rises towards a level higher than P_0 and a T-jump produces no further tension rise. During shortening (lower traces), the tension decreases to a level lower than P_0, and a T-jump produces a marked (mono-phasic) tension rise. As referred to above, in isometric muscle (middle trace), a T-jump induces a biphasic rise in tension. Detailed analyses of data from similar experiments have shown that the rate of T-jump tension rise during steady shortening increases linearly with velocity so that, near maximum shortening velocity, it could be several-fold faster than in isometric muscle (Ranatunga et al., 2010). This may well account for the difference in the speed of force generation between length-release and T-jump (or P-jump) isometric experiments. Also, the T-jump data are in keeping with the temperature-sensitivity differences seen at steady state between isometric, shortening and lengthening muscle.

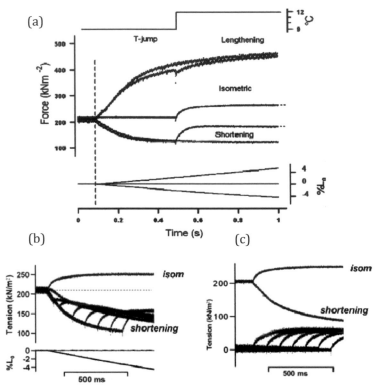

Figure 5.5   (a) A fibre held isometric was maximally Ca-activated at ~9°C and a T-jump of ~3°C applied on tension plateau (P_0) to obtain the "isometric" tension trace. Temperature was clamped again at ~9°C and a stretch at a constant velocity to obtain the two "lengthening" tension traces (with and without a T-jump). An analogous procedure was adopted to obtain "shortening" tension traces. (b) Superimposed tension records to T-jump at different times during shortening at one velocity. (c) Difference tension records—shortening tension record is subtracted from that with a T-jump (from b) (adapted from Ranatunga et al., 2007).

Figure 5.5b shows that a T-jump induces a tension rise at different stages during shortening at a certain velocity. In order to see the effect of T-jump in isolation from tension decrease by shortening, the shortening tension trace was subtracted from the T-jump plus shortening record.

Such difference tension traces in Fig. 5.5c show that the enhanced T-jump effect is seen during different times of shortening. The record at the onset of shortening indeed is biphasic like isometric, but faster; it represents T-jump force generation in crossbridges attached during isometric phase responding to negative strain, before detaching. The phase 2b rate analyses from such experiments (unpublished) show that rate could be ~10-fold higher than isometric at high shortening velocities. Thus, T-jump force generation is strain-sensitive.

5.4.2   Effects of Pi and ADP on T-Jump Force Generation

To determine the molecular step in the acto-myosin ATPase cycle that underlies endothermic force generation, force transients induced by a standard T-jump and at ~9–10°C were examined in control and in the presence of Pi or MgADP. Figure 5.6a shows a force transient induced by a T-jump in control and Fig. 5.6b shows the force transient when reactivated in the presence of 12.5 mM added Pi. It is seen that, compared to the control, the steady force before the T-jump is lower in the presence of Pi but the initial force rise (endothermic force generation) is clearly faster. Analogous experiments using MgADP showed that steady force is higher but the T-jump force rise is slower in the presence of ADP. The contrasting effects of Pi and ADP on the time course of T-jump force transient are shown in Fig. 5.6c,d. Phase 3 was not much sensitive to Pi or ADP. On the other hand, phase 2b becomes faster with increase of Pi, whereas it becomes slower with increase of ADP: In both, the time course change saturates at higher concentration levels and the relations are hyperbolic.

Pi is released earlier in the crossbridge cycle and the steady muscle force is decreased with added Pi (Cooke and Pate, 1985; Hibberd et al., 1985), but the kinetics of the approach to the new steady state were enhanced; this has been shown studies using different techniques, such as hydrostatic pressure-release (P-jump, Fortune et al., 1991), sinusoidal length oscillation (Kawai and Halvorson, 1991), Pi-jump (Dantzig et al., 1992; Tesi et al., 2000) and Pi-measurement (He et al., 1997). The unified thesis that arose from such different studies was that, in the acto-myosin ATPase cycle, force generation precedes Pi-release. Findings from T-jump experiments are consistent with that thesis and also show that this force generation is endothermic. The steady active tension was potentiated when [MgADP] is increased; the binding of MgADP to nucleotide-free crossbridges (AM) leading to accumulation of force-bearing AM-ADP states (Cooke and Pate, 1985; Dantzig et al., 1991; Lu et al., 1993, 2001; Seow and Ford, 1997), in general, may underlie the tension increase. When [MgADP] is increased, the tension rise induced by a T-jump was slower indicating that the approach to the new steady state at the post-T-jump temperature is depressed.

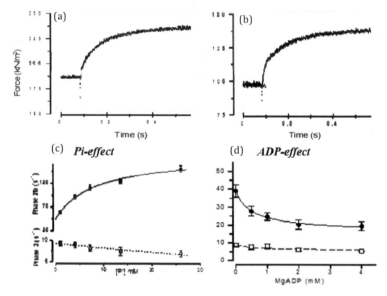

Figure 5.6 Effect of products of ATP hydrolysis, Pi and ADP. (a) Tension response to a ~3°C T-jump at ~9°C, on control maximal activation (no added Pi). (b) On reactivation with 12.5 mM Pi. (c) Phase 2b rate increases with Pi and the relation is hyperbolic. (d) Phase 2b rate decreases with ADP and the relation is hyperbolic. Phase 3 shows minimal sensitivity to Pi and ADP (adapted from Ranatunga, 1999; Coupland et al., 2005a).

5.4.3 A Minimal Crossbridge Cycle (and Modelling)

It is perhaps relevant to note that different simplified crossbridge schemes have been used to kinetically simulate the experimental findings from temperature studies on muscle, as has been done in various other studies. Although not examined here in order to account for the observations listed above, a minimal cycle of 5 steps (1–5) and 5 states (*i-v*) was necessary.

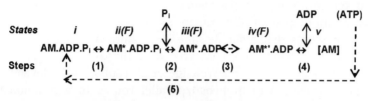

It is an extended Lymn and Taylor (1971) AM-ATPase cycle, as above, and it could accommodate and qualitatively illustrate the main findings (see Ranatunga, 2010). The key features are the following: **Step 1**, the forward rate constant is endothermic force generation (Q_{10} of ~4). **Step 2** is rapid Pi release/binding (similar to Dantzig et al., 1992). **Steps 3** and **4** represent the slow, two-step, ADP release (Dantzig et al., 1991). **Step 4** (ADP release) would be the end-crossbridge detachment on cycle-completion that is thought to determine the speed of shortening (see He et al., 1999). With milli-molar level of [ATP] under physiological conditions, the rigor [AM] state would have an insignificant life time. **Step 5** is irreversible and includes all the necessary steps after ADP release to re-prime a crossbridge for the next cycle. The overall rate in this route is low (k_{+5}, ~10 s^{-1}), probably limited by the M.ATP ↔ M.ADP.P$_i$ cleavage step after detachment (see He et al., 1999). This is shown to be endothermic in biochemical experiments on muscle proteins in solution (White and Taylor, 1976). According to the scheme, the post-stroke **states ii, iii** and **iv** are equal-force bearing states (AM*.ADP.P$_i$, AM*ADP and AM*'.ADP).

For a simplified two-state system, as convenient to readily see the sigmoidal temperature-dependence of force, all force-bearing state together would represent the high-force state (see Coupland et al., 2005a) and the pre-stroke **state i** and **state v** constitute the low-force state. Such a kinetic scheme could be used to qualitatively simulate the trends in T-jump tension responses during different ramp shortening velocities, effects of Pi and ADP (see Ranatunga et al., 2010, for details).

Kinetic modelling alone, of course, is inadequate to fully address the mechanics and energetics of active muscle contraction and a more detailed mechano-kinetic modelling would be necessary, as for example in Smith and Sleep (2004), Månsson (2010) and, recently, Offer and Ranatunga (2013); such modelling however have not been extended to include temperature-effects and temperature-jump experiments. However, kinetic modelling was useful to gain an approximate qualitative picture and, on that basis, it seems that a simple kinetic scheme can account for the main observations. This contrasts with the approach taken by Ferenczi et al. (2005) and Woledge et al. (2009), who developed complex cycles, with branched/parallel routes to accommodate T-jump data. However, phase 2a after length-release (see Section 5.2) remains unaccounted for in the simple kinetic scheme and may be a consequence of visco-elasticity in the filament compliance (Davis, 1998).

5.5 Some General Observations

5.5.1 Unresolved Issues

A conformational change of the acto-myosin crossbridge, resulting in lever arm tilting, is thought to be the molecular-structural mechanism of muscle force generation (review, Geeves and Holmes, 1999) and as mentioned above, several studies and using different techniques have concluded that this occurs before the release of inorganic phosphate in the cycle. Indeed, mechano-kinetic modelling supports this thesis that crossbridge force generation precedes Pi-release (Smith and Sleep, 2004; Sleep et al., 2005), and molecular structural modelling indicates that this is feasible (Zheng, 2010). P-jump data shows that the force generation involves a volume increase and T-jump and other data show that it is endothermic and hence is associated with increased entropy (disorder). The transition of non-stereo-specifically attached to stereo-specifically attached crossbridge states may be hydrophobic and lead to increase of entropy (Zhao and Kawai, 1994; Kawai, 2003). Davis and Epstein (2007, 2009) proposed that T-jump force generation (step 1) may represent a "protein unfolding/folding-process" within a myosin head, so that the unfolding during force generation may lead to increased entropy. Either or a combination of such molecular mechanisms is possible.

Some difficulty has been found in previous studies (see Bershitsky and Tsaturyan, 2002) in accommodating and correlating T-jump force generation with quick tension recovery after length-release. According to the experiments and formulations by Huxley and Simmons (1971), quick tension recovery (T_1–T_2 transition) by a small length-release step represents crossbridge power stroke or force generation in muscle. This quick tension recovery can be resolved in to two components (Davis and Harrington, 1993), labelled as phase 2a and 2b. It is found that the T-jump force generation in isometric muscle is much slower than tension recovery from length release. However, our observations support the thesis that phase 2 tension recovery after a length step contains a component (phase 2b) that is homologous to the endothermic force generation observed after a rapid T-jump, as originally proposed by Davis and Harrington (1993) (see also Davis, 1998). This notion also gains support from the study of Gilbert and Ford (1988) that showed experimentally that quick tension recovery from length-release is associated with heat absorption. Additionally, the rate of phase 2 tension recovery from length-release is temperature sensitive (Q_{10} of 2–3, Piazzesi et al., 2003), the temperature sensitivity being greater for phase 2b than for phase 2a component of recovery (Davis and Harrington, 1993). Thus, like the T-jump force rise, phase 2b tension recovery from length-release may represent an endothermic process. It is not exactly clear as to the kinetic basis of (fast) phase 2a.

Experimental studies on myosin-ATPase in solution (Kodama, 1985; Millar et al., 1987) have shown that the ATP cleavage step (i.e. in detached crossbridges in fibres) is endothermic and that this is due to a temperature-dependent conformational change(s) in myosin head (Werner et al., 1999; Malnasi-Csizmadia et al., 2000). Such findings need to be accommodated in a complete mechano-kinetic model of the crossbridge/A-M.ATPase cycle in muscle fibres.

Some experiments (Bershitsky and Tsaturyan, 2002) indicate that different processes underlie tension responses to T-jump and length steps; consequently, Ferenczi et al. (2005) considered apparent strain independent, but temperature-dependent, rate-limiting step prior to force generation. Also, whereas T-jump studies above required only one (or two) molecular steps for force generation, it has been argued from X-ray diffraction studies using length perturbation that, several molecular steps would be required to complete a working stroke of a crossbridge (Huxley et al., 2006). However, Knupp et al., (2009) raised some concern regarding interpretation of X-ray interference diffraction data on muscle. As considered previously (Offer and Ranatunga, 2010), the experimental findings that sarcomeric filaments in muscle are not only compliant but also their compliance may be non-linear (Edman, 2009; Nocella et al., 2013a) need to be accommodated for a fuller picture.

More recently, Offer and Ranatunga (2013) developed a model to simulate the well-known experimental mechanics of frog (Rana temporaria) muscle; the model required two force generation steps. Subsequently, the same model was used to simulate the basic effects of temperature (Offer and Ranatunga, 2015); it was found that strong endothermic ATP hydrolysis and crossbridge attachment steps (with a little endothermic character of the first force generation step) may account for the increase of active tension in isometric and shortening muscle. To what extent these simple models developed on frog muscle data can account for detailed data from mammalian muscle considered here, remains to be sorted out in the future.

5.5.2   Value of Temperature-Studies

From a detailed review on isometric muscle studies, Kawai (2003) showed that temperature studies provide useful thermodynamic parameters of muscle force generation. This was only briefly considered in this review but our analyses of isometric tension data and in relation to ATPase cycle is basically similar to Kawai's. Additionally, our studies now show that, such analyses can be extended to shortening and lengthening muscle too (Fig. 5.5). Secondly, T-jump is a useful rapid perturbation technique to probe the crossbridge/AM-ATPase cycle (Fig. 5.4) and other processes (e.g. excitation-contraction coupling process in muscle, Coupland et al., 2005b). Thirdly, the data would be relevant to understanding muscle performance in our body.

Overall, the findings briefly reviewed here provide a useful picture of the process of force and power generation in active muscle and its performance in the body. For example, accumulation of products of ATP hydrolysis, particularly Pi, is considered as contributory to moderate levels of fatigue in the in situ muscle. However, temperature studies show that, the effects on force of product accumulation, and hence relative fatigue, would be less at high physiological temperatures. Observations in support of this suggestion have been reported from fatigue-experiments at different temperatures (see Roots et al., 2009; Nocella et al., 2013b). The finding that T-jump force generation is depressed during lengthening suggests that the acto-myosin ATPase cycle is depressed/slowed. Our findings show that force during and after lengthening, in relation to isometric force, is indeed considerably less at higher physiological temperatures. The temperature studies have also shown that mechanical power output increases markedly (~20-fold) in warming from 10 to 35°C; power is significantly temperature-sensitive even within the high physiological temperature range (Q_{10} of ~2, Ranatunga, 1998). Enhanced nature of the endothermic force generation in shortening muscle, and in a velocity-dependent manner, is the underlying cause.

Acknowledgements

I wish to thank all my collaborators and colleagues for their contributions at various stages and also the publishers of various journals (the *Journal of Physiology* and the *Journal of Muscle Research and Cell Motility*) for publishing the original observations. Special thanks are extended to Professor Arthur J Buller (Bristol University), in whose laboratory I became interested in temperature-effects on muscle contraction and to Dr Gerald Offer (Bristol University) for regular discussions on molecular mechanism of muscle contraction.

References

Bershitsky SY and Tsaturyan AK (1992). Tension responses to joule temperature jump in skinned rabbit muscle fibres. *J Physiol*, **447**: 425–448.

Bershitsky SY and Tsaturyan AK (2002). The elementary force generation process probed by temperature and length perturbations in muscle fibres from the rabbit. *J Physiol,* **540**: 971–988.

Brown DES (1957). Temperature-pressure relation in muscular contraction. In *Influence of Temperature on Biological Systems* (Johnson, FH, ed.), American Physiological Society, Washington, pp. 83–100.

Clarke RSJ, Hellon RF, and Lind AR (1958). The duration of sustained contractions of the human forearm at different muscle temperatures. *J Physiol,* **143**: 454–473.

Colombini B, Nocella M, Benelli G, Cecchi G, and Bagni MA (2008). Effect of temperature on cross-bridge properties in intact frog muscle fibers. *Am J Physiol Cell Physiol,* **294**: C1113–C1117.

Cooke R and Pate E (1985). The effects of ADP and phosphate on the contraction of muscle fibers. *Biophysl J,* **48**: 789–798.

Coupland ME, Pinniger GJ, and Ranatunga KW (2005a). Endothermic force generation, temperature-jump experiments and effects of increased [MgADP] in rabbit psoas muscle fibres. *J Physiol,* **567**: 471–492.

Coupland ME, Pinniger GJ, and Ranatunga KW (2005b). Tension responses to rapid (laser) temperature-jumps during twitch contractions in intact rat muscle fibres. *J Musc Res Cell Motil,* **26**: 113–122.

Coupland ME, Puchert E, and Ranatunga KW (2001). Temperature dependence of active tension in mammalian (rabbit psoas) muscle fibres: Effect of inorganic phosphate. *J Physiol,* **536**: 879–891.

Coupland ME and Ranatunga KW (2003). Force generation induced by rapid temperature jumps in intact mammalian (rat) muscle fibres. *J Physiol,* **548**: 439–449.

Dantzig JA, Goldman YE, Millar NC, Lacktis J, and Homsher E (1992). Reversal of the cross-bridge force-generating transition by photogeneration of phosphate in rabbit psoas muscle fibres. *J Physiol,* **451**: 247–278.

Dantzig JA, Hibberd MG, Trentham DR, and Goldman YE (1991). Crossbridge kinetics in the presence of MgADP investigated by photolysis of caged ATP in rabbit psoas muscle fibres. *J Physiol,* **432**: 639–680.

Davis JS (1998). Force generation simplified. Insights from laser temperature-jump experiments on contracting muscle fibres. In *Mechanisms of Work Production and Work Absorption in Muscle* (Sugi H and Pollack GH, ed), Plenum Press, New York, pp. 343–352.

Davis JS and Epstein ND (2007). Mechanism of tension generation in muscle: An analysis of the forward and reverse rate constants. *Biophys J,* **92**: 2865–2874.

Davis JS and Epstein ND (2009). Mechanistic role of movement and strain sensitivity in muscle contraction. *Proc Nat Acad Sci,* **106**: 6140–6145.

Davis JS and Harrington W (1987). Force generation by muscle fibers in rigor: A laser temperature-jump study. *Proc Natl Acad Sci,* **84**: 975–979.

Davis JS and Harrington W (1993). A single order-disorder transition generates tension during the Huxley-Simmons phase 2 in muscle. *Biophys J,* **65**: 1886–1898.

Edman KAP (2009). Non-linear myofilament elasticity in frog intact muscle fibres. *J Exp Biol,* **212**: 1115–1119.

Elmubarak MH and Ranatunga KW (1984). Temperature sensitivity of tension development in a fast-twitch muscle of the rat. *Muscle Nerve,* **7**: 298–303.

Ferenczi MA, Bershitsky SY, Koubassova N, Siththanandan V, Helsby WI, Pannie P, Roessle M, Narayanan T, and Tsaturyan AK (2005). The 'Roll and Lock' mechanism of force generation in muscle. *Structure,* **13**: 131–141.

Fortune NS, Geeves MA, and Ranatunga KW (1991). Tension responses to rapid pressure release in glycerinated rabbit muscle fibers. *Proc Natl Acad Sci,* **88**: 7323–7327.

Galler S and Hilber K (1998). Tension/stiffness ratio of skinned rat skeletal muscle fibre types at various temperatures. *Acta Physiol Scand,* **162**: 119–126.

Geeves MA and Holmes KC (1999). Structural mechanism of muscle contraction. *Ann Rev Biochem,* **68**: 687–728.

Gilbert SH and Ford LE (1988). Heat changes during transient tension responses to small releases in active frog muscle. *Biophys J,* **54**: 611–677.

Goldman YE, Hibberd MG, and Trentham DR (1984). Relaxation of rabbit psoas muscle fibres from rigor by photochemical generation of adenosine-5'-triphosphate. *J Physiol,* **354**: 577–604.

Goldman YE, McCray JA, and Ranatunga KW (1987). Transient tension changes initiated by laser temperature jumps in rabbit psoas muscle fibres. *J Physiol,* **392**: 71–95.

Griffiths PJ, Bagni MA, Colombini B, Amenitsch H, Bernstorff S, Ashley CC, and Cecchi C (2002). Changes in myosin S1 orientation and force induced by a temperature increase. *Proc Nat Acad Sci,* **99**: 5384–5389.

Hadju S (1951). Behaviour of frog and rat muscle at higher temperatures. *Enzymologia,* **14**: 187–190.

He Z-H, Chillingworth RK, Brune M, Corrie JET, Trentham DR, Webb MR, and Ferenczi MA (1997). ATPase kinetics on activation of rabbit and frog permeabilised isometric muscle fibres: A real time phosphate assay. *J Physiol,* **501**: 125–148.

He Z-H, Chillingworth RK, Brune M, Corrie JET, Webb MR, and Ferenczi MA (1999). The efficiency of contraction in rabbit skeletal muscle fibres, determined from the rate of release of inorganic phosphate. *J Physiol,* **517**: 839–854.

Hibberd MG, Dantzig JA, Trentham DR, and Goldman YE (1985). Phosphate release and force generation in skeletal muscles fibers. *Science,* **228**: 1317–1319.

Huxley AF (1957). Muscle structure and theories of contraction. *Prog Biophys,* **7**: 285–318.

Huxley HE (1969). Mechanism of muscle contraction. *Science,* **164**: 1356–1366.

Huxley HE, Reconditi M, Stewart A, and Irving T (2006). X-ray interference studies of crossbridge action in muscle contraction: Evidence from muscles during steady shortening. *J Mol Biol,* **363**: 462–472.

Huxley AF and Simmons RM (1971). Proposed mechanism of force generation in striated muscle. *Nature,* **233**, 533–538.

Kawai M (2003). What do we learn by studying the temperature effect on isometric tension and tension transients in mammalian striated muscle fibres? *J Muscle Res Cell Motil,* **24**: 127–138.

Kawai M and Halvorson HR (1991). Two step mechanism of phosphate release and the mechanism of force generation in chemically skinned fibers of rabbit psoas muscle. *Biophys J,* **59**: 329–342.

Kawai M, Kido T, Vogel M, Fink RHA, and Ishiwata S (2006). Temperature change does not affect force between regulated actin filaments and heavy meromyosin in single molecule experiments. *J Physiol,* **574**: 877–878.

Knupp C, Offer GW, Ranatunga KW, and Squire J (2009). Probing muscle myosin motor action. X-ray (M3 and M6) interference measurements report motor domain not lever arm movement. *J Mol Biol,* **390**: 168–181.

Kodama T (1985). Thermodynamic analysis of muscle ATPase mechanisms. *Physiol Rev,* **65**: 467–551.

Linari M, Brunello E, Reconditi M, Sun Y-B, Panine P, Narayanan T, Piazzesi G, Lombardi V, and Irving M. (2005). The structural basis of the increase in isometric force production with temperature in frog skeletal muscle. *J Physiol,* **567**: 459–469.

Lu Z, Moss RL, and Walker JW (1993). Tension transients initiated by photogeneration of MgADP in skinned skeletal muscle fibers. *J Gen Physiol,* **101**: 867–888.

Lu Z, Swartz DR, Metzger JM, Moss RL, and Walker JW (2001). Regulation of force development studied by photolysis of caged ADP in rabbit skinned psoas fibers. *Biophys J,* **81**: 334–344.

Lymn RW and Taylor EW (1971). Mechanism of adenosine triphosphate hydrolysis by actomyosin. *Biochemistry,* **10**: 4617–4624.

Malnasi-Csizmadia A, Woolley RJ, and Bagshaw CR (2000). Resolution of conformational states of Dictyostelium myosin II motor domain using tryptophan (W501) mutants: Implications for the open-closed transition identified by crystallography. *Biochemistry,* **39**: 16135–16146.

Månsson A. (2010). Actomyosin-ADP states, inter-head cooperativity, and the force-velocity relation of skeletal muscle. *Biophys J,* **98**: 1237–1246.

Millar NC, Howarth JV, and Gutfreund H (1987). A transient kinetic study of enthalpy changes during the reaction of myosin subfragment 1 with ATP. *Biochem J,* **248**: 683–690.

Nocella M, Bagni MA, Cecchi G, and Colombini B (2013a). Mechanism of force enhancement during stretching of skeletal muscle fibres investigated by high time-resolved stiffness measurements. *J Muscle Res Cell Motil,* **34**: 71–81.

Nocell a M, Cecchi G, Bagni MA, and Colombini B (2013b). Effect of temperature on crossbridge force changes during fatigue in intact mouse muscle fibers. *PLOS ONE,* **8**: 1–11.

Offer G and Ranatunga KW (2010). Crossbridge and filament compliance in muscle: Implications for tension generation and lever arm swing. *J Muscle Res Cell Motil,* **31**: 245–265.

Offer G and Ranatunga KW (2013). A crossbridge cycle with two tension-generating steps simulates skeletal muscle mechanics. *Biophys J,* **105**: 928–940.

Offer G and Ranatunga KW (2015). The endothermic ATP hydrolysis and crossbridge attachment steps drive the increase of force with temperature in isometric and shortening muscle. *J Physiol,* **593**: 1997–2016.

Piazzesi G, Reconditi M, Koubassova N, Decostre V, Linari M, Lucii L, and Lombardi V. (2003). Temperature dependence of the force-generating process in single fibres from frog skeletal muscle. *J Physiol*, **549**: 93–106.

Pinniger GJ, Ranatunga KW, and Offer GW (2006). Crossbridge and non-crossbridge contributions to tension in lengthening rat muscle: Force-induced reversal of the power stroke. *J Physiol*, **573**: 627–643.

Ranatunga KW (1994). Thermal stress and Ca-independent contractile activation in mammalian skeletal muscle fibers at high temperatures. *Biophys J*, **66**: 1531–1541.

Ranatunga KW (1996). Endothermic force generation in fast and slow mammalian (rabbit) muscle fibers. *Biophys J*, **71**: 1905–1913.

Ranatunga KW (1998). Temperature dependence of mechanical power output in mammalian (rat) skeletal muscle. *Exp Physiol*, **83**: 371–376.

Ranatunga KW (1999). Effects of inorganic phosphate on endothermic force generation in muscle. *Proc R Soc B*, **266**: 1381–1385.

Ranatunga KW (2010). Force and power generating mechanism(s) in active muscle as revealed from temperature perturbation studies. *J.Physiol.*, **588**: 3657–3670.

Ranatunga KW, Coupland ME, and Mutungi G (2002). An asymmetry in the phosphate dependence of tension transients induced by length perturbation in mammalian (rabbit psoas) muscle fibres. *J Physiol*, **542**: 899–910.

Ranatunga KW, Coupland ME, Pinniger GJ, Roots H, and Offer GW (2007). Force generation examined by laser temperature-jumps in shortening and lengthening mammalian (rabbit psoas) muscle fibres. *J Physiol*, **585**: 263–277.

Ranatunga KW, Fortune NS, and Geeves MA (1990). Hydrostatic compression in glycerinated rabbit muscle fibres. *Biophys J*, **58**: 1401–1410.

Ranatunga KW, Roots H, and Offer GW (2010). Temperature jump induced force generation in rabbit muscle fibres gets faster with shortening and shows a biphasic dependence on velocity. *J Physiol*, **588**: 479–493.

Roots H, Ball G, Talbot-Ponsonby J, King M, McBeath K, and Ranatunga KW (2009). Muscle fatigue examined at different temperatures in experiments on intact mammalian (rat) muscle fibers. *J Appl Physiol*, **106**, 378–384.

Roots H, Offer GW, and Ranatunga KW (2007). Comparison of the tension responses to ramp shortening and lengthening in intact mammalian muscle fibres: Crossbridge and non-crossbridge contributions. *J Muscle Res Cell Motil,* **28**: 123–139.

Roots H, Pinniger, GJ, Offer GW, and Ranatunga KW (2012). Mechanism of force enhancement during and after lengthening of active muscle: A temperature dependence study. *J Muscle Res Cell Motil,* **33**: 313–325.

Roots H and Ranatunga KW (2008). An analysis of temperature-dependence of force, during shortening at different velocities, in (mammalian) fast muscle fibres. *J Muscle Res Cell Motil,* **29**: 9–24.

Seow CY and Ford LE (1997). Exchange of ADP on high-force cross-bridges of skinned muscle fibers. *Biophys J,* **72**: 2719–2735.

Sleep J, Irving M, and Burton K (2005). The ATP hydrolysis and phosphate release steps control the time course of force development in rabbit skeletal muscle. *J Physiol,* **563**: 671–687.

Smith DA and Sleep J (2004). Mechanokinetics of rapid tension recovery in muscle: The myosin working stroke is followed by a slower release of phosphate. *Biophys J,* **87**: 442–456.

Tesi C, Colomo F, Nencini S, Piroddi N, and Poggesi C (2000). The effect of inorganic phosphate on force generation in single myofibrils from rabbit skeletal muscle. *Biophys J,* **78**: 3081–3092.

Werner J, Urbanke C, and Wray J (1999). Fluorescence temperature-jump studies of myosin S1 structures. *Biophys J,* **76**: M-AM-G8.

White HD and Taylor EW (1976). Energetics and mechanism of actomyosin adenosine triphosphatase. *Biochemistry,* **15**: 5810–5826.

Woledge RC, Barclay CJ, and Curtin NA (2009). Temperature change as a probe of muscle crossbridge kinetics: A review and discussion *Proc R Soc B,* **276**: 2685–2695.

Zhao Y and Kawai M (1994). Kinetic and thermodynamic studies of the cross-bridge cycle in rabbit psoas muscle fibers. *Biophys J,* **67**: 1655–1668.

Zheng W (2010). Multiscale modeling of structural dynamics underlying force generation and product release in actomyosin complex. *Proteins,* **78**: 638–660.

Chapter 6

Mechanism of Force Potentiation after Stretch in Intact Mammalian Muscle

Giovanni Cecchi, Marta Nocella, Giulia Benelli, Maria Angela Bagni, and Barbara Colombini

Department of Experimental and Clinical Medicine, University of Florence, Italy

giovanni.cecchi@unifi.it

The isometric tension following a stretch applied to an active muscle is greater that the isometric tension at the same sarcomere length. This force potentiation, known as residual force enhancement (RFE), has been extensively studied; nevertheless, its mechanism remains debated. In the experiments reported here, unlike RFE studies, the excess of force after stretch, termed static tension (ST), was investigated with fast stretches (amplitude: 3–4% sarcomere length; duration: 0.6 ms) applied at low tension on the tetanus rise of FDB mouse muscle at 30°C. The measurements were made between 2.6 and 4.4 μm sarcomere length in normal and BTS-added (10 μM) Tyrode solution. ST increased with sarcomere length, reaching a peak at 3.5 μm and decreasing to zero at ~4.5 μm. At 4 μm, active force was zero but ST was still 50% of maximum. BTS reduced force by ~75% but had almost no effect on ST. Following activation, ST develops

Muscle Contraction and Cell Motility: Fundamentals and Developments
Edited by Haruo Sugi
Copyright © 2017 Pan Stanford Publishing Pte. Ltd.
ISBN 978-981-4745-16-1 (Hardcover), 978-981-4745-17-8 (eBook)
www.panstanford.com

faster than force, with a time course similar to intracellular $[Ca^{2+}]$, starting to rise 1 ms after the stimulus, at zero active force, and peaking at 3–4 ms delay after the stimulus. At 2.7 µm, for a stretch of 1% sarcomere length, ST was ~3% of tetanic force, ~7 times greater than the response of resting fibres. All these data indicate that: (1) ST has the same properties and it is equivalent to RFE, (2) it is independent of crossbridges and (3) it is likely due to the Ca^{2+}-induced stiffening of a sarcomeric structure identifiable with titin.

6.1 Introduction

Stretching of a contracting skeletal muscle induces a transient force increase followed by a period during which tension remains constantly elevated above the isometric force. This post-stretch force potentiation, termed residual force enhancement (RFE; Edman et al., 1978, 1982), is dependent on the stretch amplitude and sarcomere length, and it is independent of the stretching velocity. RFE was first demonstrated on whole muscles of the frog (Fenn, 1924; Abbott and Aubert, 1952; Hill and Howarth, 1959), and later on isolated single frog muscle fibres (Sugi, 1972; Julian and Morgan, 1979; Edman et al., 1978, 1982; Sugi and Tsuchiya, 1988; Morgan, 1990, 1994; Edman and Tsuchiya, 1996). More recently RFE was also shown on myofibrils isolated from cardiac and skeletal muscle (Rassier et al., 2003a, 2003b; Joumaa et al., 2008a, 2008b; Telley et al., 2006; Rassier and Pavlov, 2012). In spite of the extensive studies on the subject, the mechanism of RFE is still debated. Earlier studies on single fibres indicated that RFE arises from sarcomere length non-uniformity (Julian and Morgan, 1979) induced by the stretch, especially if applied on the descending limb of the length-tension relation. This was possibly associated with the recruitment of passive elastic elements (Edman and Tsuchiya, 1996). Opposing evidences suggested that RFE is due to both enhanced crossbridge force and increased passive force of the muscle fibres, attributed to a stiffening of the titin filament upon Ca^{2+} binding (Herzog and Leonard, 2002; Lee et al., 2007; Rassier, 2012; Roots et al., 2012) with a possible contribution from the titin-C protein interaction (Pinniger et al., 2006). More recently it was proposed that Ca^{2+}-induced titin stiffening could be enhanced to the level necessary to account quantitatively for RFE values found experimentally by the interaction between titin, actin and crossbridges (Leonard and Herzog, 2010; Nishikawa et al., 2012). Classically, the mechanism of RFE has been investigated by using slow stretches of 5–10% of fibre length amplitude applied at tetanus plateau, mostly on frog fibres. The results presented here were obtained on mouse muscle fibres, using a different procedure compared to the classical one (Bagni et al., 1994, 2002, 2004; Colombini et al., 2009; Nocella et al., 2012, 2014; Colombini et al., 2015) in two aspects: (1) stretches were much faster than those used for RFE studies and (2) were applied at low active force either during the tetanus rise or at tetanus plateau of fibres whose force generation was depressed by the crossbridge force inhibitor *N*-benzyl-*p*-toluene sulphonamide (BTS) (Bagni et al., 2002, 2004; Colombini et al., 2010). These studies showed that activation of the fibres greatly increased the resting sarcomere stiffness with a mechanism not involving crossbridges. In fact, stiffness increase was present also when active force was zero and it was not correlated with active force. The time course of the stiffness increase was similar to that intracellular Ca^{2+} concentration, and for this reason fibre stiffening was attributed to a Ca^{2+}-dependent stiffening of titin filaments (Bagni et al., 1994, 2004). Stretching and holding of the stiffened titin filaments gives rise to a greater steady force increase than in passive fibre, which could account for the force potentiation after stretch. Because of its characteristics, this post-stretch force potentiation was termed static tension (ST). Similarly to classical RFE, ST was almost constant for all the stimulation period, was independent of stretching velocity and proportional to stretch amplitude (Bagni et al., 2002) and increased on going from 2.2 to 2.8 µm sarcomere length (Bagni et al., 2002). These similarities suggested that ST and RFE are equivalent and attributable to the Ca^{2+}-induced titin stiffening. The experiment reported here were made to investigate the postulated equivalence of RFE and ST and to extend the analysis of force potentiation after stretch over the whole length-tension relationship up to beyond the zero myofilament overlap where crossbridge presence is not expected. The experiments were made on small intact fibre bundles from mouse flexor digitorum brevis (FDB) muscles at 30°C, close to the physiological temperature. The results showed that ST increased with sarcomere length reaching the maximum at about 3.5 µm, decreasing again at longer length and reaching zero approximately at 4.5 µm, beyond the zero overlap length, similarly to RFE in frog muscle (Edman et al., 1982). At optimal sarcomere length, the steady tension after stretch of active fibres was ~7 times greater than the passive force response to the same stretch. Our findings show the equivalence of ST and RFE and strengthen the hypothesis that they arise from a non-crossbridge Ca^{2+}-dependent increase of sarcomere stiffness, which could attributed to an increase of titin filaments stiffness following fibre activation.

6.2 Materials and Methods

6.2.1 Animals, Fibre Dissection and Measurements

This study was carried out following the EEC guidelines for animal care of The European Community Council (Directive 86/609/EEC) and the protocol was approved by the Italian Health Ministry and the Ethical Committee for Animal Experiments of the University of Florence (acceptance signed by the veterinary responsible in October 20, 2010). C57BL/6 male mice were housed at controlled temperature (21–24°C) with a 12–12 h light–dark cycle and food and water provided ad libitum. Mice (4–5 month-old) were killed by rapid cervical dislocation. Small intact fibre bundles (10–20 fibres) from the FDB muscles of the hind limb of both legs were dissected manually under a stereo-microscope with a pair of fine scissors and needles. Particular care was taken to avoid excessive stretching and to obtain bundles as clean as possible from connective tissue and debris from dead fibres. The number of animals used was minimized by using both leg muscle and by dissecting more than one bundle from the same muscle. Bundles were mounted horizontally, by means of aluminium foil clips compressed onto the tendons, between the lever arms of a capacitance force transducer (natural frequency 25–50 kHz) and a fast electromagnetic motor (minimum stretch time 100 µs) which could apply to the bundles stretches of the desired shape and amplitude. The experimental chamber (capacity 0.38 ml) was provided with a glass floor for light illumination. The motor was mounted on a micromanipulator to adjust the passive length of the bundles at selected values. Bundles were perfused permanently by means of a peristaltic pump at a rate of about 0.35 ml min^{-1} with a normal Tyrode (NT) solution of the following composition (mM): NaCl, 121; KCl, 5; CaCl$_2$, 1.8; MgCl$_2$, 0.5; NaH$_2$PO$_4$, 0.4; NaHCO$_3$, 24; glucose, 5.5; EDTA, 0.1. This solution was continuously bubbled with a mixture of 5% CO$_2$ and 95% O$_2$, which gave a pH of 7.4. Foetal calf serum (0.2%) was freshly added to the solution. In a group of experiments 10 μM *N*-benzyl-*p*-toluene sulphonamide (BTS) was added to the Tyrode solution to reduce active tension and crossbridge influence on the response to the stretches. The experiments were performed at temperature of 30°C. Stimuli of alternate polarity, 0.2–0.5 ms duration and 1.5 times threshold strength, were applied transversely to the bundles by means of two platinum-plate electrodes running parallel to them. Clip to clip fibre length including tendon (l_t), and end to end without tendons (l_0), and sarcomere length were measured using a microscope fitted with a 20× eyepieces and a 5× or 40× dry objective in the experimental chamber and checked later on digital images acquired by a video camera (Infinity camera, Lumenera Corp., Canada). The cross-sectional area was calculated as if it the section of the bundles were elliptical, with the formula $\pi ab/4$, where a and b are the smaller and the greater diameters measured along the bundles. Resting sarcomere length was measured by counting the number of consecutive sarcomeres in a calibrated scale on the pictures taken with the video camera; the desired sarcomere length was set by adjusting appropriately the resting bundle length. Mean bundle l_t and l_0 was 1161 ± 36 μm and 627 ± 12 μm ($n = 10$), respectively. After 20 min of equilibration, tetanic stimulation was applied in brief (300 ms duration) volleys using the minimum frequency necessary to obtain a just fused contraction (70–100 Hz). In general, plateau tetanic force (P_0) was stable over a period of few hours. A custom-written software (LabView, National Instruments USA) was used to drive the stimulator and the electromagnetic motor. Force and length were recorded at sampling time of 1 ms for the whole tetanus except for 30 ms, 1 ms before and 29 ms following the stretch, which were recorded at a much faster sampling time of 10 μs to obtain the appropriate time resolution.

6.2.2 Static Tension Measurements

Static tension (ST) was measured as described previously (Bagni et al., 1994, 2002; Colombini et al., 2009). Briefly, three force records were taken for each measure: (1) the response of the active fibres to the stretches and hold; (2) the response of the passive fibres to the same stretch; (3) the isometric tension. By subtracting the passive and the isometric responses from the response to fast stretches, we obtained the excess of force induced by the stretch. After a fast peak, synchronous with the stretch and caused by the stretched crossbridges, the force settled within few milliseconds to a steady value greater than the pre-stretch tension, which represents the ST. Stretches used were ramp shaped with an amplitude of 3–4% l_0 and a stretch time of 0.6–0.7 ms (corresponding to a stretching velocity greater than 70 l_0 s^{-1}) and were applied at low tension on the rise of short tetani (3 consecutive stimuli). The use of fast stretches applied at low active tension reduced to a few milliseconds the time necessary for the tension to settle to the steady level after the initial force peak, much less that than the ~2 s required when slow stretches were applied at tetanus plateau in previous RFE studies (Edman et al., 1978, 1982; Leonard and Herzog, 2010). Stretches applied at low tension also reduced the confounding contribution of crossbridges to the force transient after the stretch and avoid damaging of the fibres. The fraction of stretch applied at the sarcomere level was calculated by correcting for the fraction absorbed by the tendons using the average ratio between tendon and fibre compliances reported previously (Nocella et al., 2011, 2013). This ratio was independent of tension, therefore it was assumed for all bundles that tendon compliance was 41% of total compliance whereas the remaining 59% was attributed to the fibre independently of the tension at which the stretch was applied.

To compare different fibre response ST, in some cases, was expressed relative to P_0, for 1% l_0 stretch amplitude, representing static stiffness (SS). All tension values are reported as a fraction of P_0 and expressed as mean ± standard error.

6.3 Results

6.3.1 Static Tension

Figure 6.1 shows a typical set of records used for measuring static tension (ST). Records *b*, *c* and *d* represent the isometric tetanus plus stretch, the isometric tetanus and the passive response to the stretch, respectively. In this case the stretch was applied 3 ms after the start of stimulation. The comparison of records *b* and *c* show that the post-stretch tension remains elevated well above the isometric tension. The subtraction of the two traces, corrected for the passive tension (trace *d*), represents the excess of force (*e*), or force potentiation, induced by the stretch. It can be seen that after the initial force peak, synchronous with the stretch and caused by the stretching of the crossbridges, the excess of force settles to a constant level in 1–2 ms and remains almost constant afterwards. This steady value was taken as a measure of ST at 3 ms delay after start of stimulation.

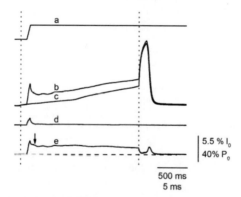

5.5 % l_0
40% P_0

500 ms
5 ms

Figure 6.1   Static tension measurements. Trace *b* shows that the stretch induces a steady force potentiation compared to the isometric record in *c*. The difference between *b* and *c* subtracted by the passive response *d* (trace *e*) represents the ST. The measure is made after the end of the fast transient, when the tension becomes steady (indicated by the arrow). Sarcomere length, 2.58 μm. Stretch, 3.2% l_0 amplitude and 630 μs duration, was applied on the rise of a short tetanus at 100 Hz, 3 ms after the first stimulus. At this time, the isometric force was 2% P_0. Sampling time is 10 μs between the vertical dotted lines and 1 ms outside the lines. Horizontal dashed line indicates zero ST.

Figure 6.2 shows the comparison between force records obtained when four stretches were applied at 1, 2, 3 and 4 ms after the start of stimulation. All the records show that (1) ST remains constant after the initial force peak in spite of the increase of active tension; (2) ST is present also when the stretch was applied 1 ms after the stimulus during the latent period when active force was still zero; (3) ST depends on the delay between the stimulus and the stretch and is it maximal when the stretch is applied 3–4 ms after the stimulus. At this time, the active tension is only a few percent of P_0. Note that the initial tension peak occurring synchronously with the stretch almost vanishes in the force record at 1 ms delay at zero active tension when there are no crossbridges attached. This confirms that the stretching of attached crossbridges is the cause of this peak.

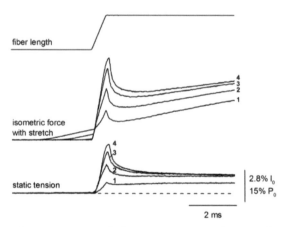

Figure 6.2 Superposition of force responses (intermediate traces) and ST (lower traces) for stretches (fibre length change, upper trace) applied 1, 2, 3 and 4 ms after the first stimulus of a short tetanus at 100 Hz. Stretch, 3.2% l_0 amplitude, duration 680 μs. Sarcomere length, 2.60 μm. Dashed horizontal line represents zero ST. Numbers near the traces are the delays of stretch application respect to the first stimulus. Note that ST is present also 1 ms after the stimulus when the fibre develops zero tension.

The ST records of Fig. 6.2 show an apparently puzzling finding: trace 1, for example, where the stretch is applied 1 ms after the stimulus, shows that ST has approximately the same value 2 ms after the stretch, whereas a stretch applied at 2 ms (trace 2), induces a much greater ST. This apparent contradiction can be easily explained by assuming that ST is caused by the stretching of a sarcomere structure whose stiffness increases with time after the stimulus. A simple increase of fibre stiffness would cause not effect on tension and it would go undetected. However, stretch applied to the fibre will reveal the increase in stiffness by showing a greater steady force response or ST. Once the responsible structure is stretched, any further change of its stiffness would cause again no change in tension and another stretch is necessary to see it. For these reasons to measure the time course of ST following the stimulus it was necessary to apply stretches at progressively increasing delay from the start of stimulation in a series of short tetani. This procedure was used to obtain the data showed in Fig. 6.3, in which ST time course was followed up to 16 ms delay in a fibre stimulated at 100 and 70 Hz.

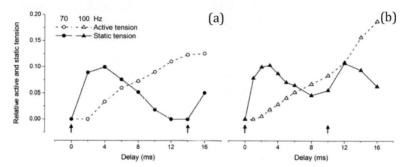

Figure 6.3 Time courses of active force (open symbols) and ST (filled symbols) development following the stimulation at two different frequency, 70 (a) and 100 (b) Hz. Note that the peak of ST was reached in both cases between 3 and 4 ms delay from the first stimulus. Active and static tension are expressed as a fraction of P_0. The arrows indicate the stimuli application. Stretch amplitude 3.1% l_0. Sarcomere length, 2.63 μm.

Note that, in agreement with previous work (Bagni et al., 2004; Colombini et al., 2009) development of ST precedes tension and it is similar to the time course of internal Ca^{2+} concentration following the stimulation (Baylor and Hollingworth, 2003). The maximum value of ST occurred when the stretch was applied at ~3 ms after the first stimulus when the active tension was only a few percent of the plateau tetanic tension. At longer delays, ST decreased in spite of the increase in active tension. ST reached a minimum and increased again after the second stimulus to reach a new maximum ~3 ms after it. Between the two peaks ST decreases similarly to Ca^{2+} concentration time course. The time necessary to reach the ST peak depends on the experimental temperature and it was shorter here at 30°C than in a previous work at 24°C (Colombini et al., 2009). The lack of correlation between active tension and ST shown in Fig. 6.3 is consistent with the idea that crossbridges are not involved in the mechanism of ST or force potentiation after stretch. Note that increasing the frequency of stimulation from 70 to 100 Hz does not alter significantly the peak of ST but increases ST at the point of minimum before the second stimulus. This finding is in agreement with the hypothesis that ST Ca^{2+} dependent. Record in literature in fact shows that Ca^{2+} concentration oscillates between maxima and minima synchronously with stimulation frequency (Baylor and Hollingworth, 2011).

The mean value of maximum ST was 7 ± 1 times ($n = 10$) greater than the tension induced by stretching the fibre under resting conditions. This means that stimulation increases by the same amount the passive stiffness of the fibre.

6.3.2 Effects of Sarcomere Length on Active and Passive Tension

It has been shown that RFE in frog fibres was dependent on sarcomere length reaching a maximum value around 2.9 μm (Edman et al., 1982). A similar investigation was made here in FDB bundles in which active, passive and ST were evaluated as function of sarcomere length between 2.6 and 4.4 μm, beyond the theoretical zero overlap length. Above 4.4 μm the bundles were usually unable to recover the original active tension when returned to the initial length and therefore no measurements were made in this range. In order to reduce the number of attached crossbridges, and their possible influence on ST, experiments were also made in 10 μM BTS solution which reduced average tetanic force at 2.7 μm by ~76% respect to experiments in NT solution. In Fig. 6.4 are shown the active tension-sarcomere length relations in NT and BTS solution. Note that the active tensions plotted were measured on the tetanus rise 3 ms after the first stimulus, when ST was maximal, and not at the tetanus plateau, as usually done for measuring the sarcomere length-tension relationship. This explains why the tension in NT solution had a maximum around 3.3 μm, and fell to zero at ~4 μm sarcomere length in contrast to the expectation of the myofilament overlap, which predicts the maximum at 2.6–2.7 μm. On the tetanus rise, in facts, in addition to myofilament overlap, force generation is modulated also by changes in myofibrillar Ca^{2+} sensitivity with sarcomere length (Williams et al., 2013). BTS strongly reduced active tension, without affecting the shape of the relationship confirming that BTS mainly reduces the number of attached crossbridges.

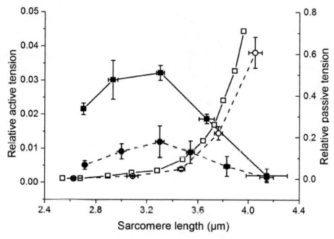

Figure 6.4 Active (filled symbols) and passive (open symbols) tension at various sarcomere lengths. Mean active tension (n = 10) in NT solution (filled square) and in 10 μM BTS solution (filled circles) measured at 3 ms after the first stimulus. Mean passive tension (n = 10) of the bundles (open squares) was near zero at 2.6 μm and increased to ~0.6 P_0 at ~4 μm sarcomere length. Open circles represent the passive tension of a single fibre from previous experiments (Colombini et al., 2009). Measurements were made under static conditions, ~20 s after the attainment of the new sarcomere length.

Figure 6.4 also shows the effects of sarcomere length on the passive tension in NT solution. The passive tension of the bundles (open squares) is not significantly different from that of single fibres (open circles) reported previously (Colombini et al., 2009) and similar to the passive curve measured in myofibrils (Leonard and Herzog, 2010). This means that passive stretch response of our preparation is not influenced significantly by the connective tissue and it is attributable mostly to titin filament. BTS had no effect on passive tension (data no showed).

6.3.3 Effects of Sarcomere Length on Static Stiffness

In order to improve the comparison of the measurements in NT and BTS solutions, in Fig. 6.5 we plotted static stiffness (SS) rather than ST. This corresponds to the ST expressed relatively to P_0 and normalized for 1% l_0 stretch amplitude. It can be seen that SS increased progressively from 2.6 µm up to a maximum value at 3.4–3.5 µm, and then started to decrease again. The sarcomere length at which SS reached zero was not directly measured, but it can be reasonably extrapolated at about 4.5 µm, well above the zero overlap length. At sarcomere length of 4.0 µm, at which the active tension was near zero, SS was still ~50% of the maximum, confirming its non-crossbridge nature. At 2.7 µm, mean

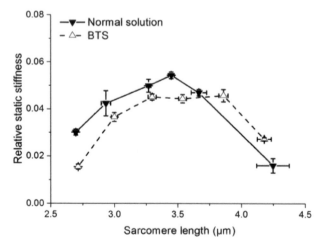

Figure 6.5 Dependence of mean SS (n = 10) from sarcomere length in both NT (filled square) and BTS solution (open square). SS measured at 3 ms after the first stimulus. Note the tendency of the SS curve in BTS to lie to the right of the curve in NT. SS is the ST, expressed as a fraction of P_0, for 1% stretch amplitude.

SS corresponded to ~3% of P_0 and increased to the maximum value of 5% P_0 at 3.5 µm to decrease to zero at ~4.5 µm. These results are not consistent with the finding that force response to stretch in activated myofibrils continues to increase progressively at sarcomere length well above 4.5 µm (Leonard and Herzog, 2010). Apart from a slight shift to the right the sarcomere length-SS curve in BTS is almost superimposed to the curve in NT. This occurred in spite of the considerable reduction of active tension (Fig. 6.4) which was about 3.5 times smaller than in NT solution at any sarcomere length.

6.4   Discussion

6.4.1   Equivalence between Residual Force Enhancement and Static Tension

A number of experimental findings indicate that RFE and ST are the same phenomenon, both measuring the excess of steady tension that follows the stretch of an active fibre. Both RFE and ST, in fact, (1) depend linearly on stretch amplitude, (2) are independent of the stretching velocity, (3) increase with sarcomere length and (4) last for the whole stimulation period (Edman et al., 1978, 1982; Bagni et al., 2002). The only difference regards the way in which the force potentiation was investigated: slow stretches applied at tetanus plateau in RFE measurements and fast stretches applied at low active tension in ST measurements. The use of fast stretches at low active tension for measuring ST has the advantage that only a small number of attached crossbridges is present at the time of the stretch and this reduced drastically the peak force synchronous with the stretch and, more importantly, the time to the attainment of the steady ST after the stretch (Bagni et al., 2002, 2004; Colombini et al., 2009). RFE and ST are also similar in amplitude (Edman et al., 1978, 1982; Cornachione and Rassier, 2012; Duvall et al., 2012). The value of 1.6% P_0 for a stretch of 1% l_0 in frog fibres for RFE reported by Edman et al. (1978) is similar to the values found for ST in frog (Bagni et al., 2002) and consistent with the value of 3% P_0 reported here at near physiological temperature in mouse muscle.

6.4.2 Dependence of Static Stiffness on Sarcomere Length

Figure 6.5 shows that static stiffness (SS) increases with sarcomere length from 2.7 µm up to a peak at 3.5 µm (optimal length), decreases again at higher lengths and reaches zero around 4.5 µm in both NT and in BTS solution, in agreement with RFE measures on frog muscle (Edman et al., 1982). The optimal length at which SS is maximal is different from that for force generation that occurs at maximum overlap near 2.6–2.7 µm (Colombini et al., 2009). An optimum sarcomere length for force potentiation does not seem present in experiments on skinned fibres, or in myofibrils in which the passive tension in activated preparation (equivalent to ST) seems to increase progressively with sarcomere length (Labeit et al., 2003; Leonard and Herzog, 2010). The reason of this difference is unclear, one possibility is that interfilamentary distance could play a role on the mechanism of force enhancement, for example by changing titin sensitivity to Ca^{2+} or by changing the Ca^{2+} promoted actin-titin interaction (Granzier and Labeit, 2007). Interfilamentary distance in fact decreases with sarcomere length in intact preparation (constant volume behaviour) whereas in skinned fibres, and presumably in myofibrils, it does not (Matsubara and Elliott, 1972).

Figure 6.5 shows that SS reaches zero at sarcomere length near 4.5 µm. This result does not seem consistent with the finding reported recently that non-crossbridge force in activated rabbit myofibrils progressively increases with sarcomere length beyond the zero overlap length up to 6 µm (Leonard and Herzog, 2010). However, given the very different conditions and procedures, myofibrils experiments are not directly comparable with the results presented here on intact fibres. Apart from the different preparation, there are two important differences: (1) myoplasmic Ca^{2+} concentration during a just fused tetanus in intact fibres is not constant but oscillates largely synchronously with the stimuli; in myofibrils, Ca^{2+} concentration was instead constant and over saturating during the whole contraction; (2) in ST study the stretches used are small (<4% l_0) and fast (stretch time of 600 µs), and were applied to fibre passively elongated to the desired sarcomere length prior the stretch. The RFE study on myofibrils, instead, employed just one very slow stretch

(stretch time of several seconds) of great amplitude (~250% l_0) spanning the whole range of sarcomere length. A further difference is that passive force in activated myofibrils was measured during the stretch itself rather than in steady conditions after it as we did.

6.4.3 BTS Effects

BTS has been shown to reduce tetanic tension and the number of attached crossbridges without affecting Ca^{2+} release (Pinniger et al., 2005). On this base, according to our hypothesis we should expect no effect of BTS on SS. In general, data in Fig. 6.5 confirm this expectation, in agreement with previous data on rat muscle (Pinniger et al., 2006). However BTS seems to shift the sarcomere length-SS relationship to the right, which would explain why SS in BTS at 2.7 μm is slightly smaller than in NT solution. The reason for this shift is unknown, but BTS, in addition to inhibiting crossbridge formation, also reduces the Ca^{2+} sensitivity of the crossbridges (Pinniger et al., 2005). It cannot be excluded that BTS has a similar effect on titin Ca^{2+} sensitivity or titin–actin interaction in a sarcomere length dependent way.

6.4.4 Independence of Static Tension from Crossbridges

Numerous findings in literature show that ST is Ca^{2+}-dependent and it is not arising from crossbridges, either force or non-force generating (Bagni et al., 1994, 2002, 2004; Colombini et al., 2009; Nocella et al., 2012, 2014; Rassier et al., 2015; Colombini et al., 2016). The results reported here at near physiological temperature in mouse muscle, are fully consistent with these findings confirming that in no conditions active tension and ST were correlated to each other making unlikely the possibility that ST could arise from stretched crossbridge. Further, the observations that (1) the sarcomere length-SS relation is very similar in NT and BTS solutions in spite of the much smaller active tension developed in BTS and (2) ST is present at sarcomere length beyond myofilament overlap where actin and myosin filaments cannot interact to form crossbridges, gives a further strong support to the hypothesis of the non-crossbridge nature of SS.

6.4.5 Residual Force Enhancement and Static Tension Mechanism

According to several studies (Julian and Morgan, 1979; Morgan, 1990, 1994; Edman, 2012), RFE is mainly consequence of sarcomere length non-uniformity developing after the stretch on the descending limb of the length tension relationship. This mechanism could be associated with recruitment of passive elasticity during force development due, for example, to misalignment of adjacent myofibrils (Edman and Tsuchiya, 1996). Non uniformity of sarcomere length would lead to fibre regions where myofilament overlap is greater than expected from the experimental sarcomere length producing the excess of tension observed. However, overlap non-uniformity does not seem a likely interpretation of our ST measurements. In fact, the sarcomere length non-uniformity and the development of passive stiffness (myofibril misalignment) upon stimulation both require the fibre to develop force and are expected to be correlated with the presence of force generating crossbridges. Without crossbridges and active force no change in overlap and no myofibrils misalignment would occur upon stretching. This is contrast with our findings showing that ST is not correlated with tension and it is present also at zero active tension either during the latent period or at sarcomere length above 4 µm. The conclusion that ST does not arise from sarcomere length non-homogeneity is consistent with the observation that ST is present even in the plateau region of the sarcomere length-tension relation, where sarcomere length non-uniformity is very small and has no effect on tension generation (Bagni et al., 2002; Lee and Herzog, 2008). A further evidence is given by the finding that force potentiation is present also in a single sarcomere (Leonard et al., 2010).

Bagni et al. (1994) postulated the possibility that Ca^{2+} could quickly stiffen some unknown sarcomeric structure, which we anticipated could be titin or nebulin. Consistently with this idea, some years later Labeit et al. (2003) made two interesting observations: they showed that (1) the elastic segment PEVK of titin, become stiffer in presence of Ca^{2+} and (2) Ca^{2+} increased the force response to slow stretch, respect to relaxing conditions, of skinned fibres previously treated to eliminate crossbridge formation. Another mechanism for titin stiffening was identified with the phosphorylation of PEVK segment (Hidalgo et al., 2009; Hudson et al., 2010). Both mechanisms seems fast enough to induce titin stiffening quickly enough, immediately after Ca^{2+} release, to be compatible with the quick development of ST after the stimulus. However, these mechanisms together were estimated to produce only ~50% of titin stiffness increases compared to resting conditions (Granzier, 2010). This figure seems too small to account for 7-fold increase of the passive stiffness upon activation found here and, in general, for passive force potentiation during stretching (Leonard and Herzog, 2010). To account for this discrepancy and to explain their results in myofibrils, Leonard and Herzog (2010) proposed that (1) in presence of Ca^{2+} titin binds to actin becoming stiffer and (2) titin–actin interaction is promoted and modulated by the crossbridge action. A similar and more elaborated hypothesis was proposed by Nishikawa et al. (2012). According to these authors crossbridges cycling could induce the rotation of thin filaments and winding of titin upon them. This will induce a great extension of the PEVK region and a noteworthy increase of its stiffness. To be effective, it was assumed that this mechanism was preceded by the Ca^{2+} induced binding of the N2 segment of titin to actin to eliminate the high compliance of proximal tandem Ig domains at low tensions (Granzier and Labeit, 2007). However, these hypotheses too require the presence of crossbridges to be effective. Force generating crossbridges are in fact necessary either to promote titin–actin interaction (Leonard and Herzog, 2010) or to induce winding of titin upon myofilaments (Nishikawa et al., 2012). Consequently, RFE or ST amplitude is expected to be directly linked to the force developed by the fibre. As pointed out before, this is not the case since our previous and present findings show that force potentiation is not correlated with the tension developed by the fibre. Any mechanism of force potentiation based on crossbridges seems to be ruled out by the absence of correlation between ST (or RFE) and the force developed by the fibre. Thus, force enhancement post stretch is mainly attributable to stretching of an elastic sarcomeric structure, non-correlated to crossbridge action, whose stiffness greatly increases upon Ca^{2+} concentration increase, which is likely identified with titin. As pointed out above, stiffening of titin by Ca^{2+} does not seem great enough to explain the ST increase of the fibre upon stimulation (Granzier, 2010). There are some aspects, however, which could explain at least partially this discrepancy. The experiments on skinned fibre by Labeit et al. (2003) from which the stiffening of titin upon Ca^{2+} binding was derived, were made in soleus muscle, but soleus has a SS of about 1/5 of EDL and 1/2 of FDB (Colombini et al., 2009; Nocella et al., 2012). This effect would reduce the discrepancy. In addition, in skinned fibres and myofibrils, passive tension in resting and Ca^{2+} activated fibres was found to increase monotonically with sarcomere length up to 3.5 μm and beyond in myofibrils. Experiments in intact preparations from both frog and mouse muscle, showed instead the presence of an optimum sarcomere length at which ST was maximal (Edman et al., 1982; Bagni et al., 2002; Colombini et al., 2009). This different result could be correlated with the constant volume behaviour of intact fibres not present in skinned fibres. In intact fibres the myofilament lattice spacing decreases when the sarcomere length increases due to the constant volume behaviour of the lattice. This effect might influence the interaction between Ca^{2+} and PEVK titin segments. A further possibility is that PEVK segment could have an optimum elongation for Ca^{2+} sensitivity, which could be attained at sarcomere length around 3.5 μm. This could explain great increase of ST with Ca^{2+} and the effects of sarcomere length on it. Finally is should be considered that increase of titin stiffness could be contributed by the stiffness increase of the I27 region upon Ca^{2+} binding (Leonard et al., 2010). It is interesting that Nishikawa et al. (2012) reported that activation increases the passive stiffness of titin by about 2.5 times in soleus. Considering that ST of soleus is about one-half of that of FDB, this would correspond to an increase of 5 times in FDB sarcomere stiffness, in relatively good agreement with increment of 7 times found in this study.

6.4.6 Conclusions

The data reported here show that stimulation, in addition to promoting crossbridge interaction, increases the sarcomere stiffness in a non-crossbridge Ca^{2+}-dependent way. This stiffness increase, upon a stretch, gives rise to a constant force level, which adds to active force resulting in a force potentiation that correspond to the RFE or ST found in skeletal muscle. Although not directly demonstrated, all our data indicate that passive sarcomere stiffness increase upon activation is due to a Ca^{2+} -based titin stiffening, possibly associated with a Ca^{2+}-dependent titin–actin interaction. The postulated titin stiffness increase upon stimulation could have a significant physiological role as a way to increase the sarcomere stiffness, which would help in maintaining the uniformity of the sarcomere length. ST development anticipates the force and reaches the maximum increase 3–4 ms after the stimulus. This seems well appropriate to counteract the sarcomere instability that may occur during the early phases of contraction when different sarcomeres in series could develop different forces due to the likely non-uniform Ca^{2+} distribution among sarcomeres.

Grants

This study was supported by grant from Ministero dell'Istruzione, dell'Università e della Ricerca (PRIN 2010R8JK2X_002), from the University of Florence and from Ente Cassa di Risparmio di Firenze (CRF 2011.0302), Italy. The funders had no role in study design, data collection and analysis, decision to publish, or preparation of the manuscript.

References

Abbott BC, Aubert XM (1952) The force exerted by active striated muscle during and after change of length. *J Physiol*, **117**: 77–86.

Bagni MA, Cecchi G, Colombini B, Colomo F (2002). A non-cross-bridge stiffness in activated frog muscle fibres. *Biophys J*, **82**: 3118–3127.

Bagni MA, Cecchi G, Colomo F, Garzella P (1994). Development of stiffness precedes cross–bridge attachment during the early tension rise in single frog muscle fibres. *J Physiol*, **481**: 273–278.

Bagni MA, Colombini B, Geiger P, Berlinguer Palmini R, Cecchi G (2004). Non-cross-bridge calcium dependent stiffness in frog muscle fibres. *Am J Physiol Cell Physiol*, **286**: C1353–C1357.

Baylor SM, Hollingworth S (2003). Sarcoplasmic reticulum calcium release compared in slow-twitch and fast-twitch fibres of mouse muscle. *J Physiol*, **551**: 125–138.

Baylor SM, Hollingworth S (2011). Calcium indicators and calcium signalling in skeletal muscle fibres during excitation-contraction coupling. *Prog Biophys Mol Biol*, **105**: 162–179.

Colombini B, Benelli G, Nocella M, Musarò A, Cecchi G, Bagni MA (2009). Mechanical properties of intact single fibres from wild-type and MLC/mIgf-1 transgenic mouse muscle. *J Muscle Res Cell Motil*, **30**: 199–207.

Colombini B, Nocella M, Bagni MA (2016). Non-crossbridge stiffness in active muscle fibres. *J Exp Biol*, **219**: 153–160.

Colombini B, Nocella M, Bagni MA, Griffiths PJ, Cecchi G (2010). Is the cross-bridge stiffness proportional to tension during muscle fiber activation? *Biophys J*, **98**: 2582–2590.

Cornachione AS, Rassier DE (2012). A non-cross-bridge, static tension is present in permeabilized skeletal muscle fibers after active force inhibition or actin extraction. *Am J Physiol Cell Physiol*, **302**: C566–C574.

Duvall MM, Gifford JL, Amrein M, Herzog W (2012). Altered mechanical properties of titin immunoglobulin domain 27 in the presence of calcium. *Eur Biophys J*, **42**: 301–307.

Edman KA (2012). Residual force enhancement after stretch in striated muscle. A consequence of increased myofilament overlap? *J Physiol*, **590**: 1339–1345.

Edman KA, Elzinga G, Noble MI (1978). Enhancement of mechanical performance by stretch during tetanic contractions of vertebrate skeletal muscle fibres. *J Physiol*, **281**: 139–155.

Edman KA, Elzinga G, Noble MI (1982). Residual force enhancement after stretch of contracting frog single muscle fibers. *J Gen Physiol*, **80**: 769–784.

Edman KA, Tsuchiya T (1996). Strain of passive elements during force enhancement by stretch in frog muscle fibres. *J Physiol*, **490**: 191–205.

Fenn WO (1924). The relation between the work performed and the energy liberated in muscular contraction. *J Physiol*, **58**: 373–395.

Granzier HL (2010). Activation and stretch-induced passive force enhancement—are you pulling my chain? Focus on "Regulation of muscle force in the absence of actin-myosin-based cross-bridge interaction". *Am J Physiol Cell Physiol*, **299**: C11–C13.

Granzier H, Labeit S (2007). Structure-function relations of the giant elastic protein titin in striated and smooth muscle cells. *Muscle Nerve*, **36**: 740–755.

Herzog W, Leonard TR (2002). Force enhancement following stretching of skeletal muscle: A new mechanism. *J Exp Biol*, **205**: 1275–1283.

Hidalgo C, Hudson B, Bogomolovas J, Zhu Y, Anderson B, Greaser M, Labeit S, Granzier H (2009). PKC phosphorylation of titin's PEVK element: A novel and conserved pathway for modulating myocardial stiffness. *Circ Res*, **105**: 631–638.

Hill AV, Howarth JV (1959). The reversal of chemical reactions in contracting muscle during an applied stretch. *Proc R Soc Lond B Biol Sci*, **151**: 169–193.

Hudson BD, Hidalgo CG, Gotthardt M, Granzier HL (2010). Excision of titin's cardiac PEVK spring element abolishes PKCalpha-induced increases in myocardial stiffness. *J Mol Cell Cardiol*, **48**: 972–978.

Joumaa V, Leonard TR, Herzog W (2008a). Residual force enhancement in myofibrils and sarcomeres *Proc R Soc B*, **275**: 1411–1419.

Joumaa V, Rassier DE, Leonard TR, Herzog W (2008b). The origin of passive force enhancement in skeletal muscle. *Am J Physiol Cell Physiol*, **294**: C74–C78.

Julian FJ, Morgan DL (1979). The effect on tension of non-uniform distribution of length changes applied to frog muscle fibres. *J Physiol*, **293**: 379–392.

Labeit D, Watanabe K, Witt C, Fujita H, Wu Y, Lahmers S, Funck T, Labeit S, Granzier H (2003). Calcium-dependent molecular spring elements in the giant protein titin. *Proc Natl Acad Sci*, **100**: 13716–13721.

Lee EJ, Herzog W (2008). Residual force enhancement exceeds the isometric force at optimal sarcomere length for optimized stretch conditions. *J Appl Physiol*, **105**: 457–462.

Lee EJ, Joumaa V, Herzog W (2007). New insights into the passive force enhancement in skeletal muscles. *J Biomech*, **40**: 719–727.

Leonard TR, DuVall M, Herzog W (2010). Force enhancement following stretch in a single sarcomere. *Am J Physiol Cell Physiol*, **299**: C1398–C1401.

Leonard TR, Herzog W (2010). Regulation of muscle force in the absence of actin-myosin-based cross-bridge interaction. *Am J Physiol Cell Physiol*, **299**: C14–C20.

Matsubara I, Elliott GF (1972). X-ray diffraction studies on skinned single fibres of frog skeletal muscle. *J Mol Biol*, **72**: 657–669.

Morgan DL (1990). New insights into the behavior of muscle during active lengthening. *Biophys J*, **57**: 209–221.

Morgan DL (1994). An explanation for residual increased tension in striated muscle after stretch during contraction. *Exp Physiol*, **79**: 831–838.

Nishikawa KC, Monroy JA, Uyeno TE, Yeo SH, Pai DK, Lindstedt SL (2012). Is titin a 'winding filament'? A new twist on muscle contraction. *Proc R Soc B*, **279**: 981–990.

Nocella M, Cecchi G, Bagni MA, Colombini B (2013). Effect of temperature on crossbridge force changes during fatigue and recovery in intact mouse muscle fibers. *PLoS One*, **8**: e78918.

Nocella M, Cecchi G, Bagni MA, Colombini B (2014). Force enhancement after stretch in mammalian muscle fiber: No evidence of cross-bridge involvement. *Am J Physiol Cell Physiol*, 307: C1123–C1129.

Nocella M, Colombini B, Bagni MA, Bruton J, Cecchi G (2012). Non-crossbridge calcium-dependent stiffness in slow and fast skeletal fibres from mouse muscle. *J Muscle Res Cell Motil*, **32**: 403–409.

Nocella M, Colombini B, Benelli G, Cecchi G, Bagni MA, Bruton J (2011). Force decline during fatigue is due to both a decrease in the force per individual cross-bridge and the number of cross-bridges. *J Physiol*, **589**: 3371–3381.

Pinniger GJ, Bruton JD, Westerblad H, Ranatunga KW (2005). Effects of a myosin-II inhibitor (N-benzyl-p-toluene sulphonamide, BTS) on contractile characteristics of intact fast-twitch mammalian muscle fibres. *J Muscle Res Cell Motil*, **26**: 135–141.

Pinniger GJ, Ranatunga KW, Offer GW (2006). Crossbridge and non-crossbridge contributions to tension in lengthening rat muscle: Force-induced reversal of the power stroke. *J Physiol*, **573**: 627–643.

Rassier DE (2012). The mechanisms of the residual force enhancement after stretch of skeletal muscle: Non-uniformity in half-sarcomeres and stiffness of titin. *Proc R Soc B*, **279**: 2705–2713.

Rassier DE, Herzog W, Pollack GH (2003a). Dynamics of individual sarcomeres during and after stretch in activated single myofibrils. *Proc R Soc Lond Biol Sci*, **270**: 1735–1740.

Rassier DE, Herzog W, Pollack GH (2003b). Stretch-induced force enhancement and stability of skeletal muscle myofibrils. *Adv Exp Med Biol*, **538**: 501–515.

Rassier DE, Leite FS, Nocella M, Cornachione AS, Colombini B, Bagni MA (2015). Non-crossbridge forces in activated striated muscles: A titin dependent mechanism of regulation? *J Muscle Res Cell Motil*, **36**: 37–45.

Rassier DE, Pavlov I (2012). Force produced by isolated sarcomeres and half-sarcomeres after imposed stretch. *Am J Physiol Cell Physiol*, **302**: C240–C248.

Roots H, Pinniger GJ, Offer GW, Ranatunga KW (2012). Mechanism of force enhancement during and after lengthening of active muscle: A temperature dependence study. *J Muscle Res Cell Motil*, **33**: 313–325.

Sugi H (1972). Tension changes during and after stretch in frog muscle fibres. *J Physiol*, **225**: 237–253.

Sugi H, Tsuchiya T (1988). Stiffness changes during enhancement and deficit of isometric force by slow length changes in frog skeletal muscle fibres. *J Physiol*, **407**: 215–229.

Telley IA, Stehle R, Ranatunga KW, Pfizer G, Stüssi E, Denoth J (2006). Dynamic behaviour of half-sarcomeres during and after stretch in activated rabbit psoas myofibrils: Sarcomere asymmetry but no 'sarcomere popping'. *J Physiol*, **573**: 173–185.

Williams CD, Salcedo MK, Irving TC, Regnier M, Daniel TL (2013). The length-tension curve in muscle depends on lattice spacing. *Proc Biol Sci B*, **280**: 20130697.

Chapter 7

The Static Tension in Skeletal Muscles and Its Regulation by Titin

Dilson E. Rassier,[a,b] Anabelle S. Cornachione,[c] Felipe S. Leite,[a] Marta Nocella,[d] Barbara Colombini,[d] and Maria Angela Bagni[d]

[a]*Department of Kinesiology and Physical Education McGill University, Canada*
[b]*Departments of Physiology and Physics, McGill University, Canada*
[c]*University of São Paulo, Brazil*
[d]*Department of Experimental and Clinical Medicine, University of Florence, Italy*

dilson.rassier@mcgill.ca

Skeletal muscles present a static tension: an increase in force after stretch that is dependent on Ca^{2+} but independent of myosin–actin interactions. This chapter presents the main characteristics of the static tension and discusses the proposed mechanisms responsible for this phenomenon. Evidence will be presented showing that the static tension is caused by a Ca^{2+}-induced increase in the stiffness of titin during activation and stretch. Such increase in titin stiffness increases the overall sarcomere stiffness for as long as muscle activation persists. This form of Ca^{2+} regulation has important implications for our understating of the basic mechanisms of muscle contraction.

Muscle Contraction and Cell Motility: Fundamentals and Developments
Edited by Haruo Sugi
Copyright © 2017 Pan Stanford Publishing Pte. Ltd.
ISBN 978-981-4745-16-1 (Hardcover), 978-981-4745-17-8 (eBook)
www.panstanford.com

7.1 Introduction

Skeletal muscles present a static tension: a Ca^{2+}-dependent increase in force in response to stretch that is not associated with myosin–actin interactions and cross-bridge formation, and that persists through activation. The static tension was observed for the first time in intact muscle fibres from the frog (Bagni et al., 2004, 2005; Rassier et al., 2005) and it has also been shown in intact and permeabilized fibres from mammalians (Cornachione and Rassier, 2012; Nocella et al., 2012; Cornachione et al., 2015). The mechanisms responsible for the static tension are still not known, but recent studies suggest that this phenomenon is directly associated with the properties of titin molecules, that may become stiffer upon muscle activation (Labeit et al., 2003; Bagni et al., 2004, 2005; Rassier et al., 2005). This hypothesis is appealing, as titin is the main structure responsible for passive forces and sarcomere stiffness in striated muscles when they are stretched.

Static tension may have important implications for the understanding of muscle functions, including the long-lasting force enhancement following stretch of skeletal muscles (residual force enhancement), the length dependence of muscle activation, and the near-instantaneous increase stiffness of muscle fibres when activated. The static tension is also associated with a shift in the passive force–sarcomere length relation towards higher forces in the presence of Ca^{2+} and in the absence of active force (Labeit et al., 2003; Cornachione and Rassier, 2012), a phenomenon with implications for the understanding of the entire force–length relation and muscle contraction.

This chapter reviews the characteristics of the static tension in skeletal muscles, providing insights into potential mechanisms by which titin may change its characteristics upon activation and stretch. It shows evidence that the static tension is an important characteristic of skeletal muscle contraction.

7.2 Characteristics of the Static Tension

The fact that muscle fibres that are electrically stimulated present a measureable increase in fibre stiffness that precedes force development is not new—it has been observed since the early 1980s (Ford et al., 1981; Cecchi et al., 1982). Such early increase in stiffness is established during the period lagging between muscle stimulation and force development, and it continues throughout the entire tension rise during twitch or tetanic contractions. Early studies suggested that cross-bridges formed early during muscle stimulation were responsible for the early increase in stiffness. These cross-bridges would operate in a special state that contributes to stiffness but not to tension (Cecchi et al., 1982; Bagni et al., 1988), and would start developing force after a significant delay, explaining the time difference between stiffness and force. This hypothesis was advanced by studies suggesting that such cross-bridges would operate in a special state weakly attached to actin: the "weakly binding cross-bridges" (Chalovich et al., 1981; Brenner et al., 1982). The weakly binding cross-bridges would be characterized by a fast equilibrium between attached and detached states. Weakly binding bridges might result in a fibre stiffness increase without a corresponding increase in force, and thus could explain the stiffness presence earlier than force early during tension development.

Subsequent studies aimed at investigating the details of the weakly binding bridges were unable to confirm their existence. In these studies, fast ramp stretches were imposed on muscle fibres at rest, during the latent period, or at various times during twitch tension development, and the mechanical response was measured to detect the presence of weakly binding bridges (Bagni et al., 1992, 1994) (Fig. 7.1). In both rested and activated fibres, the force response was composed by a fast phase followed by a slower phase. A particularly interesting effect of the activation observed in these studies was that the force produced at the end of the stretch did not decrease quickly (as it would in passive fibres), but it was followed by a period during which the tension remained at an approximately constant value. The amplitude of the stretches applied in these studies was greater than the elastic limit of cross-bridges extension, but stretches applied during the latent period did not show signs of cross-bridges detachment in the force response (Fig. 7.1A) while a stretch applied later during the development of force showed the detachment effect (Fig. 7.1B). Static tension was always observed in these studies (Bagni et al., 1994). The static tension after the stretch was investigated in details in a series of subsequent studies, using a variety of experimental approaches and muscle fibres isolated from the frog and the mouse (Colombini et al., 2009; Nocella et al., 2012).

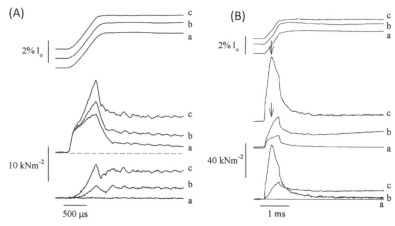

Figure 7.1   Force responses to fast stretches imposed at rest and during activation in a single fibre isolated from the *lumbricalis digiti IV* muscle from the frog. Sarcomere length traces (upper), force responses (middle), and force responses after subtraction of the twitch tension and the passive force from the twitch tension with stretch (lower) are shown. (A) Stretches (amplitude: 3.6% sarcomere length (l_0), velocity: 60 l_0 s^{-1}, sarcomere length: 2.13 μm) were applied at rest (a) and during the latent period, 3 ms (b) and 4 ms (c) after a single stimulus. (B) Stretches (amplitude: 2.6% l_0, velocity: 51 l_0 s^{-1}, sarcomere length: 2.15 μm) were applied at rest (a), 8 ms (b) and 25 ms (c) after a single stimulus. Arrows on the force transient, 8 and 25 ms depict the point of cross-bridge detachment (force "gives"). Note that, although the "give" effect is absent in the force records at 3 and 4 ms after the stimulation, the static tension is present, indicating that it is not related to cross-bridge formation.

The time course of the static tension was investigated also in fibres isolated from the frog during tetanic contractions in the presence of 2,3-butanedione monoxime (BDM), a chemical that inhibits myosin–actin interactions (Bagni et al., 2002). BDM allows the investigation of muscles in a situation in which there is Ca^{2+} activation of the contractile system without myosin–actin interactions. The static tension was maintained with BDM. Interestingly, the static tension was depressed when the investigators used chemicals that result in an inhibition of Ca^{2+}

release from the sarcoplasmic reticulum (dantrolene sodium, deuterium oxide (D₂O) and methoxyverapamil (D600) which inhibits the voltage sensor activator). These studies strengthened the fact that the static tension is regulated by Ca^{2+} release from the sarcoplasmic reticulum (Bagni et al., 2004), and not cross-bridge kinetics. Along the same line of investigation, Campbell and Moss (2002) studied permeabilized rat soleus fibres and observed that the initial tension of a non-cross-bridge structure is significantly greater in pCa 4.5 than pCa 9.0, suggesting the action of a Ca^{2+}-sensitive parallel elastic element in skeletal muscles. The presence of non-cross-bridge components in the persistent increase in tension when skeletal muscles are stretched has also been confirmed in muscle from the rat (Roots et al., 2007).

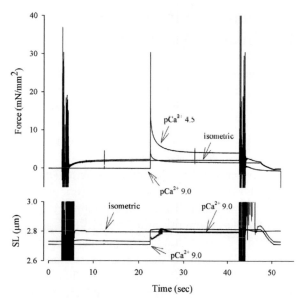

Figure 7.2 Superimposed contractions produced by an activated fibre, before and after stretch in pCa 4.5, and after stretch in pCa 9.0. The fibre was depleted from TnC, thin filaments, and treated with blebbistatin. The isometric force was inhibited after these treatments, but during stretch the force increased substantially. The steady-state force obtained after stretch in pCa 4.5 was higher than the force obtained after stretch in pCa 9.0 and during an isometric contraction at a similar length.

Finally, two studies performed with permeabilized fibres and myofibrils from the rabbit confirmed the presence of the static tension in several conditions that eliminate the possibility of cross-bridge involvement (Cornachione and Rassier, 2012). In one of these studies, the authors determined the static tension in fibres isolated from the psoas muscle from the rabbit following (i) depletion from troponin C (TnC) which eliminates the switch-on mechanism of actomyosin interactions, (ii) depletion of thin filaments, and (iii) treatment with blebbistatin, a powerful, specific myosin II inhibitor. The static tension, induced by stretching a fibre in an experimental bath with a high Ca^{2+} concentration (pCa 4.5), was present at levels similar to what has been observed in previous studies with intact fibres (Fig. 7.2).

7.3 Mechanisms of Increase in Non-Cross-Bridge Forces

It has been suggested that elastic structures, and most specifically titin molecules are responsible for the static tension (Labeit et al., 2003; Bagni et al., 2004, 2005; Rassier et al., 2005). Titin is a multifunctional protein that covers the half-sarcomere, spanning from the Z-disk to the M-line in parallel with the contractile filaments. Titin regulates the development of passive force in response to increases in the sarcomere length. The regulation of passive force rises from the work of different elastic segments arranged in series within the titin I-band region, which contains two immunoglobulin-like segments and one PEVK segment (Proline-, Glutamate-, Valine-, and Lysine-rich segment). Differential splicing of titin creates PEVK segments that vary in size among the different isoforms and additional titin's I-band specific domains. Cardiac muscles express two titin isoforms: N2B, which contains a spring-like segment called N2B, and N2BA, which contains two spring-like domains, N2B and N2A. Skeletal muscles express N2A isoforms, but slow twitch skeletal muscle express a longer N2A isoform than fast twitch muscles (Neagoe et al., 2003). The spring-like elements define the extensibility and compliance of titin. The shorter isoforms of titin start to produce passive forces at shorter sarcomere lengths, as the number and length of spring elements determine the passive tension that titin produces upon muscle stretch.

The amplitude and time course of the static tension are different in muscles containing different titin isoforms. The ratio between the static tension and the stretch amplitude (static stiffness) is ~5 times greater in EDL than in soleus fibres (Nocella et al., 2012). Most conclusively, a recent study developed with isolated myofibrils treated for depletion of myosin–actin interactions showed that the static tension varied in a direct relation with the titin isoforms: the static tension was higher in psoas than in soleus myofibrils, and it was not present in ventricle myofibrils (Cornachione et al., 2015).

Finally, studies performed with intact fibres from the frog (Rassier et al., 2005) and permeabilized fibres from the rabbit (Labeit et al., 2003; Cornachione and Rassier, 2012) have shown that an increasing Ca^{2+} concentration causes an upward shift in the sarcomere length-passive force relationship, that is independent of myosin–actin interactions. In Fig. 7.3, a typical result from these studies is shown. A permeabilized muscle fibre with no cross-bridge formation was stretched consecutively in a variety of sarcomere lengths, in pCa 9.0 and pCa 4.5. There was an upward shift in the sarcomere length–force relation when the experiment was conducted in pCa 4.5.

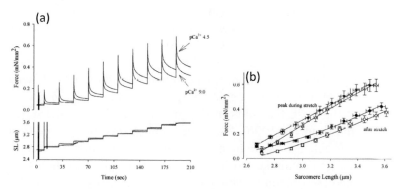

Figure 7.3   (a) Consecutive stretches performed on a muscle fibre after depletion of TnC, thin filaments, and treatment with blebbistatin. The force is higher when the fibre is stretched in pCa 4.5 than in pCa 9.0, showing that Ca^{2+} increases the non-cross-bridge force. (b) The force–sarcomere length relation for fibres experimented in pCa 4.5 and pCa 9.0. The forces obtained during the peak of the stretches (triangles), and after the stretches (circles) are shown in the graph, in experiments conducted in pCa 4.5 (filled symbols) and pCa 9.0 (open symbols).

Mechanisms. The mechanism by which Ca^{2+} regulates mechanisms and the static tension is unknown. There are two major mechanisms that have been proposed in the literature: (i) an effect of Ca^{2+} on titin interactions with actin that could increase the overall sarcomere stiffness, and (ii) a direct effect of Ca^{2+} on titin stiffness.

(i) **Ca^{2+}-induced increase in titin–actin interactions.** It has been proposed that Ca^{2+} may strengthen the interactions between titin and actin, increasing the overall sarcomere stiffness. The high malleability of the PEVK domain of titin allows the molecule to transit among different conformational states (Ma and Wang, 2003) and binds F-actin (Kulke et al., 2001; Yamasaki et al., 2001; Linke et al., 2002; Nagy et al., 2005). Interestingly, it has been shown that the binding of the PEVK domain of titin to actin can be modulated by S100A1, a member of the S100 family of EF-hand Ca^{2+} binding proteins (Yamasaki et al., 2001), which is present in high concentrations in striated muscles (Kato and Kimura, 1985).

Although this hypothesis is attractive due to the proximity between titin and actin filament in the I-band of the sarcomeres (Kellermayer and Granzier, 1996; Kellermayer and Granzier, 1996; Linke et al., 1997; Trombitas et al., 1997), most evidence suggest the contrary. A study investigating in-vitro motility assays for myosin-driven actin motility showed that titin inhibited significantly, or even blocked, the sliding of the actin filaments in the presence of Ca^{2+} (Kellermayer and Granzier, 1996). Most importantly, this inhibitory effect was enhanced with increased concentrations of Ca^{2+}. Studies using recombinant titin fragments also failed to detect an increased binding between different PEVK segments of titin and actin as a direct result of Ca^{2+} (Kulke et al., 2001; Yamasaki et al., 2001). Kulke et al. (2001) found that the PEVK-induced inhibition of actin filament sliding over myosin was reversed with a high Ca^{2+} concentration, and Yamasaki et al. (2001) suggested that S100A1-PEVK binding alleviates the PEVK-based inhibition of F-actin motility, inhibiting PEVK–actin interaction and providing the sarcomere with a mechanism to free the thin filament from titin before contraction.

Finally, binding of PEVK fragments of titin to actin was inhibited with Ca^{2+} in a study conducted by Stuyvers et al. (1998) with cardiac muscles; the titin–actin-based stiffness of the rat cardiac trabeculae increased when Ca^{2+} levels were lowered during muscle relaxation.

(ii) **Ca^{2+}-induced increase in titin stiffness**. An increase in the intracellular Ca^{2+} concentration can increase the stiffness of different regions of titin, allowing the static tension to be developed upon stretch. In fact, since the early 1990s studies (Takahashi et al., 1992; Tatsumi et al., 1996) have shown that titin (then called α-connectin) has an affinity for Ca^{2+} ions, containing binding from the N2A segment to the M-line (then called β-connectin portion). These studies suggested that the main Ca^{2+} binding region of titin was the PEVK segment, which has a strong negative net charge at physiological pH of 5.0 (Kolmerer et al., 1996), and thus amenable for positively charged ions. These findings were confirmed by the observation that circular dichroic spectra of a 400 kDa fragment which constitutes the N-terminal elastic region of β-connectin—in the PEVK region—were changed by the binding of Ca^{2+} ions, suggesting that titin alter its structure (Tatsumi et al., 2001).

In an elegant study using different fragments of titin tested with atomic force microscopy, Labeit et al. (2003) observed that Ca^{2+} binding to the PEVK region of the molecules caused a decrease in its persistence length. A decrease in the persistence length is associated with an increase in stiffness. The authors also showed that the minimal titin fragment that responded to Ca^{2+} ions contained a central E-rich domain with glutamates flanked by PEVK repeats. Since skeletal muscle titin isoforms contain a variable number of PEVK repeats and E-rich motifs (Bang et al., 2001), their result strengthens the hypothesis that Ca^{2+} affects the conformation of the PEVK segment. Since the glutamate-rich (E-rich) motif is essential for the titin response to Ca^{2+}, the result has important implications for muscle regulation.

There is one study using molecular dynamics simulation suggesting that binding of Ca^{2+} ions to the molecule could also regulate the Ig domains of titin molecules, also responsible for regulation of passive forces (Lu et al., 1998). However, an experimental study performed by Watanabe et al. (2002) investigated the potential effect of Ca^{2+} on differentially spliced (I65–70) and constitutive (I91–98) regions from Ig domains of titin. The authors observed that the average domain unfolding force in I91–98 and the persistence length of the unfolded I91–98 chain were not different when experiments were conducted in pCa 9.0 and pCa 3.0, suggesting that Ca^{2+} does not change the properties of titin through an increased stiffness of Ig domains.

Experimental evidence strongly points towards a mechanism by which Ca^{2+} affects titin by increasing the molecule stiffness, and especially on the PEVK domain containing glutamate-rich (E-rich) motifs. Ca^{2+} does not seem to increase the titin–actin interaction—in fat it may decrease the titin–actin interaction upon a rise in Ca^{2+} concentration.

7.4 Conclusion and Physiological Implications

Evidence shows that an increase in intracellular Ca^{2+} concentration during muscle activation induces an increase in the sarcomere stiffness that is dissociated from cross-bridges. Instead, activation leads to Ca^{2+} binding to the PEVK domains of titin, and specifically the glutamate-rich (E-rich) motifs. The binding of Ca^{2+} to PEVK domains decreases the persistence length of titin and increase the stiffness of the molecule, and as a result the stiffness of the sarcomere. When muscles fibres are stretched, a stiffer titin molecule contributes to an increased force that persists for as long as the activation continues.

The Ca^{2+}-induced increase in titin stiffness may have important physiological roles. First, the increase in the stiffness of titin upon muscle activation must be important to balance unequal and increasing forces in half-sarcomeres due to myosin–actin interactions. Such unequal forces may cause movements of the thick filaments from the centre towards the edges of the sarcomeres, which would lead to mechanical instabilities. Recently, a study conducted with mechanically isolated sarcomeres showed a close relation between force production and A-band displacements; beyond lengths of 2.2–2.4 μm, the A-band displacements deceased linearly with increasing sarcomere lengths in a region in which titin is stiffer (Pavlov et al., 2009). In this way, Ca^{2+} regulation of titin can also provide a powerful mechanism for the stable behaviour of sarcomeres when they are activated along the descending limb of the force–length relationship (Pavlov et al., 2009; Rassier and Pavlov, 2010). Several studies conducted with isolated myofibrils have shown that sarcomeres do not present large instabilities despite a decrease in active force when activated at long lengths (Pavlov et al., 2009; Rassier and Pavlov, 2010). Such relative stability prevents large sarcomere length non-uniformities and prevents overextension of the sarcomeres.

A Ca^{2+}-dependent increase in the stiffness of titin may also explain the so-called "residual force enhancement" (Rassier, 2012; Nocella et al., 2014)—a phenomenon observed for more than 50 years (Abbott and Aubert, 1952) but with mechanisms that are still elusive. The force after stretch of activated muscles is higher than the isometric force produced at similar lengths, a phenomenon that cannot be explained by the overlap between myosin and actin filaments. A recent study suggests that the residual force enhancement is caused by half-sarcomere non-uniformities and a stiffening of titin molecules, in a mechanism similar to the one described in this paper to explain the static tension (Rassier and Pavlov, 2012). The characteristics of static tension and the residual force enhancement are striking (Cornachione and Rassier, 2012), and likely underlie the same phenomenon caused by a titin-dependent mechanism of regulation.

Acknowledgements

This research was supported by the Canadian Institutes of Health Research (CIHR), the Natural Science and Engineering Research Council of Canada (NSERC) and University of Florence, PRIN2010R8JK2X_002.

References

Abbott, B. C., and X. Aubert (1952). The force exerted by active striated muscle during and after change of length. *J. Physiol.*, **117**(1): 77–86.

Bagni, M. A., G. Cecchi, B. Colombini, and F. Colomo (2002). A non-cross-bridge stiffness in activated frog muscle fibers. *Biophys. J.*, **82**(6): 3118–3127.

Bagni, M. A., G. Cecchi, F. Colomo, and P. Garzella (1992). Are weakly binding bridges present in resting intact muscle fibers? *Biophys. J.*, **63**(5): 1412–1415.

Bagni, M. A., G. Cecchi, F. Colomo, and P. Garzella (1994). Development of stiffness precedes cross-bridge attachment during the early tension rise in single frog muscle fibres. *J. Physiol.*, **481**(Pt 2): 273–278.

Bagni, M. A., G. Cecchi, and M. Schoenberg (1988). A model of force production that explains the lag between crossbridge attachment and force after electrical stimulation of striated muscle fibers. *Biophys. J.*, **54**(6): 1105–1114.

Bagni, M. A., B. Colombini, F. Colomo, R. B. Palmini, and G. Cecchi (2005). Non cross-bridge stiffness in skeletal muscle fibres at rest and during activity. *Adv. Exp. Med. Biol.*, **565**: 141–154.

Bagni, M. A., B. Colombini, P. Geiger, P. R. Berlinguer, and G. Cecchi (2004). Non-cross-bridge calcium-dependent stiffness in frog muscle fibers. *Am. J. Physiol. Cell Physiol.*, **286**(6): C1353–C1357.

Bang, M. L., T. Centner, F. Fornoff, A. J. Geach, M. Gotthardt, M. McNabb, C. C. Witt, D. Labeit, C. C. Gregorio, H. Granzier, and S. Labeit (2001). The complete gene sequence of titin, expression of an unusual approximately 700-kDa titin isoform, and its interaction with obscurin identify a novel Z-line to I-band linking system. *Circ. Res.*, **89**(11): 1065–1072.

Brenner, B., M. Schoenberg, J. M. Chalovich, L. E. Greene, and E. Eisenberg (1982). Evidence for cross-bridge attachment in relaxed muscle at low ionic strength. *Proc. Natl. Acad. Sci. U. S. A.*, **79**(23): 7288–7291.

Campbell, K. S., and R. L. Moss (2002). History-dependent mechanical properties of permeabilized rat soleus muscle fibers. *Biophys. J.*, **82**(2): 929–943.

Cecchi, G., P. J. Griffiths, and S. Taylor (1982). Muscular contraction: Kinetics of crossbridge attachment studied by high-frequency stiffness measurements. *Science*, **217**(4554): 70–72.

Chalovich, J. M., P. B. Chock, and E. Eisenberg (1981). Mechanism of action of troponin. tropomyosin. Inhibition of actomyosin ATPase activity without inhibition of myosin binding to actin. *J. Biol. Chem.*, **256**(2): 575–578.

Colombini, B., G. Benelli, M. Nocella, A. Musaro, G. Cecchi, and M. A. Bagni (2009). Mechanical properties of intact single fibres from wild-type and MLC/mIgf-1 transgenic mouse muscle. *J. Muscle Res. Cell Motil.*, **30**(5–6):199–207.

Cornachione, A. S., F. S. Leite, M. A. Bagni, and D. E. Rassier (2015). The increase in non-crossbridge forces after stretch of activated striated muscle is related to titin isoforms. *Am. J. Physiol. Cell Physiol.,* **310**(1): C19–C26..

Cornachione, A. S., and D. E. Rassier (2012). A non-cross-bridge, static tension is present in permeabilized skeletal muscle fibers after active force inhibition or actin extraction. *Am. J. Physiol. Cell Physiol.,* **302**(3): C566–C574.

Ford, L. E., A. F. Huxley, and R. M. Simmons (1981). The relation between stiffness and filament overlap in stimulated frog muscle fibres. *J. Physiol.,* **311**: 219–249.

Kato, K., and S. Kimura (1985). S100ao (alpha alpha) protein is mainly located in the heart and striated muscles. *Biochim. Biophys. Acta,* **842**(2–3): 146–150.

Kellermayer, M. S., and H. L. Granzier (1996a). Calcium-dependent inhibition of in vitro thin-filament motility by native titin. *FEBS Lett.,* **380**(3): 281–286.

Kellermayer, M. S., and H. L. Granzier (1996b). Elastic properties of single titin molecules made visible through fluorescent F-actin binding. *Biochem. Biophys. Res. Commun.,* **221**(3): 491–497.

Kolmerer, B., N. Olivieri, C. C. Witt, B. G. Herrmann, and S. Labeit (1996). Genomic organization of M line titin and its tissue-specific expression in two distinct isoforms. *J. Mol. Biol.,* **256**(3): 556–563.

Kulke, M., S. Fujita-Becker, F. Rostkova, C. Neagoe, D. Labeit, D. J. Manstein, M. Gautel, and W. A. Linke (2001). Interaction between PEVK-titin and actin filaments: Origin of a viscous force component in cardiac myofibrils. *Circ. Res.,* **89**(10): 874–881.

Labeit, D., K. Watanabe, C. Witt, H. Fujita, Y. Wu, S. Lahmers, T. Funck, S. Labeit, and H. Granzier (2003). Calcium-dependent molecular spring elements in the giant protein titin. *Proc. Natl. Acad. Sci. U. S. A.,* **100**(23): 13716–13721.

Linke, W. A., M. Ivemeyer, S. Labeit, H. Hinssen, J. C. Ruegg, and M. Gautel (1997). Actin-titin interaction in cardiac myofibrils: Probing a physiological role. *Biophys. J.,* **73**(2): 905–919.

Linke, W. A., M. Kulke, H. Li, S. Fujita-Becker, C. Neagoe, D. J. Manstein, M. Gautel, and J. M. Fernandez (2002). PEVK domain of titin: An entropic spring with actin-binding properties. *J. Struct. Biol.,* **137**(1–2): 194–205.

Lu, H., B. Isralewitz, A. Krammer, V. Vogel, and K. Schulten (1998). Unfolding of titin immunoglobulin domains by steered molecular dynamics simulation. *Biophys. J.,* **75**(2): 662–671.

Ma, K., and K. Wang (2003). Malleable conformation of the elastic PEVK segment of titin: Non-co-operative interconversion of polyproline II helix, beta-turn and unordered structures. *Biochem. J.,* **374**(Pt 3): 687–695.

Nagy, A., L. Grama, T. Huber, P. Bianco, K. Trombitas, H. L. Granzier, and M. S. Kellermayer (2005). Hierarchical extensibility in the PEVK domain of skeletal-muscle titin. *Biophys. J.,* **89**(1): 329–336.

Neagoe, C., C. A. Opitz, I. Makarenko, and W. A. Linke (2003). Gigantic variety: Expression patterns of titin isoforms in striated muscles and consequences for myofibrillar passive stiffness. *J. Muscle Res. Cell Motil.,* **24**(2–3): 175–189.

Nocella, M., G. Cecchi, M. A. Bagni, and B. Colombini (2014). Force enhancement after stretch in mammalian muscle fiber: No evidence of cross-bridge involvement. *Am. J. Physiol. Cell Physiol.,* **307**(12): C1123–C1129.

Nocella, M., B. Colombini, M. A. Bagni, J. Bruton, and G. Cecchi (2012). Non-crossbridge calcium-dependent stiffness in slow and fast skeletal fibres from mouse muscle. *J. Muscle Res. Cell Motil.,* **32**(6): 403–409.

Pavlov, I., R. Novinger, and D. E. Rassier (2009a). The mechanical behavior of individual sarcomeres of myofibrils isolated from rabbit psoas muscle. *Am. J. Physiol. Cell Physiol.,* **297**(5): C1211–C1219.

Pavlov, I., R. Novinger, and D. E. Rassier (2009b). Sarcomere dynamics in skeletal muscle myofibrils during isometric contractions. *J. Biomech.,* **42**(16): 2808–2812.

Rassier, D. E. (2012a). The mechanisms of the residual force enhancement after stretch of skeletal muscle: Non-uniformity in half-sarcomeres and stiffness of titin. *Proc. Biol. Sci.,* **279**(1739): 2705–2713.

Rassier, D. E. (2012b). Residual force enhancement in skeletal muscles: One sarcomere after the other. *J. Muscle Res. Cell Motil.,* **33**(3–4): 155–165.

Rassier, D. E., E. J. Lee, and W. Herzog (2005). Modulation of passive force in single skeletal muscle fibres. *Biol. Lett.,* **1**(3): 342–345.

Rassier, D. E., and I. Pavlov (2010). Contractile characteristics of sarcomeres arranged in series or mechanically isolated from myofibrils. *Adv. Exp. Med. Biol.,* **682**: 123–140.

Rassier, D. E., and I. Pavlov (2012). Force produced by isolated sarcomeres and half-sarcomeres after an imposed stretch. *Am. J. Physiol. Cell Physiol.*, **302**(1): C240–C248.

Roots, H., G. W. Offer, and K. W. Ranatunga (2007). Comparison of the tension responses to ramp shortening and lengthening in intact mammalian muscle fibres: Crossbridge and non-crossbridge contributions. *J. Muscle Res. Cell Motil.*, **28**(2–3): 123–139.

Stuyvers, B. D., M. Miura, J. P. Jin, and H. E. ter Keurs (1998). $Ca^{(2+)}$-dependence of diastolic properties of cardiac sarcomeres: Involvement of titin. *Prog. Biophys. Mol. Biol.*, **69**(2–3): 425–443.

Takahashi, K., A. Hattori, R. Tatsumi, and K. Takai (1992). Calcium-induced splitting of connectin filaments into beta-connectin and a 1,200-kDa subfragment. *J. Biochem.*, **111**(6): 778–782.

Tatsumi, R., A. Hattori, and K. Takahashi (1996). Splitting of connectin/titin filaments into beta-connectin/T2 and a 1,200-kDa subfragment by 0.1 mM calcium ions. *Adv. Biophys.*, **33**: 65–77.

Tatsumi, R., K. Maeda, A. Hattori, and K. Takahashi (2001). Calcium binding to an elastic portion of connectin/titin filaments. *J. Muscle Res. Cell Motil.*, **22**(2): 149–162.

Trombitas, K., M. L. Greaser, and G. H. Pollack (1997). Interaction between titin and thin filaments in intact cardiac muscle. *J. Muscle Res. Cell Motil.*, **18**(3): 345–351.

Watanabe, K., C. Muhle-Goll, M. S. Kellermayer, S. Labeit, and H. Granzier (2002). Different molecular mechanics displayed by titin's constitutively and differentially expressed tandem Ig segments. *J. Struct. Biol.*, **137**(1–2): 248–258.

Yamasaki, R., M. Berri, Y. Wu, K. Trombitas, M. McNabb, M. S. Kellermayer, C. Witt, D. Labeit, S. Labeit, M. Greaser, and H. Granzier (2001). Titin-actin interaction in mouse myocardium: Passive tension modulation and its regulation by calcium/S100A1. *Biophys. J.*, **81**(4): 2297–2313.

Chapter 8

Stiffness of Contracting Human Muscle Measured with Supersonic Shear Imaging

Kazushige Sasaki[a] and Naokata Ishii[b]

[a]*Faculty of Human Sciences and Design, Japan Women's University, Tokyo 112-8681, Japan*
[b]*Department of Life Sciences, Graduate School of Arts and Sciences, The University of Tokyo, Tokyo 153-8902, Japan*

sasakik@fc.jwu.ac.jp, ishii@idaten.c.u-tokyo.ac.jp

Recently, an ultrasound-based elastographic technique called supersonic shear imaging (SSI) has been developed and used to measure stiffness (shear modulus) of in vivo muscles. This review describes the theoretical background of SSI, summarizes some basic observations on the shear modulus of contracting human muscles, and presents the latest experimental findings. It is well documented that the muscle shear modulus increases with increasing intensity of contraction. A linear association has been found between the muscle shear modulus and motor unit activity assessed with surface electromyography. Moreover, we have demonstrated both the length-dependent changes in shear

Muscle Contraction and Cell Motility: Fundamentals and Developments
Edited by Haruo Sugi
Copyright © 2017 Pan Stanford Publishing Pte. Ltd.
ISBN 978-981-4745-16-1 (Hardcover), 978-981-4745-17-8 (eBook)
www.panstanford.com

modulus and the association of shear modulus with contractile force, even when the motor unit activity is controlled by direct electric stimulation of muscle. These findings provide strong evidence that the muscle shear modulus measured with SSI can be a useful indicator of muscle activation level or contractile force in a variety of conditions. While the structures and mechanisms determining muscle stiffness in vivo are not fully understood, the result of our pilot study suggests that the shear modulus of contracting muscle may reflect both the single-fiber stiffness (cross-bridge kinetics) and the motor unit recruitment, i.e., the number of activated muscle fibers.

8.1   Introduction

In studies of muscle mechanics, stiffness of contracting single fibers has been used as a measure of the number of attached cross-bridges at any instance. It has usually been quantified by measuring force responses to small (<1% of fiber length) sinusoidal length changes given to contracting fibers. Muscle contraction involves several exponential processes associated with cross-bridge cycling, so that stiffness of contracting fibers is "dynamic" in nature and varies depending on the frequency of length oscillation. Sinusoidal analyses with skinned fibers from rabbit muscle have shown that the dynamic stiffness of contracting fibers involves three viscous (exponential) components, and length oscillation at a frequency much higher than ~100 Hz (e.g., ~1 kHz) can be used to measure the series elasticity representing the number of cross-bridges attached at either "rigor state" or "power stroke" in their cyclic reaction (Kawai, 1979).

During both force-developing phase and steady state of isometric contractions, the stiffness of skinned single fibers is directly proportional to the contractile force (Fig. 8.1; Rüegg et al., 1979). In steady-state contractions, the stiffness decreases in a linear fashion with increasing sarcomere length beyond the optimal length for force generation (L_o), indicating that it is proportional to the amount of overlap between thick and thin filaments (Fig. 8.2; Rüegg et al., 1979). For isotonic contractions, Tsuchiya et al. (1979) have shown that the stiffness linearly increases with force and reaches a maximum under maximal isometric force (Fig. 8.3). Alternatively, the stiffness is inversely related to the shortening velocity, suggesting that the probability of interaction between actin and myosin molecules decreases with increasing the sliding velocity between thick and thin filaments, as proposed by Huxley (1957).

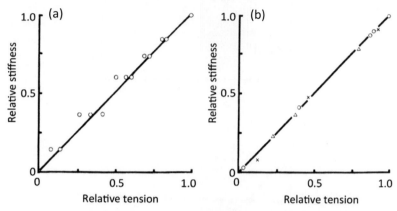

Figure 8.1   Relations between contractile tension and stiffness in skinned frog muscle fibers. (a) Stiffness measured during the tension rising phase after "calcium jump." (b) Stiffness measured during steady-state tension in contractions at varied Ca^{2+} concentrations (modified from Rüegg et al., 1979).

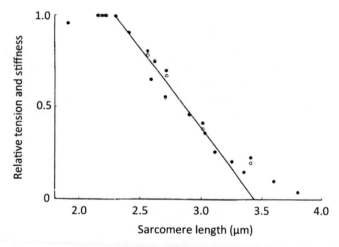

Figure 8.2   Dependence of active tension (filled circles) and stiffness (open circles) on sarcomere length in skinned frog muscle fibers, showing that both are proportional to the overlap between thick and thin filaments (modified from Rüegg et al., 1979).

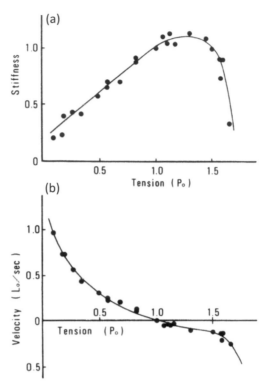

Figure 8.3 Dependence of relative stiffness on isotonic load (a) and the force–velocity relation (b) obtained from the same preparation of frog single muscle fiber. Stiffness was determined by measuring length changes of fibers after quick changes in isotonic loading. Tension is expressed relative to the maximal isometric tension (P_0). Negative velocity represents forced lengthening under the load $>P_0$ (adapted from Tsuchiya et al., 1979).

Measuring stiffness of contracting human muscles in vivo is also of great physiological significance, because it may provide us with information about the force-generating capacity of muscle fibers, which is determined by the relation between sarcomere length and contractile force (length–force relation). The length–force characteristics of muscle can be estimated in vivo by measuring maximal voluntary torques at varied joint angles. However, obtained relation between joint torque and joint angle may be considerably truncated from the original length–force relation of muscle, due mainly to changes in effective moment-arm length with joint angles (Maganaris, 2001; Sasaki et al., 2014). It can also be influenced by activation of synergistic and antagonistic muscle groups. Therefore, direct determination of the relation between muscle length and stiffness (length–stiffness relation) is regarded as highly effective to predict the length–force relations of a variety of muscles in the body, even without measurements of joint torques.

However, application of length oscillation with small amplitude and high frequency to muscles in vivo is substantially impossible, due to the presence of a large amount of series elasticity and intervening soft tissues. A recently developed ultrasound-based elastographic technique, "supersonic shear imaging" (SSI; Bercoff et al., 2004) can overcome this problem and potentially be useful for in vivo measurements of stiffness in contracting muscle. Also, in place of its poor time resolution due to complicated image processing, SSI can visualize changes in regional stiffness within muscle during steady-state contractions. Among other things, it may provide us with an insight into the localization of recruited fibers or motor units in a variety of conditions, e.g., in contractions at varied voluntary activation level, during sustained exertion of small contractile force, during the course of muscle fatigue, etc.

This review lists some recent studies on stiffness of contracting human muscles, with special reference to the effects of muscle activation level, muscle length, and contraction types.

8.2 Methods and Materials

8.2.1 Theoretical Basis of Supersonic Shear Imaging

SSI is based on the B-mode ultrasound imaging that has widely been used in research and clinical diagnosis. In addition to usual scanning supersonic waves for image acquisition, SSI projects another strong supersonic beam that is focused on and hits given portions within a tissue subjected to observation. There, it gives rise to a shear deformation that then propagates three dimensionally as shear wave. In a linearly elastic and transversely isotropic material, its shear elastic modulus (G) is a function of the propagation velocity of shear wave (V_s) as described by the following equation:

$$G = \rho V_s^2, \tag{8.1}$$

where ρ is the density of muscle (generally assumed to be 1,000 kg/m³). Therefore, regional stiffness can be estimated by processing the reflected ultrasound signals and measuring the propagation velocity of shear waves.

When muscle is subjected to measurements, shear deformations produced at given portions of muscle fibers can also propagate three dimensionally. Thus, observations of longitudinal plane should provide regional shear elastic modulus along the fiber axis. In general, shear elastic modulus (G) of a rod-shaped cantilever is proportional to Young's modulus (E) as described by the following equation:

$$G = E/2(1 + v), \tag{8.2}$$

where v is the Poisson ratio. Therefore, measured value of shear elasticity presumably represents Young's modulus averaged for muscle fibers included in the region of interest.

Standing on the above theoretical basis, the SSI scanner (Aixplorer, SuperSonic Imagine, France) implements an ultrafast (up to 20 kHz) echographic imaging of the shear wave propagation to calculate the shear wave velocity along the principal axis of ultrasound probe in less than 20 ms (Bercoff et al., 2004; Hug et al., 2015). Such a short acquisition time minimizes the influence of any motion artifacts (Gennisson et al., 2010).

At present, the short acquisition time is a critical advantage of SSI over the other techniques such as magnetic resonance elastography. Although magnetic resonance elastography can provide three-dimensional shear elasticity map with an excellent spatial resolution, the long acquisition time (several minutes even for two-dimensional measurements) (Bensamoun et al., 2008) limits its application to relatively static organs/conditions. Therefore, SSI opens a new possibility for assessing elastic properties of in vivo human muscles during forceful but brief contractions. Moreover, the SSI scanner is portable and requires no external vibrator, so that the measurement can be free from various experimental constraints.

In 2010, some researchers presented preliminary data on the stiffness of in vivo human muscles determined by SSI (Gennisson et al., 2010; Nordez and Hug, 2010; Shinohara et al.,

2010). Since then, this technique has drawn increasing attention in the field of human skeletal muscle physiology and biomechanics.

8.2.2 Some Technical Issues

Typical examples of shear elasticity imaging using SSI are shown in Fig. 8.4. The muscle shear modulus obtained with a resolution of 1 × 1 mm is spatially filtered and color-coded, comprising a two-dimensional map superimposed on a B-mode ultrasound image. To obtain a representative value, the shear modulus is generally averaged over a selected region of interest (ROI) using bundled software of the SSI scanner or custom-designed computer program (Bouillard et al., 2011, 2012a).

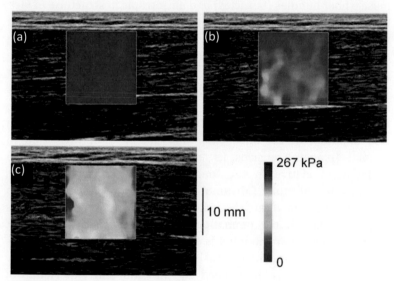

Figure 8.4   Examples of shear modulus distribution superimposed on longitudinal ultrasound image of the biceps brachii muscle at rest (a) and during contractions at 10% (b) and 40% (c) of maximal voluntary contraction. The shear modulus typically increases with increasing contraction intensity.

While it has been well demonstrated that the shear modulus measurement using SSI is highly accurate and reliable (Bouillard et al., 2011; Eby et al., 2013; Koo et al., 2013; Lacourpaille et al., 2012; Yoshitake et al., 2014), there are some technical issues that require careful consideration. First, the upper limit of shear elasticity measurement is currently 266.6 kPa (equivalent shear wave velocity of 16.3 m/s). Despite large inter-muscle and inter-individual differences (Sasaki et al., 2014), this limit is generally insufficient for assessing the muscle shear modulus during maximal contractions. Second, a time resolution of 1 Hz in the current SSI scanner precludes researchers from studying the muscle stiffness changes during ballistic (quick and explosive) contractions or fast movements. A recent study, however, suggests that the above two limitations can be overcome by both hardware and software improvements in the near future (Ateş et al., 2015). Third, the orientation of ultrasound probe greatly influences the measured shear modulus, because skeletal muscle is composed of muscle fiber bundles (fascicles) and anisotropic in structure. In fact, Gennisson et al. (2010) showed that in the human biceps brachii muscle, the shear wave velocity was highest when propagating along the muscle fascicles, and decreased with increasing the probe angle relative to the fascicles. This finding suggests that the ultrasound probe should be placed parallel to the fascicles for the accurate measurement of muscle shear modulus. The dependence of shear wave velocity on the probe orientation also implies that the shear modulus can be underestimated in pennate (pinnate) muscles, i.e., muscles with oblique orientation of fascicles relative to the longitudinal axis of whole muscle, though a recent study (Miyamoto et al., 2015) on resting human muscles suggests that the magnitude of underestimation is negligibly small if the pennation angle is less than 20°. Finally, the measured shear modulus is more or less associated with the clarity of ultrasound image, so that the accuracy and reliability of measurement are influenced by the skill and experience of operator (Hug et al., 2015).

8.3 Muscle Activation Level and Stiffness

8.3.1 Association of Shear Modulus with Joint Torque

A simple and practical way of associating muscle stiffness with activation level is to examine the shear modulus at several different contraction intensities. In general, contraction intensity is defined as a contraction-induced muscle force generation relative to that during maximal voluntary contraction (MVC). Because of the difficulty to directly measure individual muscle force in vivo, most of the studies on human muscles use the torque around the relevant joint axis (joint torque) as a global measure of muscle force generation.

Nordez and Hug (2010) investigated the shear modulus of the human biceps brachii muscle and its association with elbow flexion torque using SSI. Although they employed only low contraction intensities (ramp contraction of up to 30%MVC) because of the limited range (0–100 kPa) of shear modulus measurement in the earlier version of SSI scanner, a curvilinear relation between the shear modulus and contraction intensity was observed. Namely, they reported a relatively sharp increase in shear modulus preceded by little change at very low contraction intensities. The same group of authors subsequently performed another experiment (Bouillard et al., 2012b) in which the shear modulus was measured in elbow flexor synergists (the short and long heads of biceps brachii, brachialis, and brachioradialis muscles). The result indicated that the non-linear shear modulus–torque relation of the biceps brachii muscle (Nordez and Hug, 2010) could be explained by the change in relative contribution of elbow flexor synergists to joint torque as a function of contraction intensity. By contrast, Yoshitake et al. (2014) studied the biceps brachii muscle with a broader range of contraction intensities (up to 60%MVC) and found a linear association of the shear modulus with elbow flexion torque. A linear association of the biceps brachii stiffness and elbow flexion torque was also demonstrated by Dresner et al. (2001) using magnetic resonance elastography.

Bouillard et al. (2011, 2012a) have studied the association of shear modulus with joint torque in human finger muscles (the first dorsal interosseous and the abductor digiti minimi). During isometric ramp contractions with linearly increasing joint torque, the shear modulus increased linearly in both muscles. As these muscles are considered the single agonist for abduction of index finger and little finger, respectively, the individual muscle force can be directly inferred from the measurement of joint torque, assuming a negligible change in moment arm during contraction (Hug et al., 2015). Therefore, these results provide evidence that the shear modulus determined by SSI is a measure of contractile force produced by the muscle of interest.

8.3.2 Association of Shear Modulus with Motor Unit Activity

Since the shear modulus determined by SSI represents a regional stiffness of target tissue, it is likely that the muscle shear modulus is related more to motor unit activity within a single muscle rather than to joint torque that represents a net effect of all synergistic and antagonistic muscles crossing the joint. In fact, several studies have investigated the association of muscle shear modulus with motor unit activity in addition to joint torque. In human muscle studies, motor unit activity is commonly examined by surface electromyography (EMG).

With regard to the relation between EMG and muscle mechanical activity, it has been frequently observed that surface EMG amplitude in large limb muscles increases non-linearly with joint torque (Bouillard et al., 2012b; Lawrence and De Luca, 1983; Nordez and Hug, 2010; Sasaki and Ishii, 2005; Watanabe and Akima, 2009). Several physiological and technical reasons may account for the non-linearity, including motor unit recruitment strategy (Fuglevand et al., 1993; Lawrence and De Luca, 1983), inhomogeneous muscle activity (van Zuylen et al., 1988), mixed muscle fiber composition (Woods and Bigland-Ritchie, 1983), and amplitude cancellation (Keenan et al., 2005). Apart from these explanations, the above-mentioned study (Bouillard et al., 2012b) on the shear modulus of human elbow flexor muscle synergists raised an intriguing possibility that the changes in load sharing, i.e., relative contribution to joint torque, between synergists partly explain the non-linear EMG–torque relation of the biceps brachii muscle. In fact, several studies have consistently shown that the shear modulus can be linearly related to EMG amplitude in the biceps brachii muscle (Lapole et al., 2015; Nordez and Hug, 2010; Yoshitake et al., 2014). The linear association also holds true for other muscles including small hand muscles where both shear modulus and EMG are linearly related to joint torque (Bouillard et al., 2011, 2012a).

8.3.3 Usefulness as a Measure of Muscle Activation Level

The linear association of shear modulus with surface EMG amplitude observed in many human muscles implies that muscle shear modulus can be used as a valid alternative to surface EMG for evaluating muscle activation. In fact, shear modulus measurement has several features that may be advantageous over surface EMG. First, the measurement is unlikely to be affected by cross talk from adjacent muscles or signal cancellation due to action potential overlap (Bouillard et al., 2012b). Rather, the ROI can be manually and precisely selected in terms of the corresponding anatomical structures imaged by B-mode ultrasonography (see Fig. 8.4). Second, the measurement is potentially applied to deep muscles and relatively deep regions of superficial muscles, although there is currently a depth limit of approximately 30 mm from the probe surface, within which the shear modulus can be accurately measured (Miyamoto et al., 2015). Finally, the muscle shear modulus at a given contraction intensity was shown to be insensitive to neuromuscular fatigue (Bouillard et al., 2012a). This is presumably explained by the fact that the shear modulus represents mechanical rather than electrical activity of the muscle examined. A simultaneous measurement of muscle shear modulus and EMG may thus provide a deeper insight into the mechanisms of neuromuscular fatigue and increased stiffness during muscle contractions in vivo.

8.4 Relations between Length, Force, and Stiffness

The key findings of Bouillard et al. (2011, 2012a) that the shear modulus represents the mechanical activity, i.e., contractile force, of individual muscle have been obtained mainly by measuring the shear modulus during submaximal voluntary muscle contractions with varied intensities. During submaximal voluntary contractions, however, contractile force is modulated by changes in motor unit activity, namely the number and average firing rate of motor units (or muscle fibers) recruited. Therefore, it remains unclear whether the muscle stiffness changes with force-generating capacity of muscle fibers even without changes in the motor unit activity. To address this issue, we conducted an experiment with the human tibialis anterior muscle and investigated the effects of muscle length on both force and shear modulus (Sasaki et al., 2014).

8.4.1 Length-Dependent Changes in Shear Modulus

In the experiment, percutaneous electrical stimulation with an 80-Hz train of 0.25-ms rectangular pulses was used to induce a 5-s tetanic contraction while controlling the motor unit activity. Stimulus intensity was determined on an individual basis, being set to the maximal tolerable level. Using a custom-designed ankle dynamometer (Sasaki and Ishii, 2005, 2010), the ankle joint torque and shear modulus were measured concurrently during tetanic contractions at five different ankle joint angles (from 15° of dorsiflexion to 25° of plantar flexion), while the corresponding muscle fascicle length and pennation angle were determined by analyzing B-mode ultrasound images captured by the SSI scanner. Muscle force, defined as the contractile force acting parallel to the muscle fiber orientation, was calculated from joint torque, tendon moment arm length (determined by another experiment), and pennation angle.

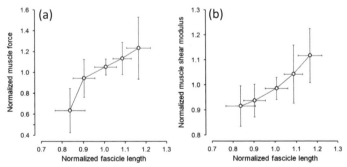

Figure 8.5 Length–force (a) and length–shear modulus (b) relations of the tetanized tibialis anterior muscle. Data are normalized to the average of five different joint positions in each participant and expressed as means and SD ($n = 9$). Regression analysis revealed significant positive associations of muscle force ($R^2 = 0.51$, $n = 45$, $P < 0.001$) and shear modulus ($R^2 = 0.42$, $n = 45$, $P < 0.001$) with fascicle length (adapted from Sasaki et al., 2014).

Figure 8.5a shows length–force relation, whereas Fig. 8.5b shows length–shear modulus relation of the tetanized tibialis anterior muscle. These results indicate that in vivo human tibialis anterior muscle mainly operates in the "ascending limb," which is consistent with the finding of Maganaris (2001), and that the shear modulus is also length-dependent despite a relatively constant motor unit activity.

8.4.2 Linear Association of Force and Shear Modulus

As both muscle force and shear modulus showed similar length-dependent changes, the association of these variables was then explored. Figure 8.6 shows a significant linear association of shear modulus with contractile force (R^2 = 0.52, P < 0.001). This result is in line with the close link between force and stiffness in contracting muscle fibers, both of which represent the number of attached cross-bridges (Ford et al., 1981), and also supports the view that the muscle shear modulus serves as an indirect estimate of individual muscle force (Bouillard et al., 2011, 2012a).

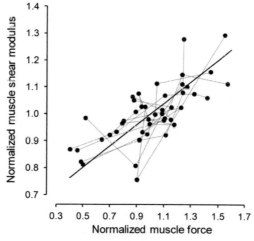

Figure 8.6   Association between muscle force and shear modulus of the tetanized tibialis anterior muscle. Data are normalized to the average of five different joint positions in each participant and are shown as individual line plots. Regression analysis revealed a significant positive association of muscle force with shear modulus (R^2 = 0.52, n = 45, P < 0.001) (adapted from Sasaki et al., 2014).

It should be noted, however, that in the ascending limb of length–force relation, the stiffness of single muscle fibers may not be necessarily proportional to the number of attached cross-bridges or contractile force because of the filament compliance (Julian and Morgan, 1981). In fact, our result showed that the length-dependent changes in shear modulus were small in magnitude compared to the corresponding changes in muscle force, as illustrated in Fig. 8.6. Accordingly, the changes in shear modulus with contractile force during tetanic contractions with different muscle length may not be fully accounted for by the changes in muscle-fiber stiffness.

8.4.3 Difference between Tetanic and Voluntary Contractions

While the percutaneous electrical stimulation was assumed to activate the tibialis anterior muscle selectively, such selective activation can be rarely seen in human voluntary movements. Thus we sought to determine the shear modulus of the tibialis anterior during MVC and compare the length–shear modulus relation of voluntarily activated muscle with that of the tetanized muscle. Figure 8.7 shows the difference in the length–shear modulus

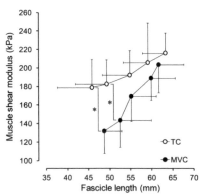

Figure 8.7 Comparison of length–shear modulus relations of the tibialis anterior muscle during tetanic contractions (TC, open circles) and maximal voluntary contractions (MVC, filled circles). Data are means and SD (n = 9). *Significant difference between the two contraction modes (P < 0.05, paired t-test with the false discovery rate procedure) (adapted from Sasaki et al., 2014).

relations between electrically evoked tetanic contractions and MVC. Although the muscle shear modulus measured during MVC increased with fascicle length, the slope of length–shear modulus relation was much steeper in MVC than in tetanic contractions. Statistical analysis revealed significant differences in the shear modulus measured at short fascicle lengths (dorsiflexed positions). These differences are probably due to relatively low motor unit firing rates during MVC, which would lead to greater attenuation of muscle force at shorter muscle lengths (Balnave and Allen, 1996; Marsh et al., 1981). In fact, the average motor unit firing rates in the tibialis anterior muscle during voluntary contractions has been shown to be 5–30 Hz (De Luca and Hostage, 2010), which is considerably lower than the stimulation frequency used to induce tetanic contractions (80 Hz).

8.5 Stiffness Measured during Dynamic Contractions

As mentioned earlier, a low time resolution (1 Hz) of the current technology confines the application of SSI to static muscle contractions. However, the shear modulus measurement during dynamic muscle contractions is worth challenging, leading not only to a better understanding of how in vivo muscle stiffness is determined during contractions but also to several important applications such as an analysis of neural and mechanical control of dynamic human movements. This section presents the results of our pilot study on the shear modulus in the biceps brachii muscle during isometric, shortening, and lengthening contractions against a given load.

8.5.1 Differences in Shear Modulus among Contraction Types

Using an custom-designed arm dynamometer (Sasaki et al., 2011), the muscle shear modulus, elbow flexion force, elbow joint angle, and motor unit activities of the biceps brachii and triceps brachii muscles (monitored by surface EMG) were concurrently measured during voluntary muscle contractions that were performed by holding (isometric), lifting (shortening), or lowering (lengthening) a weight load corresponding to 30%, 40%, and 50% of MVC. During isometric contractions, the weight was held as steady as possible at elbow joint angles of 50°, 70°, and 90° (0° represents full extension). During shortening and lengthening contractions, the elbow was flexed and extended, respectively, at a very slow speed (~10°/s) within a range of 40° to 100° of elbow flexion. The data obtained from the isometric contraction were time-averaged and presented as a mean of the three contractions at different joint angles, i.e., 50°, 70°, and 90°. The data obtained from the shortening and lengthening contractions were time-averaged from 50 to 90° of elbow flexion.

Figure 8.8 shows the differences in shear modulus and EMG amplitude (relative to MVC) in the biceps brachii muscle among the three different types of contraction. Similar results were obtained with the three load conditions, so that only the results at 40%MVC are presented here. The muscle shear modulus was significantly lower in lengthening contraction than in the other two contraction types, while no significant difference was found between isometric and shortening contractions (Fig. 8.8a). In agreement with previous observations (Altenburg et al., 2008; Bigland and Lippold, 1954; Moritani et al., 1987; Nakazawa et al., 1993), the EMG amplitude was significantly different among the three contraction types. Specifically, it was highest in shortening contraction, and lowest in lengthening contraction (Fig. 8.8b).

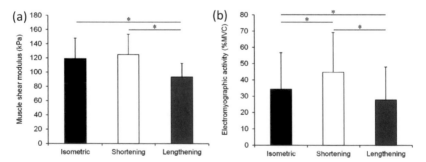

Figure 8.8 Differences in shear modulus (a) and electromyographic activity (b) of the biceps brachii muscle among contraction types. Data are expressed as means and SD (n = 9). MVC, maximal voluntary contraction. *Significantly different ($P <$ 0.05, paired t-test with false discovery rate procedure).

8.5.2   Putative Mechanisms

It is well documented that the stiffness of contracting muscle fiber decreases with increasing shortening velocity (Ford et al., 1985; Griffiths et al., 1993; Julian and Sollins, 1975; Sugi and Tsuchiya, 1988; Tsuchiya et al., 1979), primarily reflecting the change in the number of attached cross-bridges (Ford et al., 1985; Piazzesi et al., 2007). Contrary to this, our result showed that the muscle shear modulus was similar between isometric and shortening contractions. To interpret this discrepancy properly, it should be kept in mind that in our experiment, the muscle sheer modulus was measured during submaximal voluntary contractions where not all the motor units (or muscle fibers) were activated. In fact, the EMG amplitude, an index of motor unit activity, was different among the three contraction types despite the nearly identical elbow flexion force. Thus the shear modulus in shortening contraction is likely to represent a competing effect of the decrease in single fiber stiffness (due to muscle shortening) and the increase in the number of activated muscle fibers (suggested by the large EMG amplitude) compared to isometric contraction. Admittedly, however, the contraction velocity was kept very low in this experiment because of the low time resolution (1 Hz) of shear elasticity measurement. Therefore, the possibility cannot be excluded that the muscle shear modulus decreases at higher shortening velocities, as suggested by single-fiber studies.

The assumption that the shear modulus is influenced by both of the average stiffness and number of activated fibers within the ROI may also explain the shear modulus in the actively lengthening muscle. We observed the decrease in EMG amplitude, which suggests the decrease in the number of activated muscle fibers, during lengthening contraction compared to isometric contraction. Furthermore, there were a few observations that even after the completion of stretch, the stiffness of contracting muscle fiber remained almost unchanged (Sugi and Tsuchiya, 1988) or increased to a lesser extent than did the contractile force (Rassier and Herzog, 2005). These findings suggest that the lower shear modulus in lengthening contraction may be accounted for by the decrease in the number of activated muscle fibers without increasing the muscle fiber stiffness, compared to isometric contraction. Also, the same contractile force (isotonic loading) with the reduced shear modulus of the lengthening muscle implies that larger force is generated by each cross-bridge in lengthening contraction than in isometric contraction.

8.6   General Conclusions and Perspectives

The studies briefly reviewed in this chapter provide strong evidence that the stiffness (shear modulus) of contracting muscle measured with SSI can be a useful indicator of muscle activation level or contractile force in a variety of conditions. With further technical improvements expected in the near future, this approach will become a more powerful tool for the study of human neuromuscular function. However, there are some limitations and unsolved issues that should be addressed for future research.

First, the ROI in which the shear modulus can be instantaneously measured is currently limited ($\sim$1.5 × 1.5 cm). The measured data are typically averaged over the ROI on the assumption that the average value serves as a representative of the whole muscle. However, this assumption has not been tested rigorously. In fact, even within a small area, relatively large variations in muscle shear modulus have been observed even in low-intensity contractions (Fig. 8.4). Moreover, studies using surface EMG and magnetic resonance imaging have provided evidence that the muscle activation is three dimensionally heterogeneous within an individual muscle (Damon et al., 2008; Kinugasa et al., 2011; Watanabe et al., 2014). The spatial variability in fiber-type distribution (Dahmane et al., 2005; Johnson et al., 1973) and the possible fiber-type difference in stiffness (Metzger and Moss, 1990; Petit et al., 1990) may introduce even greater spatial variations in muscle shear modulus.

Second, as noted above, the spatial variations in shear modulus observed within a small ROI have been overlooked in previous studies. Since skeletal muscle is composed not only of muscle fibers but also of collagenous connective tissues that surround and bind muscle fibers into small bundles (fascicles), the spatial variations in shear modulus are partly attributable to the difference in elastic properties between muscle fibers and intramuscular connective tissues. In addition, there is a possibility that the spatial variability in motor unit activity or mechanical load is reflected by the variations in shear modulus. Although the variability in motor unit activity can be studied by intramuscular EMG technique, several advantages of the SSI (e.g., non-invasiveness, construction of a two-dimensional map, and applicability to relatively deep tissues) may allow a more comprehensive and sensitive measurement. Given a mean muscle fiber diameter of 50 μm in humans (Maier and Bornemann, 1999), the spatial resolution of shear elasticity measurement (currently 1 × 1 mm) implies that the shear modulus in each pixel represents a mean stiffness of approximately 20 muscle fibers. This is much smaller than the average innervation number of motor units estimated in the human first dorsal interosseous muscle (300–400 fibers; Enoka and Fuglevand, 2001). Therefore, the current technology may have the potential to visualize and quantify the activation of a few or even a single motor unit, although the muscle fibers belonging to the same motor unit are scattered over a broad region of the muscle (Fuglevand and Segal, 1997).

Third, Hug et al. (2015) have provided a line of evidence (Bouillard et al., 2011, 2012a; Maïsetti et al., 2012) that the muscle shear modulus can be used as a reliable measure of force or torque produced by an individual muscle. For a more direct estimation of individual muscle force, however, information of moment arm (the perpendicular distance from the joint center of rotation to the muscle action line) and physiological cross-sectional area (the total cross-sectional area perpendicular to muscle fibers) is necessary (for details, see Hug et al., 2015). In addition, our preliminary data suggest that the slope of force–shear modulus relation may be different among contraction types (Fig. 8.8a), because of the possible competing effect of the average stiffness and number of activated muscle fibers within the ROI. Further systematic studies are thus needed to test whether the estimation of individual muscle force is also feasible during dynamic contractions.

Finally, while we and other researchers have consistently observed the contraction-induced increase in muscle shear modulus, the structures and mechanisms underlying this phenomenon are not fully understood. The experimental data (Bouillard et al., 2012a; Sasaki et al., 2014) suggest that the shear modulus is determined, at least in part, by mechanism(s) independent of motor unit activity, i.e., the number and firing rate of motor units or muscle fibers activated. In fact, the shear modulus in a resting muscle has been shown to increase with increasing passive force (Koo et al., 2013, 2014; Maïsetti et al., 2012). One possible mechanism is the biaxial (longitudinal and transverse) stretch of interfascicular connective tissue during contraction, by which its longitudinal stiffness changes dynamically (Azizi and Roberts, 2009). Another mechanism lies in a recently proposed three-filament model of muscle force generation (Herzog et al., 2015; Schappacher-Tilp et al., 2015), where the structural protein titin plays an essential role in muscle force regulation. According to this model, titin alters its spring stiffness not only when being stretched but also upon muscle activation through binding of calcium ions to its specific sites and/or by binding its proximal region to actin filament. While the model is developed to explain the phenomenon known as residual force enhancement (the increase in steady-state isometric force following an active muscle stretch), it may provide a unified explanation for changes in muscle shear modulus with both active and passive forces.

Acknowledgments

We gratefully acknowledge the invaluable contribution of our colleagues (at the University of Tokyo), especially Sho Toyama, Daisuke Tsushima, Gen Yamamoto, and Shota Narimatsu, to the experiments and data analyses.

References

Altenburg TM, de Ruiter CJ, Verdijk PW, van Mechelen W, de Haan A (2008). Vastus lateralis surface and single motor unit EMG following submaximal shortening and lengthening contractions. *Appl Physiol Nutr Metab,* **33**: 1086–1095.

Ateş F, Hug F, Bouillard K, Jubeau M, Frappart T, Couade M, Bercoff J, Nordez A (2015). Muscle shear elastic modulus is linearly related to muscle torque over the entire range of isometric contraction intensity. *J Electromyogr Kinesiol,* **25**: 703–708.

Azizi E, Roberts TJ (2009). Biaxial strain and variable stiffness in aponeuroses. *J Physiol,* **587**: 4309–4318.

Balnave CD, Allen DG (1996). The effect of muscle length on intracellular calcium and force in single fibres from mouse skeletal muscle. *J Physiol,* **492**: 705–713.

Bensamoun SF, Glaser KJ, Ringleb SI, Chen Q, Ehman RL, An KN (2008). Rapid magnetic resonance elastography of muscle using one-dimensional projection. *J Magn Reson Imaging*, **27**: 1083–1088.

Bercoff J, Tanter M, Fink M (2004). Supersonic shear imaging: A new technique for soft tissue elasticity mapping. *IEEE Trans Ultrason Ferroelectr Freq Control*, **51**: 396–409.

Bigland B, Lippold OC (1954). The relation between force, velocity and integrated electrical activity in human muscles. *J Physiol*, **123**: 214–224.

Bouillard K, Hug F, Guével A, Nordez A (2012a). Shear elastic modulus can be used to estimate an index of individual muscle force during a submaximal isometric fatiguing contraction. *J Appl Physiol*, **113**: 1353–1361.

Bouillard K, Nordez A, Hodges PW, Cornu C, Hug F (2012b). Evidence of changes in load sharing during isometric elbow flexion with ramped torque. *J Biomech*, **45**: 1424–1429.

Bouillard K, Nordez A, Hug F (2011). Estimation of individual muscle force using elastography. *PLoS One*, **6**: e29261.

Dahmane R, Djordjevič S, Šimunič B, Valenčič V (2005). Spatial fiber type distribution in normal human muscle: Histochemical and tensiomyographical evaluation. *J Biomech*, **38**: 2451–2459.

Damon BM, Wadington MC, Lansdown DA, Hornberger JL (2008). Spatial heterogeneity in the muscle functional MRI signal intensity time course: Effect of exercise intensity. *Magn Reson Imaging*, **26**: 1114–1121.

De Luca CJ, Hostage EC (2010). Relationship between firing rate and recruitment threshold of motoneurons in voluntary isometric contractions. *J Neurophysiol*, **104**: 1034–1046.

Dresner MA, Rose GH, Rossman PJ, Muthupillai R, Manduca A, Ehman RL (2001). Magnetic resonance elastography of skeletal muscle. *J Magn Reson Imaging*, **13**: 269–276.

Eby SF, Song P, Chen S, Chen Q, Greenleaf JF, An KN (2013). Validation of shear wave elastography in skeletal muscle. *J Biomech*, **46**: 2381–2387.

Enoka RM, Fuglevand AJ (2001). Motor unit physiology: Some unresolved issues. *Muscle Nerve*, **24**: 4–17.

Ford LE, Huxley AF, Simmons RM (1981). The relation between stiffness and filament overlap in stimulated frog muscle fibres. *J Physiol*, **311**: 219–249.

Ford LE, Huxley AF, Simmons RM (1985). Tension transients during steady shortening of frog muscle fibres. *J Physiol,* **361**: 131–150.

Fuglevand AJ, Segal SS (1997). Simulation of motor unit recruitment and microvascular unit perfusion: Spatial considerations. *J Appl Physiol,* **83**: 1223–1234.

Fuglevand AJ, Winter DA, Patla AE (1993). Models of recruitment and rate coding organization in motor-unit pools. *J Neurophysiol,* **70**: 2470–2488.

Gennisson JL, Deffieux T, Macé E, Montaldo G, Fink M, Tanter M (2010). Viscoelastic and anisotropic mechanical properties of in vivo muscle tissue assessed by supersonic shear imaging. *Ultrasound Med Biol,* **36**: 789–801.

Griffiths PJ, Ashley CC, Bagni MA, Maéda Y, Cecchi G (1993). Cross-bridge attachment and stiffness during isotonic shortening of intact single muscle fibers. *Biophys J,* **64**: 1150–1160.

Herzog W, Powers K, Johnston K, Duvall M (2015). A new paradigm for muscle contraction. *Front Physiol,* **6**: 174.

Hug F, Tucker K, Gennisson JL, Tanter M, Nordez A (2015). Elastography for muscle biomechanics: Toward the estimation of individual muscle force. *Exerc Sport Sci Rev,* **43**: 125–133.

Huxley AF (1957). Muscle structure and theories of contraction. *Prog Biophys Biophys Chem,* **7**: 255–318.

Johnson MA, Polgar J, Weightman D, Appleton D (1973). Data on the distribution of fibre types in thirty-six human muscles: An autopsy study. *J Neurol Sci,* **18**: 111–129.

Julian FJ, Morgan DL (1981). Tension, stiffness, unloaded shortening speed and potentiation of frog muscle fibres at sarcomere lengths below optimum. *J Physiol,* **319**: 205–217.

Julian FJ, Sollins MR (1975). Variation of muscle stiffness with force at increasing speeds of shortening. *J Gen Physiol,* **66**: 287–302.

Kawai M (1979). Effect of MgATP on cross-bridge kinetics in chemically skinned rabbit psoas fibers as measured by sinusoidal analysis technique. In *Cross-Bridge Mechanism in Muscle Contraction* (Sugi H, Pollack GH, ed), University of Tokyo Press, Tokyo, pp. 149–169.

Keenan KG, Farina D, Maluf KS, Merletti R, Enoka RM (2005). Influence of amplitude cancellation on the simulated surface electromyogram. *J Appl Physiol,* **98**: 120–131.

Kinugasa R, Kawakami Y, Sinha S, Fukunaga T (2011). Unique spatial distribution of in vivo human muscle activation. *Exp Physiol,* **96**: 938–948.

Koo TK, Guo JY, Cohen JH, Parker KJ (2013). Relationship between shear elastic modulus and passive muscle force: An ex-vivo study. *J Biomech,* **46**: 2053–2059.

Koo TK, Guo JY, Cohen JH, Parker KJ (2014). Quantifying the passive stretching response of human tibialis anterior muscle using shear wave elastography. *Clin Biomech,* **29**: 33–39.

Lacourpaille L, Hug F, Bouillard K, Hogrel JY, Nordez A (2012). Supersonic shear imaging provides a reliable measurement of resting muscle shear elastic modulus. *Physiol Meas,* **33**: N19–N28.

Lapole T, Tindel J, Galy R, Nordez A (2015). Contracting biceps brachii elastic properties can be reliably characterized using supersonic shear imaging. *Eur J Appl Physiol,* **115**: 497–505.

Lawrence JH, De Luca CJ (1983). Myoelectric signal versus force relationship in different human muscles. *J Appl Physiol,* **54**: 1653–1659.

Maganaris CN (2001). Force-length characteristics of in vivo human skeletal muscle. *Acta Physiol Scand,* **172**: 279–285.

Maier F, Bornemann A (1999). Comparison of the muscle fiber diameter and satellite cell frequency in human muscle biopsies. *Muscle Nerve,* **22**: 578–583.

Maïsetti O, Hug F, Bouillard K, Nordez A (2012). Characterization of passive elastic properties of the human medial gastrocnemius muscle belly using supersonic shear imaging. *J Biomech,* **45**: 978–984.

Marsh E, Sale D, McComas AJ, Quinlan J (1981). Influence of joint position on ankle dorsiflexion in humans. *J Appl Physiol,* **51**: 160–167.

Metzger JM, Moss RL (1990). Effects of tension and stiffness due to reduced pH in mammalian fast- and slow-twitch skinned skeletal muscle fibres. *J Physiol,* **428**: 737–750.

Miyamoto N, Hirata K, Kanehisa H, Yoshitake Y (2015). Validity of measurement of shear modulus by ultrasound shear wave elastography in human pennate muscle. *PLoS One,* **10**: e0124311.

Moritani T, Muramatsu S, Muro M (1987). Activity of motor units during concentric and eccentric contractions. *Am J Phys Med,* **66**: 338–350.

Nakazawa K, Kawakami Y, Fukunaga T, Yano H, Miyashita M (1993). Differences in activation patterns in elbow flexor muscles during isometric, concentric and eccentric contractions. *Eur J Appl Physiol,* **66**: 214–220.

Nordez A, Hug F (2010). Muscle shear elastic modulus measured using supersonic shear imaging is highly related to muscle activity level. *J Appl Physiol,* **108**: 1389–1394.

Petit J, Filippi GM, Emonet-Denand F, Hunt CC, Laporte Y (1990). Changes in muscle stiffness produced by motor units of different types in peroneus longus muscle of cat. *J Neurophysiol,* **63**: 190–197.

Piazzesi G, Reconditi M, Linari M, Lucii L, Bianco P, Brunello E, Decostre V, Stewart A, Gore DB, Irving TC, Irving M, Lombardi V (2007). Skeletal muscle performance determined by modulation of number of myosin motors rather than motor force or stroke size. *Cell,* **131**: 784–795.

Rassier DE, Herzog W (2005). Relationship between force and stiffness in muscle fibers after stretch. *J Appl Physiol,* **99**: 1769–1775.

Rüegg JC, Güth K, Kuhn HJ, Herzig JW, Griffiths PJ, Yamamoto T (1979). Muscle stiffness in relation to tension development of skinned striated muscle fibres. In *Cross-Bridge Mechanism in Muscle Contraction* (Sugi H, Pollack GH, ed), University of Tokyo Press, Tokyo, pp. 125–143.

Sasaki K, Ishii N (2005). Shortening velocity of human triceps surae muscle measured with the slack test in vivo. *J Physiol,* **567**: 1047–1056.

Sasaki K, Ishii N (2010). Unloaded shortening velocity of voluntarily and electrically activated human dorsiflexor muscles in vivo. *PLoS One,* **5**: e13043.

Sasaki K, Sasaki T, Ishii N (2011). Acceleration and force reveal different mechanisms of electromechanical delay. *Med Sci Sports Exerc,* **43**: 1200–1206.

Sasaki K, Toyama S, Ishii N (2014). Length-force characteristics of in vivo human muscle reflected by supersonic shear imaging. *J Appl Physiol,* **117**: 153–162.

Schappacher-Tilp G, Leonard T, Desch G, Herzog W (2015). A novel three-filament model of force generation in eccentric contraction of skeletal muscles. *PLoS One,* **10**: e0117634.

Shinohara M, Sabra K, Gennisson JL, Fink M, Tanter M (2010). Real-time visualization of muscle stiffness distribution with ultrasound shear wave imaging during muscle contraction. *Muscle Nerve,* **42**: 438–441.

Sugi H, Tsuchiya T (1988). Stiffness changes during enhancement and deficit of isometric force by slow length changes in frog skeletal muscle fibres. *J Physiol,* **407**: 215–229.

Tsuchiya T, Sugi H, Kometani K (1979). Isotonic velocity transients and enhancement of mechanical performance in frog skeletal muscle fibers after quick increases in load. In *Cross-Bridge Mechanism in Muscle Contraction* (Sugi H, Pollack GH, ed), University of Tokyo Press, Tokyo, pp. 225–240.

van Zuylen EJ, Gielen CC, Denier van der Gon JJ (1988). Coordination and inhomogeneous activation of human arm muscles during isometric torques. *J Neurophysiol,* **60**: 1523–1548.

Watanabe K, Akima H (2009). Normalized EMG to normalized torque relationship of vastus intermedius muscle during isometric knee extension. *Eur J Appl Physiol,* **106**: 665–673.

Watanabe K, Kouzaki M, Moritani T (2014). Non-uniform surface electromyographic responses to change in joint angle within rectus femoris muscle. *Muscle Nerve,* **50**: 794–802.

Woods JJ, Bigland-Ritchie B (1983). Linear and non-linear surface EMG/force relationships in human muscles: An anatomical/functional argument for the existence of both. *Am J Phys Med,* **62**: 287–299.

Yoshitake Y, Takai Y, Kanehisa H, Shinohara M (2014). Muscle shear modulus measured with ultrasound shear-wave elastography across a wide range of contraction intensity. *Muscle Nerve,* **50**: 103–113.

Chapter 9

Effect of DTT on Force and Stiffness during Recovery from Fatigue in Mouse Muscle Fibres

Barbara Colombini,[a] Marta Nocella,[a] Joseph D. Bruton,[b] Maria Angela Bagni,[a] and Giovanni Cecchi[a]

[a]Department of Experimental and Clinical Medicine, University of Florence, Italy
[b]Department of Physiology and Pharmacology, Karolinska Institutet, Sweden

barbara.colombini@unifi.it

Intense muscle activity can result in fatigue, a state where tetanic force remains depressed for a considerable period after the end of activity. At the level of interaction between myosin and actin, the force loss might reflect either a decrease in the number of force-generating crossbridges or a decrease in the mean force generated by single crossbridge. The cause of these changes is unclear but one recurrent suggestion is that free radicals or reactive oxygen species (ROS) have modified the contractile proteins. The present experiments investigated this point using single fibres or small fibre bundles isolated from the mouse flexor digitorum brevis muscle at 22–24°C. Fibres were repetitively stimulated to induce fatigue and then force and stiffness recovery were followed during

Muscle Contraction and Cell Motility: Fundamentals and Developments
Edited by Haruo Sugi
Copyright © 2017 Pan Stanford Publishing Pte. Ltd.
ISBN 978-981-4745-16-1 (Hardcover), 978-981-4745-17-8 (eBook)
www.panstanford.com

exposure to normal Tyrode solution or Tyrode solution to which 1 mM dithiothreitol (DTT) had been added. Force and fibre stiffness were measured before fatigue and during recovery from fatigue during 30 and 120 Hz test tetani. During the whole recovery from fatigue, force was slightly though significantly depressed in DTT with respect to Tyrode solution, whereas fibre stiffness remained unchanged. Our findings suggest that during recovery from fatigue, the impaired force production of crossbridges is not easily reversed or modified by a powerful reducing agent. Since force reduction by DTT occurred without alteration of fibre stiffness, our results suggest that force reduction is caused mainly by a mechanism which does not reduce crossbridge number, such as a reduction of the mean crossbridge force.

9.1 Introduction

Following periods of intense exercise, a state of fatigue persists and an individual cannot generate the force or power output that was possible before exercise started. It has long been known that this force loss is more marked at low compared to high fibre recruitment or stimulation frequencies (Edwards et al., 1977). Similar results were found in isolated skeletal muscles that were induced to contract repeatedly and where force was monitored and subsequently followed during recovery. The causes and the mechanisms underlying this force loss after fatigue are incompletely understood. Several hypotheses have been advanced and one that is currently receiving much attention suggests that the oxidation-reduction status of the muscle has been altered by increased production of reactive oxygen species (ROS), including the superoxide anion (Allen et al., 2008; Bruton et al., 2008; Lamb and Westerblad, 2011).

Production of ROS is generally agreed to increase with exercise. ROS induces modifications of both actin and myosin filaments (Fedorova et al., 2010). Recently, it was demonstrated that a variety of anti-oxidant agents alone or in combination were able to restore tetanic calcium transients but were unable to reverse the force depression associated with fatigue (Cheng et al., 2015). However, in the presence of saturating $[Ca^{2+}]$, high concentrations of oxidising agents have been shown to markedly alter force in intact muscle (Andrade et al., 1998) and in skinned muscle fibres (Plant et al., 2000). It has also been demonstrated that dithiothreitol (DTT) can radically affect the oxidation status of muscle and can reverse the force depression induced by H_2O_2 (Andrade at al., 1998; Dutka et al., 2012; Posterino et al., 2003). Thus, we were interested to determine the effect of DTT on crossbridge properties during recovery from fatigue.

9.2   Methods

9.2.1   Fibre Dissection and Measurements

Male mice (C57BL/6 strain, 3–6 months old) were housed at controlled temperature (21–24°C) with a 12–12 h light–dark cycle. Food and water were provided ad libitum. Mice were killed by rapid cervical dislocation, according to the procedure suggested by the Ethical Committee for Animal Experiments of the University of Florence and the EEC guidelines for animal care of the European Community Council (Directive 86/609/EEC). All efforts were made to minimize animal suffering and to use only the number of mice necessary to obtain reliable data. Both flexor digitorum brevis (FDB) muscles were removed and placed in oxygenated Tyrode solution: one FDB was used for DTT treatment and the contralateral muscle was used as control. Single intact fibres or small bundles of 2 to 10 fibres were dissected as described previously (Colombini et al., 2009). Aluminium clips were attached to tendons as close as possible to the end of the fibre preparations and used to mount the fibres horizontally in an experimental chamber (capacity 0.38 ml) between the lever arms of a capacitance force transducer (resonance frequency, 16–20 kHz) and an electromagnetic motor that was used to change fibre length. Fibres were perfused continuously at a rate of about 0.35 ml min^{-1} with a normal Tyrode (NT) solution of the following composition (mM): NaCl, 121; KCl, 5; $CaCl_2$, 1.8; $MgCl_2$, 0.5; NaH_2PO_4, 0.4; $NaHCO_3$, 24; glucose, 5.5; EDTA, 0.1 and bubbled with 5% CO_2–95% O_2 (pH of 7.4). Foetal calf serum (0.2%) was routinely added to the solution. Tyrode solution containing 1 mM dithiothreitol (DTT) was prepared fresh before each experiment following the procedure of Andrade et al. (1998).

Experiments were performed at room temperature (22–24°C). Bipolar stimuli (0.5 ms duration and 1.5 times threshold strength) were applied to the fibre using two platinum-plate electrodes mounted parallel to the fibre. Preparations were stretched to the length at which tetanic force was maximal. The resting preparation's length (l_0), largest and smallest diameters of fibres and the resting sarcomere length were set using a microscope fitted with 20× eyepieces and a 5× or 40× objective lens over the experimental chamber and rechecked later on digital images obtained with a video camera (Infinity Camera, Lumenera Corp., Canada). The cross-sectional area of the single fibres or bundles was calculated as $a*b*\pi/4$ where a and b are the average values of the width and the vertical height (measured with fine focusing) of the preparation, respectively, measured at several points along the preparation. Sarcomere length was measured by counting the number of sarcomeres present in a fixed segment of a calibrated scale on the acquired images. Stimulation and preparation length changes were controlled by a custom-written software (LabView, National Instruments, USA) which was also used to record force and length at sampling speed of up to 200 kHz.

9.2.2 Force and Stiffness Measurements

Control tetanic contractions (300 ms duration, 30 and 120 Hz stimulation frequencies) were obtained at intervals of 90 s. Only those preparations in which plateau tetanic force (P_0) was stable (P_0 decreased by <10%) were used. Fatigue was induced using protocol consisting of a series of isometric tetani (240 ms duration, 120 Hz frequency) applied at 1.5 s intervals. Fatigue data were collected in NT from 14 fibres and the recovery was followed in NT ($n = 7$) and in DTT ($n = 7$) solution. Test tetani, applied at regular intervals of 2 min (except the first two points applied every 1 min) during the recovery period, were composed by an initial part of 160 ms at 30 Hz followed by a second part of 160 ms at 120 Hz. This allowed to test force recovery at high and low stimulation frequency.

The relative changes of attached crossbridges number during recovery from fatigue was estimated as described previously by measuring fibre stiffness (Cecchi et al., 1982, 1986; Ford et al., 1977; Nocella et al., 2011). This was done by applying small

(~0.15% l_0 peak-to-peak amplitude) 6.5 kHz sinusoidal length changes (*dl*) to one end of the activated fibres and measuring the resulting force oscillations (*dP*) at the other end. Stiffness was then calculated as the ratio $(dP/P_0)/(dl/l_0)$. Considering that *dl* was maintained constant during a given experiment, relative stiffness changes were measure simply by *dP*. The high oscillation frequency used ensures that stiffness is not affected by the quick force recovery mechanism (Ford et al., 1977), whereas the short length of the preparation avoids the effects of mechanical fibre resonance. This was confirmed by the absence of any measurable phase shift between force and length during sinusoidal oscillations similarly to our previous work (Nocella et al., 2011). These stiffness measurements are influenced by tendons stiffness and cannot be used directly to assess crossbridge number; however, since tendon stiffness is not affected by fatigue (Nocella et al., 2011), changes occurring during fatigue or recovery are attributable exclusively to changes of crossbridge stiffness that is proportional to crossbridge number. Fibre stiffness was not corrected for passive and static stiffness (Nocella et al., 2012, 2014) which were considered negligibly small.

The results from small bundles were indistinguishable from those of single fibres; then the data were grouped together. Force and stiffness data were always expressed relatively to the control data before fatigue. Values are shown as mean ± SEM. Statistical significance was tested by two sample *t*-test; values were considered to be statistically significantly different for *P* value <0.05.

9.3 Results

Figure 9.1 shows typical examples of tetanic records taken during an experiment: superimposed 30 and 120 Hz tetani before the induction of fatigue (Fig. 9.1a), the last 120 Hz tetanus of the fatiguing protocol (Fig. 9.1b) and combined 30–120 Hz test tetani recorded at 6 and 20 min after the end of fatigue (Fig. 9.1c,d), respectively, during recovery either in NT (upper row) or in 1 mM DTT (lower row) solution. The black band at the top of the tetanic records represents the peak-to-peak amplitude of force oscillations at 6.5 kHz. Since the length change amplitude *dl* remained constant throughout the experiment, the amplitude of the black band is proportional to fibre stiffness (Nocella et al.,

2011). Stiffness and force are clearly reduced at the end of fatigue (Fig. 9.1a,b) and the subsequent recovery after fatigue was smaller in DTT than in NT solution (Fig. 9.1c,d).

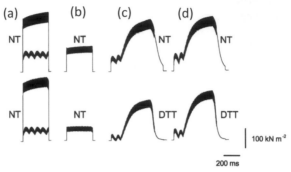

Figure 9.1   Typical records of tetanic contractions with superimposed high frequency sinusoidal length changes during the first (a) and the last (b) tetani of a fatiguing series of tetanic contractions in normal Tyrode (NT) solution and at 6 min (c) and 20 min (d) of recovery from fatigue either in NT (upper row) or 1 mM DTT (lower row) solution. Records are from two separate fibres. The thicker black band on the tetanic records represent the superimposed oscillations at 6.5 kHz not time resolved in this figure. In c and d the tetanic contractions during recovery were composed of an initial 160 ms at 30 Hz followed by 160 ms at 120 Hz. Sampling time of 1 ms/point was used at the start of all the records followed by a period at 5 μs/point and successively by a slow speed sampling again. Double sampling time was necessary to resolve both the slow time course of tetanic tension and the fast changes during force oscillations. The apparent longer duration of tetanic contraction and much slower tetanus rise at 120 Hz in c and d compared to a and b are caused by the different duration and start of fast sampling time. In this particular experiment, tension at the end of fatigue is smaller in the group tested with DTT than in the control group. However, the mean force fall at the end of fatigue was the same in both groups.

Figure 9.2 is a bar chart contrasting the tetanic force and fibre stiffness in the absence or presence of DTT at 6, 14 and 20 min during recovery after the end of fatigue. In agreement with previous data (Nocella et al., 2013), at the end of fatigue mean force decreased to ~30%, whereas the stiffness decreased much less than force to ~60% indicating that part of the force drop occurs by a mechanism that did not alter the fibre stiffness, i.e. a reduction in the mean crossbridge force (Nocella et al., 2013). Mean force recovery in NT at 20 min was greater at 120 Hz (92 ± 6% of control) than at 30 Hz (70 ± 6%). Similar to tension, stiffness recovered more at 120 Hz (90 ± 2%) than at 30 Hz (81 ± 8%). Bathing the fibres with DTT solution reduced force and stiffness recovery with a pattern similar to NT solution. At 20 min force recovered to 76 ± 7% at 120 Hz and 61 ± 9% at 30 Hz and stiffness recovered to 90 ± 3% at 120 Hz and 78 ± 3% at 30 Hz. The only

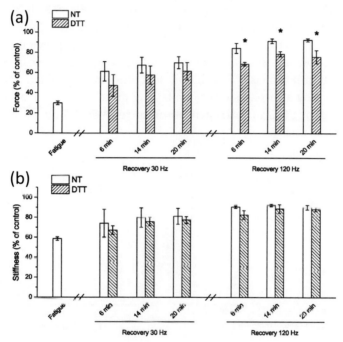

Figure 9.2   Mean tetanic force (a) and stiffness (b) in the last tetanus of fatigue and in tetani at 30 and 120 Hz frequency stimulation, at different times during recovery, in NT (empty bars) and in 1 mM DTT (dashed bars) solution. Both force and stiffness recovery at 120 and 30 Hz are smaller in DTT; however, only force at 120 Hz was significantly reduced by DTT. Data are presented with respect to that measured in control contractions before fatigue. Values are mean ± SEM (n = 7, except for the data of stiffness recovery at 30 Hz where n = 5, and at end of fatigue where n = 14). Asterisks indicate statistically significant changes (P < 0.05) respect to NT solution.

change induced by DTT statistically significant was a reduction of the force recovery at 120 Hz, whereas stiffness changes were not significant. This finding suggests that DTT reduces the force during recovery mainly by reducing the mean crossbridge force without affecting crossbridge number. The same force depressing effect of DTT was also present at 30 Hz frequency; however, very likely due the relatively high data scattering, the difference with NT solution was not statistically significant.

The whole time courses of force and stiffness during recovery in both NT and DTT solutions are shown in Fig. 9.3. It can be seen that inhibition of force recovery by DTT occurs throughout the whole recovery period. In agreement with previous data (Nocella et al., 2013), force and stiffness recovery occurred in two different phases: an initial fast one lasting about 2 min followed by a second slower one lasting up to 20 min. On average, more than 50% of force and stiffness recovery occurred during the initial phase.

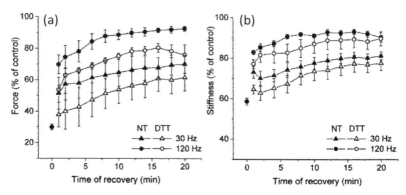

Figure 9.3 Mean time courses of tetanic force (a) and stiffness (b) during recovery from fatigue in NT or in 1 mM DTT solution. The filled circle at time zero represents force and stiffness at the end of fatigue at 120 Hz in NT solution. Both force and stiffness show the greatest degree of recovery during the first 2 min in both NT and DTT solution, and thereafter both recover much more slowly. Tension and stiffness data are presented with respect to that measured in control contractions before fatigue. Values are mean ± SEM (n = 7, except for the data of stiffness recovery at 30 Hz, where n = 5, and at end of fatigue, where n = 14).

9.4 Discussion

The novel findings of this study are that when the oxidation/ reduction status of muscle contractile proteins was changed by the addition of a reducing agent as DTT, force recovery after fatigue was depressed compared to NT and this effect was greater at 120 Hz than at 30 Hz. The negative effect of DTT was significant during the whole time course of the recovery, amounting to ~19% of tension loss compared to NT solution after 20 min of recovery. In principle, changes in tetanic force can be due to a reduction of crossbridge number or reduction of the mean crossbridge force or both. The reduction of tension recovery in DTT occurred without any significant reduction of stiffness recovery. Considering that tendon stiffness is not altered during fatigue (Nocella et al., 2011), this means that crossbridge stiffness, and therefore crossbridge number, was not reduced by DTT. Thus the reduction of tetanic force caused by DTT is mainly attributable to a mechanism operating throughout a reduction of the mean crossbridge force. DTT had a similar depressing effect on recovery from fatigue also at 30 Hz; however, likely due to greater scattering of data, the depression was not statistically significant. These results are in agreement with data in literature at higher temperature (32°C) which show no positive effect of ROS/RNS-neutralizing compounds on force production during induction of fatigue or in the sub-sequent recovery period (Cheng et al., 2015).

A slower recovery of force at sub-maximal levels after strenuous fatigue has been described in both humans (Allman and Rice, 2001; Edwards et al., 1977) and in isolated rodent muscle (Allen et al., 2008). It has been demonstrated in isolated fibres that while during the recovery period, the tetanic calcium transient is depressed, frequently the ensuing force is less than would be expected from the force-calcium relationship (Cheng et al., 2015; Westerblad and Allen, 1993). This suggests that the number of crossbridges or average force generated by a crossbridge is decreased. Recently, we addressed this question and found that the force recovery after fatigue occurred in two different phases: an initial one caused by the increase of the force per crossbridge, followed by a second part due to the increase of the attached crossbridge number (Nocella et al., 2013; Germinario et al., 2016).

Previous studies indicated that in rested fibres, DTT had either no effect in skinned fibres (Posterino et al., 2003) or a depressant effect on tetanic force in intact fibres, at least on contractions evoked at 40 Hz (Andrade et al., 1998). These results suggest that there are sites on contractile proteins in intact muscle fibres that are sensitive to reducing agents under resting conditions. In the present study, we found that when muscle fibres experienced increased ROS production during repeated contractions resulting to fatigue, force recovery was less in the presence than in the absence of DTT and force recovered less than stiffness, suggesting that DTT depresses mainly the force per crossbridge rather than crossbridge number. In agreement with our results, a depressant effect of DTT was described earlier by Andrade et al. (1998). However, these authors attributed the depressant effect of DTT to its capacity to reduce the myofibrillar Ca^{2+} sensitivity, a mechanism consistent with a reduction of crossbridge number rather than the force per crossbridge. It should be noted that the conditions under which DTT was used were different in the present study from earlier studies (Andrade et al., 1998; Posterino et al., 2003). Here, DTT was applied to fatigued fibres during 20 min of recovery following a fatiguing protocol, whereas Andrade et al. (1998) used DTT for a shorter period in rested fibres and Posterino et al. (2003) tested DTT on skinned fibres. It is likely that rested fibres have a different REDOX state compared to that existing after fatigue. Immediately after fatigue, in fact the ROS arising from mitochondrial and cytosolic enzymes have the potential to modify accessible sites on the contractile proteins. Thus in fibres recovering from fatigue, these modified sites on contractile proteins will respond differently to DTT. In conclusion, according to our results, the alteration of the oxidation/reduction state of the contractile proteins induced by DTT depresses significantly the recovery of force after fatigue by reducing the recovery of both crossbridge number and crossbridge individual force.

Grants

This study was supported by grant from Ministero dell'Istruzione, dell'Università e della Ricerca (PRIN 2010R8JK2X_002) and from the University of Florence. The funders had no role in study design, data collection and analysis, decision to publish, or preparation of the manuscript.

References

Allen DG, Lamb GD, Westerblad H (2008). Skeletal muscle fatigue: Cellular mechanisms. *Physiol Rev,* **88**: 287–332.

Allman BL, Rice CL (2001). Incomplete recovery of voluntary isometric force after fatigue is not affected by old age. *Muscle Nerve,* **24**(9): 1156–1167.

Andrade FH, Reid MB, Allen DG, Westerblad H (1998). Effect of hydrogen peroxide and dithiothreitol on contractile function of single skeletal muscle fibres from the mouse. *J Physiol,* **509**: 565–575.

Bruton JD, Place N, Yamada T, Silva JP, Andrade FH, Dahlstedt AJ, Zhang SJ, Katz A, Larsson NG, Westerblad H (2008). Reactive oxygen species and fatigue-induced prolonged low-frequency force depression in skeletal muscle fibres of rats, mice and SOD2 overexpressing mice. *J Physiol,* **586**: 175–184.

Cecchi G, Griffiths PJ, Taylor S (1982). Muscular contraction: Kinetics of crossbridge attachment studied by high-frequency stiffness measurements. *Science,* **217**: 70–72.

Cecchi G, Griffiths PJ, Taylor S (1986). Stiffness and force in activated frog skeletal muscle fibers. *Biophys J,* **49**: 437–451.

Cheng AJ, Bruton JD, Lanner JT, Westerblad H (2015). Antioxidant treatments do not improve force recovery after fatiguing stimulation of mouse skeletal muscle fibres. *J Physiol,* **593**: 457–472.

Colombini B, Benelli G, Nocella M, Musaro A, Cecchi G, Bagni MA (2009). Mechanical properties of intact single fibres from wild-type and MLC/mIgf-1 transgenic mouse muscle. *J Muscle Res Cell Motil,* **30**: 199–207.

Dutka TL, Verburg E, Larkins N, Hortemo KH, Lunde PK, Sejersted OM, Lamb GD (2012). ROS-mediated decline in maximum Ca^{2+}-activated force in rat skeletal muscle fibers following in vitro and in vivo stimulation. *PLoS One,* **7**: e35226.

Edwards RH, Hill DK, Jones DA, Merton PA (1977). Fatigue of long duration in human skeletal muscle after exercise. *J Physiol,* **272**: 769–778.

Fedorova M, Kuleva N, Hoffmann R (2010). Identification of cysteine, methionine and tryptophan residues of actin oxidized in vivo during oxidative stress. *J Proteome Res,* **9**: 1598–1609.

Ford LE, Huxley AF, Simmons RM (1977). Tension responses to sudden length change in stimulated frog muscle fibres near slack length. *J Physiol,* **269**: 441–515.

Germinario E, Bondì M, Cencetti F, Donati C, Nocella M, Colombini B, Betto R, Bruni P, Bagni MA, Danieli-Betto D (2016). S1P3 receptor influences key physiological properties of fast-twitch extensor digitorum longus muscle. *J Appl Physiol,* **120**: 1288–1230.

Lamb GD, Westerblad H (2011). Acute effects of reactive oxygen and nitrogen species on the contractile function of skeletal muscle. *J Physiol,* **589**: 2119–2127.

Nocella M, Cecchi G, Bagni MA, Colombini B (2013). Effect of temperature on crossbridge force changes during fatigue and recovery in intact mouse muscle fibers. *PLoS One,* **8**: e78918.

Nocella M, Cecchi G, Bagni MA, Colombini B (2014). Force enhancement after stretch in mammalian muscle fiber: No evidence of cross-bridge involvement. *Am J Physiol Cell Physiol,* **307**: C1123–C1129.

Nocella M, Colombini B, Bagni MA, Bruton J, Cecchi G (2012). Non-crossbridge calcium-dependent stiffness in slow and fast skeletal fibres from mouse muscle. *J Muscle Res Cell Motil,* **32**: 403–409.

Nocella M, Colombini B, Benelli G, Cecchi G, Bagni MA, Bruton J (2011). Force decline during fatigue is due to both a decrease in the force per individual cross-bridge and the number of cross-bridges. *J Physiol,* **589**: 3371–3381.

Plant DR, Lynch GS, Williams DA (2000). Hydrogen peroxide modulates Ca^{2+}-activation of single permeabilized fibres from fast- and slow-twitch skeletal muscles of rats. *J Muscle Res Cell Motil,* **21**: 747–752.

Posterino GS, Cellini MA, Lamb GD (2003). Effects of oxidation and cytosolic redox conditions on excitation-contraction coupling in rat skeletal muscle. *J Physiol,* **547**: 807–823.

Westerblad H, Allen DG (1993). The contribution of $[Ca^{2+}]i$ to the slowing of relaxation in fatigued single fibres from mouse skeletal muscle. *J Physiol,* **468**: 729–740.

PART II
CARDIAC AND SMOOTH MUSCLE

Chapter 10

ATP Utilization in Skeletal and Cardiac Muscle: Economy and Efficiency

G. J. M. Stienen

Department of Physiology, Institute for Cardiovascular Research,
VU University Medical Center, Amsterdam, the Netherlands and Faculty of Science,
Department of Physics and Astronomy, VU University, Amsterdam, the Netherlands

g.stienen@vumc.nl

During muscle contraction, chemical energy in the form of ATP is continuously converted into mechanical work and heat. Energy turnover in muscle is studied not only to understand the energetic costs of locomotion of the body but also to understand the fundamental properties of the contractile mechanism itself. In this review, an overview will be given of the dependence of the rate of ATP utilization on the level of activity, fiber type and species in permeabilized skeletal and cardiac muscle preparations from amphibian (frog and Xenopus laevis) and mammalian (rat, rabbit, guinea pig and human). Special attention will be paid to the definitions of—and the distinction between— economy and efficiency of muscle contraction. Results in particular with respect to the increase in energy turnover during muscle shortening (the Fenn effect) and the future perspectives of

---

Muscle Contraction and Cell Motility: Fundamentals and Developments
Edited by Haruo Sugi
Copyright © 2017 Pan Stanford Publishing Pte. Ltd.
ISBN 978-981-4745-16-1 (Hardcover), 978-981-4745-17-8 (eBook)
www.panstanford.com

studies on muscle energetics will be discussed in terms of a simplified model for crossbridge interaction.

10.1 Introduction

During muscle contraction, chemical energy (E) in the form of ATP is continuously converted into mechanical work (W) and heat (H). In skeletal and cardiac muscle, ATP utilization in quiescent tissue (basal metabolism) is rather low. During muscle activity, it rises considerably, by up to a factor of 100 at maximal activation (Barclay et al., 1993; Balaban, 2002). To meet this high energy demand, muscle is also equipped with very robust ATP recycling systems, regenerating ATP from its hydrolysis products ADP and inorganic phosphate (Pi).

Three different modi operandi can be distinguished. The first mode concerns the basal metabolic activity, required to maintain cellular integrity (ion homeostasis, protein synthesis, etc.). The second mode of action concerns isometrically contracting cells, in which an action potential at the surface membrane of the cell elicits a sudden rise in the intracellular free Ca^{2+} concentration. This activates the contractile machinery and results in tension (force divided by cross-sectional area) generation in the muscle. When the muscle is kept isometrically (at constant length), the contractile units inside the cell (the sarcomeres) do not shorten. Accordingly, in an isometrically contracting muscle as well as in quiescent muscle no external work is being performed and ultimately all chemical energy required for basal metabolism, Ca^{2+} handling and force generation is liberated as heat. The net efficiency of the motor (W/(H + W)) is zero. During an isometric contraction, the relevant energetic parameter is tension cost: the rate of ATP utilization (in moles per muscle volume per second or $mol \cdot m^{-3} \cdot s^{-1}$) divided by the tension generated by the muscle (in $N \cdot m^{-2}$). It should be noted that whereas efficiency is dimensionless (and usually is expressed as a percentage), tension cost is expressed in $mol \cdot N^{-1} \cdot m^{-1} \cdot s^{-1}$ (or equivalent). The inverse of tension cost, *economy*, is also frequently used, but this term should not be confused with the thermodynamically rigorously defined term *efficiency*. Isometric contractions in slow muscle are more economically than in fast muscle because a similar amount of tension is maintained at the expense of a lower rate of ATP utilization. The third mode of action pertains to activated muscle cells which either shorten (concentric contractions) and external work is generated or lengthen (eccentric contractions) in which external work is imposed and the muscle acts as a brake. During concentric contractions efficiency is positive, but during eccentric contraction efficiency can be negative.

Energy turnover in muscle is studied not only to understand the energetic costs of locomotion of the body but also to understand the fundamental properties of the contractile mechanism itself. Evidently, the efficiency of energy conversion at the level of the whole body will be less than the efficiency of muscle contraction at the cellular level, because of friction in joints, opposing forces of antagonists required for joint stabilization, etc. Interesting studies have been performed regarding the storage of elastic energy in tendons (e.g., de Haan et al., 1989) and the effects of anatomy and training (e.g., Maloiy et al., 1986). The second focus of the study of muscle energetics, directed towards the implications of muscle energetics for the contractile mechanism is the main topic of this review. For a related review focused on the alterations in of force development during heart failure, see Stienen (2015).

Energy transduction at the molecular level (the crossbridge) will be described and an overview will be given of the dependence of the rate of ATP utilization on the level of activity, fiber type and species. It is well known for almost a century from the work of W. O. Fenn (Fenn, 1923; Rall, 1982) that the energetic cost of contraction increases with an increase in power output of the muscle. It is very surprisingly that a theoretical explanation of this so-called Fenn effect is still lacking. A postulate in conventional models for muscle contraction is that the individual myosin heads interact independently with actin. Recent studies, however, suggest that the two heads of the myosin molecule do not act independently during stretch (Brunello et al., 2007). Moreover, evidence has been provided suggesting that myosin heads not only bind cooperatively along the thin filament but also that local differences may exist in attachment or detachment probability, regulated by myosin binding protein C (e.g., Moss et al., 2015), titin (e.g., Anderson and Granzier, 2012) and filament compliance (e.g., Huxley et al., 1994; Brunello et al., 2014). These "confounding" factors will be discussed briefly in the final part of this review but still need to be incorporated in order to arrive at a comprehensive model of muscle contraction.

10.2   The Crossbridge Cycle

Force is developed in the contractile unit (the half sarcomere) which consists of a hexagonal array of myosin an actin filaments upon binding of a myosin head which extends from the thick (myosin) filament to a myosin binding site on the thin (actin) filament (Huxley, 1957). This is a cyclic process in which the myosin head binds, subsequently performs a power stroke and thereafter detaches from actin (Huxley and Simmons, 1971). The power stroke results in force production and/or sliding of the actin and myosin filament relative to each other, resulting in sarcomere and muscle shortening. Single molecule studies indicate that each attached myosin head (or crossbridge) generates a force of approximately 10 pN and that the size of the working stroke amounts to approximately 10 nm (Finer et al., 1994; Takagi et al., 2006; Sugi et al., 2008). Mechanical studies in intact single muscle fibers and cardiac trabeculae have shown that the size and speed of the power stroke are load dependent (Reconditi et al., 2004; Caremani et al., 2016)).

In a simplified two-state scheme of the crossbridge cycle, based on the original Huxley (1957) model, the attachment process takes place at a rate f_{app} and the detachment process takes place with a rate g_{app} (Brenner, 1988). This implies that the number of myosin heads attached equals the total number of available myosin heads (N), multiplied by the fraction of attached crossbridges ($f_{app}/(f_{app} + g_{app})$). The number of force generating crossbridges thus depends on f_{app} and g_{app}.

The rate constant governing the initial tension development in the cardiomyocyte is determined by the speed of the activation process, i.e., membrane depolarization and the rise of the intracellular Ca^{2+} concentration and the speed of tension development at the crossbridge level. The latter rate can be determined during the quick release and restretch maneuver, in which an activated muscle cell is abruptly shortened until it becomes slack, and thereafter quickly restretched to its original length (Brenner and Eisenberg, 1986). In the simplified two-state scheme of the crossbridge cycle, the rate of tension redevelopment (k_{tr}) is determined by the sum of the apparent attachment and detachment rates, i.e., $f_{app} + g_{app}$.

Each crossbridge undergoes a cycle of attachment and detachment during which one molecule ATP is hydrolyzed. In an isometrically contracting muscle cell in steady state, the rate of crossbridge cycling is given by $f_{app} \cdot g_{app} / (f_{app} + g_{app})$. Detachment is generally considered to be the rate-limiting step within the cycle ($f_{app} \gg g_{app}$). In this case the rate of crossbridge cycling and thus the rate of ATP utilization equals g_{app}. A measure of the economy of isometric contraction is tension cost, i.e., force divided by the rate of contraction related ATP utilization which is proportional to g_{app}, even when g_{app} would not be rate limiting.

Measurements of energy turnover in muscle revealed that the mechanical efficiency of muscle contraction, i.e., external work/energy input from ATP hydrolysis, is optimal at intermediate velocities of muscle shortening. A typical figure for mechanical crossbridge efficiency is ~40%. However, overall cardiac muscle efficiency is ~22% because energy is required for other processes within the cardiomyocyte such as SR Ca^{2+}-uptake and mitochondrial oxidative phosphorylation (Smith et al., 2005; Barclay and Widén, 2010).

10.3 Dependence of ATP Utilization on Activity, Fiber Type and Species

Stimulated by—and in part in collaboration with—experts on muscle energetics such as Elzinga, Curtin, Ferenczi and Woledge (Woledge et al., 1985; Buschman et al., 1996; He et al., 1998), the work in my group for many years has been committed to measurements of the rate of ATP utilization in permeabilized muscle tissue. The methods employed in these studies were based on the enzymatic coupling of ATP resynthesis using phosphoenol pyruvate (PEP) with the breakdown of NADH, pioneered by, amongst others, Sleep (Glyn and Sleep, 1985) and Güth (Güth and Wojciechowski, 1986) and has been used to resolve a plethora of research questions (e.g., Kobayashi et al., 2004).

The overall reaction scheme is depicted in Table 10.1. In this way it is possible to measure ATP utilization, while the ATP concentration remains constant. The breakdown of NADH was measured photometrically from the change in NADH absorbance at 340 nm in the measuring chamber just underneath the preparation, rather than the change in NADH fluorescence inside the preparation, in order to prevent motion artifacts and to facilitate calibration of the signal.

Table 10.1 Principle of the measurement of ATP consumption in permeabilized preparations

ATP → ADP + Pi	ATP hydrolysis
ADP + PEP → ATP + Pyruvate	ATP resynthesis, catalyzed by pyruvate kinase
Pyruvate + NADH → NAD⁺ + Lactate	Breakdown of NADH, catalyzed by lactate dehydrogenase

The initial studies were performed using a measuring chamber with a volume of around 70 µL. In later studies this volume was reduced to approximately 4 µL, to quantitate the picomolar increase in ATP utilization during rapid shortening (Potma and Stienen, 1996). An overview of our studies in many different preparations and species, as well as of related studies of colleagues in this field, is given in Table 10.2.

Many of the initial studies were performed on fast muscle fibers frog skeletal muscle at low temperature (~4°C). Later studies were performed on single muscle fibers from the iliofibularis muscle from Xenopus laevis (the South African claw toad; an ugly beast, according to A. V. Hill (Hill, 1970)), which has the advantage that it consists of different fiber types, which can easily recognized by eye during dissection (Lännergren and Hoh, 1984). A second advantage was that as in frog muscle (fibers), measurements of heat production could be used for comparison.

It can be clearly seen that the rate of ATP utilization in frog muscle fibers is rather high in comparison with mammalian muscle fibers, despite the fact that the measurements were performed at low temperature. Frog leg fibers are clearly designed for speed of contraction. Experiments performed in saponin or mechanically skinned fibers, in which the sarcoplasmic reticulum (SR) is still intact, allowed us to study the calcium and temperature dependence of the SR calcium ATPase (SERCA2a), and its relative contribution to the total ATP utilization during isometric contraction. One striking observation was that the SR-ATPase was more sensitive to calcium in comparison to the actomyosin (AM-)ATPase (Stienen et al., 1995), which explains why overall muscle efficiency is reduced at submaximal activation.

Table 10.2 Overview of measurements of ATPase and force in permeabilized preparations

Species	Skinning method	Tempera- ture (°C)	ATPase (mM·s⁻¹)	ATPase rate (s⁻¹)	Tension (kN·m⁻²)	Tension cost (µmol. N⁻¹·m⁻¹·s⁻¹)	Additional variable	Remarks	References
Amphibian									
Frog (anterior tibialis)	triton	4	0.34 ± 0.05 n = 7						(Stienen et al., 1990)
Frog (semitendino- sus)	mech+ lubrol	4	See remark	See remark			SL, Ca	Force: 121 mg ATPase: 0.094 ± 0.027 µm/min/ mg protein	(Levy et al., 1976)
Xenopus laevis Fast (type 1 + 2) Slow (type 3)	triton	4.3	0.2 0.1				Temperature, BDM	SR ATPase (saponin and mech) Q10	(Stienen et al., 1995)
Xenopus type 1 Xenopus type 3		20		10.6 ± 0.9 3.6 ± 0.4				Economy/heat	(Buschman et al., 1996)
Rabbit									
Adductor. magnus								Fast twitch	(Levy et al., 1976)
							ΔL: 1.9×		(Arata et al., 1979)

(Contiued)

Table 10.2 (*Continued*)

Species	Skinning method	Tempera-ture (°C)	ATPase (mM·s⁻¹)	ATPase rate (s⁻¹)	Tension (kN·m⁻²)	Tension cost (μmol·N⁻¹·m⁻¹·s⁻¹)	Additional variable	Remarks	References
Psoas	Glycerol	10		1.27 ± 0.12	13.2 ± 0.9		pH. Pi	pH 7.0, 3 mM Pi	(Cooke et al., 1988)
Psoas	Glycerol	15	0.46 ± 0.01 $n = 28$	2.30 ± 0.05			ΔL: +	[S1] = 0.2 mM	(Potma et al., 1994a)
Psoas	Glycerol	15		1.78 ± 0.09 $n = 6$					(Glyn and Sleep, 1985)
Psoas	Glycerol	20	See remark	3				1.25 nmol s⁻¹ m⁻¹ fiber⁻¹	(Kawai et al., 1987)
Psoas	Glycerol + triton	15	0.43 ± 0.02 $n = 21$	2.1 ± 0.1 $n = 21$	136 ± 5 $n = 21$		pH = 7.3 [S1] = 0.2 mM		(Potma et al., 1994b)
Psoas	Glycerol + triton	15	0.056 ± 0.004 $n = 14$	0.28 ± 0.02 $n = 14$	115 ± 6 $n = 14$		pH = 7.3 [S1] = 0.2 mM		(Potma et al., 1994b)
Psoas soleus		15	0.41 ± 0.03 0.050 ± 0.003 $n = 14$	2.0 ± 0.2 0.25 ± 0.02	130 ± 5 110 ± 5		pH, Pi	[S1] = 0.2 mM	(Potma et al., 1995)
Psoas		15	0.42 ± 0.03	2.1 ± 0.1	141 ± 6		ΔL: 2.6 μm	High resolution during shortening	(Potma and Stienen, 1996)
Psoas soleus		15	4.7 ± 0.2 0.57 ± 0.05	31.5 3.8	230 ± 11 194 ± 23		Pi mob	Pi release	(He et al., 1998)

Species	Skinning method	Temperature (°C)	ATPase (mM·s⁻¹)	ATPase rate (s⁻¹)	Tension (kN·m⁻²)	Tension cost (μmol·N⁻¹·m⁻¹·s⁻¹)	Additional variable	Remarks	References
Rat									
EDL	Mech	21–22		3.80 ± 0.53 n = 8					(Stephenson et al., 1989)
IIX	Triton 2 h	12	0.23 ± 0.01		81.8 ± 3.7	2.9 ± 0.09	Light chain composition		(Bottinelli et al., 1994)
IIB			0.178 ± 0.023		95.0 ± 11.3	1.89 ± 0.22			
IIA			0.168 ± 0.026		111.4 ±	1.52 ± 0.13			
I			0.045 ± 0.006		15.2	0.66 ± 0.004			
					57.6 ± 4.2				
Human									
IIB		20	0.41 ± 0.06		171 ± 7				(Stienen et al., 1996)
IIA			0.27 ± 0.02		136 ± 5				
I			0.10 ± 0.01		114 ± 7				
IIB	Glycerol/ saponin/ triton	20	0.108 ± 0.013 (n = 2)			0.169 ± 0.042		Force and AM ATPase lower but not tension cost because diameter measured before triton	(Szentesi et al., 2001)
IIA			0.139 ± 0.014 (n = 6)			0.069 ± 0.008 (SR)			
I			0.046 ± 0.004 (n = 5)			0.019 ± 0.002			
IIX	Freeze	20	0.325 ± 0.032 (n = 7)				Chronic heart failure		
IIA			0.248						

(Contiued)

Table 10.2 *(Continued)*

Species	Skinning method	Temperature (°C)	ATPase (mM·s⁻¹)	ATPase rate (s⁻¹)	Tension (kN·m⁻²)	Tension cost (µmol. N⁻¹·m⁻¹·s⁻¹)	Additional variable	Remarks	References
I	dried + triton		± 0.017 (n = 14) 0.134 ± 0.024 (n = 11)						
Mice									
Tibialis cranialis		20	~0.35		84 ± 4	4.4		Bundles mainly neonatal MHC	(Chandra et al., 2009)
Cardiac rat									
RV trabecula	Triton	20	0.48 ± 0.05 n = 10	3.2	64 ± 3 n = 10		ΔL: +162%	Pig: 0.48 s⁻¹ (Kuhn et al., 1990) [S1] = 0.15 mM	(Stienen et al., 1993)
RV trabecula	Triton	21	0.56 ± 0.05 n = 6	3.3	58.3 ± 0.5 n = 6		SL		(Kentish and Stienen, 1994)
RV trabecula	Triton	20	0.43 ± 0.02 n = 23	2.7 ± 0.1 n = ?	51 ± 3 n = 23		Pi, pH, ΔL: +152%		(Ebus et al., 1994)
RV trabecula	Triton		0.41	2.7	52.8 ± 8.8		PKA		(de Tombe and Stienen, 1995)
RV trabecula	Triton	20	0.348 ± 0.023 0.48 ± 0.04		39 ± 1 53 ± 3		ΔL, BDM	Concurrent ATPases	(Ebus and Stienen, 1996a) (Ebus and Stienen, 1996b)

Species	Skinning method	Tempera-ture (°C)	ATPase (mM·s⁻¹)	ATPase rate (s⁻¹)	Tension (kN·m⁻²)	Tension cost (µmol·N⁻¹·m⁻¹·s⁻¹)	Additional variable	Remarks	References
Cardiac guinea pig and human									
RV trabecula	Triton	12	0.063 ± 0.005	0.4	38 ± 2			ADP formation [S1] = 157 µM	(Barsotti and Ferenczi, 1988)
α-MHC		20	0.39 ± 0.06	41.2 ± 1.2	10.95 ± 1.17			LV + RV	(van der Velden et al., 1998)
β-MHC			0.10 ± 0.01		2.28 ± 0.13				
β-MHC		20	0.10 ± 0.01		41 ± 3	2.4 ± 0.2	Minoxidil	Control 96% β-MHC	(van der Velden et al., 1999)
Atrium		20	0.260 ± 0.025		13.7 ± 1.0	11.4 ± 1.4	MHC	Tension not corrected for myofibrillar density	(Narolska et al., 2005b)
ventricle			0.051 ± 0.011		18.6 ± 2.3	2.4 ± 0.3			

Note: Different permeabilization methods were used: chemical treatment using detergents such as Triton X-100 (triton), lubrol, saponin or prolonged storage in glycerol (at around −20°C); mechanical skinning (mech) according to Natori (Natori, 1954) and freeze-drying (for further details see references and Stienen (2000)).

Abbreviations: AM: actomyosin; ATPase rate: rate of ATP utilization expressed per myosin head; BDM: butanedione monoxime; ΔL: changes in length; EDL: m. extensor digitorum longus; MHC: myosin heavy chain; Pi: inorganic phosphate; PKA: protein kinase A; RV and LV: right and left ventricle from the heart; S1: myosin subfragment-S1; S_L: sarcomere length; SR: sarcoplasmic reticulum; Q10: temperature coefficient; I and II: slow and fast skeletal muscle fiber types I and II.

In subsequent studies we turned to (fast) rabbit psoas muscle fibers, which are frequently used in biochemical studies, and their slow counterparts in soleus muscle. Part of these studies were designed to understand crossbridge interactions associated with the power stroke. A surprising finding was the divergent decline in isometric force and ATP consumption with an increase in inorganic phosphate (Pi) concentration in fast fibers and the parallel decline in both in slow muscle fibers. Moreover, the studies in which both Pi and pH were varied provided interesting insights in the process of muscle fatigue. Comparison with literature values in rabbit (and rat) showed good agreement between the data from different groups (c.f. Levy et al., 1976; Arata et al., 1979; Glyn and Sleep, 1985; Kawai et al., 1987; Cooke et al., 1988).

The experiments in rat muscle fibers and the subsequent studies in human fibers (Stienen et al., 1996; Szentcsi et al., 2001) permit a general comparison between fibers of the same type in rabbit, rat, and man. In addition, the study of the temperature dependence of the ATP consumption in human fibers (Stienen et al., 1996) allows the extrapolation of the findings to body temperature. The general trend that can be observed in the results summarized in Table 10.2 is—even in fibers of the corresponding type (identified by biochemical analysis of MHC composition)—that tension cost decreases with body weight of the species, i.e., the contractions become more economically, likely at the expense of speed (c.f. Rome et al., 1999).

10.4   ATP Utilization in Cardiac Muscle

Measurements of ATP utilization in cardiac muscle were carried out in part in collaboration with Papp, Kentish and de Tombe (Stienen et al., 1993; Kentish and Stienen, 1994; de Tombe and Stienen, 2007). The initial studies were mainly performed on thin right ventricular (RV) trabeculae from the rat which consists of a mixture of predominantly (~75%) fast (α-)MHC and ~25% slow (β-)MHC. In some studies RV trabeculae from guinea pig were used which mainly consist of β-MHC, and therefore are considered to resemble human myocardium more closely. These measurements provided insight in the effect of length changes (Stienen et al., 1993), β-adrenergic stimulation (mimicked by in vitro protein kinase A treatment) (de Tombe and Stienen, 1995), the relation between ATP utilization and sarcomere length (Kentish and Stienen, 1994) and the effect of temperature (de Tombe and Stienen, 2007). Moreover, by using a combination of permeabilization protocols and pharmacological interventions we studied the partitioning of the different ATP utilizing processes in cardiac muscle (Ebus and Stienen, 1996a). In addition, we investigated the influence of changes in the free calcium and inorganic phosphate (Pi) concentration and in pH (Ebus et al., 1994) to describe the alterations in energy requirement of the tissue during cardiac ischemia. The results from these studies were largely in agreement with the observations in fast skeletal muscle where the effects of muscle fatigue were addressed.

The effects of changes in sarcomere length and temperature are noteworthy for a variety of reasons. It appeared that at short sarcomere length (1.6–2.2 μm), force decreased more than ATP utilization, probably as a result of restoring forces originating from connective tissue and titin. The effects of temperature and calcium concentration were studied not only on ATP utilization but—in parallel experiments—also on the rate of tension regeneration (k_{tr}). The picture that emerges from this study (de Tombe and Stienen, 2007) which is consistent with the simplified crossbridge cycle described above is that at low calcium concentration k_{tr} is dominated by the apparent rate of crossbridge detachment (g_{app}), with a temperature coefficient very similar to that of tension cost, while at high calcium concentration ATP utilization and k_{tr} are dominated by the apparent rate of crossbridge attachment (f_{app}) plus g_{app}, with a temperature coefficient larger than that of tension cost and of isometric tension.

In guinea pig the MHC isoform changes with age (van der Velden et al., 1998) and disease (van der Velden et al., 1999). This allowed us to study the relation between ATP utilization and tension cost with the pure β- and α-MHC isoforms. It appeared that the slow β-MHC isoform is about fivefold more economical than the fast α-MHC isoform.

Next we studied force and ATP utilization in human myocardium obtained during open heart or cardiac transplant surgery. Atrial biopsies are composed of a mixture of fast and slow MHC isoforms, and ventricular biopsies are almost exclusively composed of the slow MHC isoform. These studies (Narolska et al., 2005a; Narolska et al., 2005b) indicated that also in human tissue the slow β-MHC isoform is about fivefold more economical than the fast α-MHC isoform. Comparison of the data on β-MHC in guinea pig and human tissue confirm the decline in ATP utilization and the increase in economy with body weight observed in skeletal muscle. More recently these measurements were performed (and are still ongoing) in muscle strips isolated from human myectomy samples from the septum. Myectomy is performed to remove the LV outflow tract obstruction in patients with familial hypertrophic cardiomyopathy (Witjas-Paalberends et al., 2013; Witjas-Paalberends et al., 2014). These studies yield information on the energetic consequences of sarcomeric mutations, which will enhance our understanding of the primary and secondary effects of sarcomeric mutations in familial hypertrophic cardiomyopathy, which occur rather frequently (1:500) in the human population.

10.5 The Fenn Effect

During the ejection phase of the cardiac contraction, the cardiomyocytes shorten and the rate of crossbridge cycling (g_{app}) increases in proportion to the externally developed mechanical work, a property known as the Fenn effect (Fenn, 1923). The Fenn effect is a paragon of the difference between muscle efficiency and economy. Our first experiments to study the Fenn effect were conducted in cardiac trabeculae (Stienen et al., 1993). In subsequent studies, rabbit psoas muscle fibers were used to investigate the effect of muscle shortening on ATP consumption (Potma et al., 1994a; Potma et al., 1994b; Potma et al., 1995; Potma and Stienen, 1996). In view of the limited resolution in the early experiments we first looked at the effects of repetitive pulse-shaped changes in length. Previous heat measurements (Abbott et al., 1951) suggested that the increase in ATP consumption observed was largely caused by the acceleration of crossbridge cycling during shortening whereas energy turnover during lengthening was expected to be similar to that observed during isometric contraction. Therefore we reasoned that the net effects of pulse-shaped changes in length would resemble those that would occur during staircase shortening. Changes in either amplitude of frequency were used to study the influence of differences in "effective" speed of shortening during the staircase. This implicit assumption was bolstered experimentally by using ramp-like wave forms in which the speeds during the shortening and lengthening phases were varied. In the final series of experiments of this type using the 4 μL measuring chamber, conducted in collaboration with Reggiani and Bottinelli, we managed to record the increase in ATP consumption during and after steady shortening at different speeds (Reggiani et al., 1997) in different fiber types from rat (at 12°C). These experiments revealed that the difference in efficiency in fast and slow muscle fibers was rather small. They also provided insight in the modulatory role of myosin light chain composition on muscle energetics, on top of the dominant role of the myosin heavy chain composition (Reggiani et al., 1997).

The effect of shortening (or external load) on the rate of ATP utilization has also been studied at high time resolution (Sun et al., 2001) using an NADH-linked assay and by Sugi and coworkers (Sugi et al., 1998; Sugi et al., 2003) from the amount of work performed during auxotonic shortening when the amount of ATP available was sufficient for a single crossbridge power stroke. It is clear from these studies that the rate of ATP utilization depends on the mechanical conditions but that a comprehensive crossbridge model which takes all experimental data with regards of crossbridge and filament compliance, number of crossbridges attached, and the distribution of crossbridges among the various biochemical states into account is still lacking.

Probably the most intriguing observation regarding the effects of length changes on ATP utilization was the reversal of the Fenn effect at low calcium concentration (Stienen et al., 1993). We observed that length changes that at high calcium concentration resulted in an increase in the rate of ATP utilization above the corresponding isometric value, at low calcium concentration resulted in a decrease in the rate of ATP utilization relative to the corresponding isometric value (Fig. 10.1). The most straightforward explanation of this finding—also consistent with the reduction in the average force level during the length changes—appeared to be an acceleration of both the apparent rate of attachment and detachment during shortening. It has been discussed that the magnitude of the Fenn effect in cardiac tissue is smaller than in frog skeletal muscle (Rall, 1982). This might at least partly be caused by the submaximal level of the free cytosolic Ca^{2+} concentration in cardiac myocytes because indications have been obtained that the Fenn effect may be reversed in cardiac tissue at very low Ca^{2+} concentrations (Stienen et al., 1993).

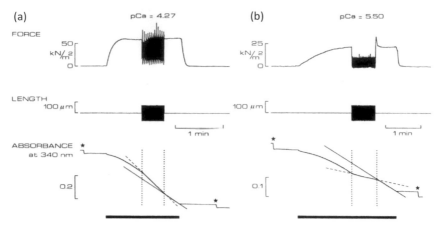

Figure 10.1 Reversal of the Fenn effect. Recordings of production of force per cross-sectional area (tension) (*upper traces*), displacement of the motor to change fiber length (*middle traces*) and NADH absorbance (*lower traces*) at saturating (a) and submaximal (b) Ca^{2+} concentration. The rate of ATP utilization was determined from the decline in NADH absorbance from the slopes of the straight lines fitted to selected parts of the signal. Solid bars indicate when the trabeculae from rat was immersed in the measuring chamber, in which the activating solution was present. The length of the preparation was varied with a square wave at 23 Hz by ±2.5% of the initial length (corresponding to a sarcomere length of 2.2 μm) for approximately 30 s. At (*) 0.5 nmol ADP was injected into the measuring chamber to calibrate the absorbance signal. Note that at saturating Ca^{2+} concentration, the repetitive length changes result in an increase in the rate of ATP utilization (interrupted line) relative to the isometric value (solid line), whereas at submaximal Ca^{2+} concentration the slope was decreased. (Stienen et al., 1993). Figure reproduced with kind permission from Springer Science + Business Media.

10.6   Future Perspectives

The explanation of the economy of contraction, i.e., the ATP utilization required for maintenance of isometric tension in terms of a model of crossbridge interaction appears to be fairly straightforward. The main parameters are the fraction of attached crossbridges, which depends on the f_{app} and g_{app}, the average force per crossbridge and the rate of crossbridge turnover, which is dominated by g_{app} and the free energy which comes available from the hydrolysis of ATP. Tension cost does not depend on the free calcium concentration. Hence, under isometric conditions the free calcium concentration appears to influence solely the fraction of attached crossbridges. The concentration of other cytosolic ingredients such as Pi, ADP, and H^+ influence tension development and ATP utilization in a more complex manner, but even a simplified version of a crossbridge model as described by Pate and Cooke (1989) would take these factors into account, c.f. Papp et al. (2002).

The explanation of the efficiency of the contractile process is far more complex and the available crossbridge models to date only account for a partial description of the mechanical and energetical properties of active muscle during shortening and lengthening. Factors that complicate the general picture are (actin and myosin) filament and end compliance (tendons), the complex 3D geometry of the whole heart, and of whole muscles, which consist of a mixture of fiber types. Filament (and end) compliance require the need of a distributed crossbridge model simulation, which should be tailored to the prevailing experimental conditions and configuration. Additional factors complicating the general picture are cooperative interaction of myosin heads with the actin filament and the possible interaction between the two myosin heads of a single myosin molecule (Chaen et al., 1988). These factors as well as the role of myosin binding protein C and titin are still incompletely understood.

What would be an appropriate approach to resolve these issues? Clearly, considerable progress has been made in our understanding of the individual constituents involved in the energy transduction process in muscle but a "super" model which takes all currently known factors into account appears to be beyond our reach. Understanding of the Fenn effect in terms of the strain dependency of the rate constants governing the predominant crossbridge states appears a logical first step. The experimental approach of choice could be the study of phosphate release (with ms-time resolution) during load clamps which eliminate the influence of filament and end compliance (Linari et al., 2009). This could lead to an energy-transduction-crossbridge module which could be integrated into a more general comprehensive theoretical framework of muscle contraction.

References

Abbott BC, Aubert XM, Hill AV (1951) The absorption of work by a muscle stretched during a single twitch or a short tetanus. *Proc R Soc London Ser B Biol Sci*, 139: 86–104.

Anderson BR, Granzier HL (2012) Titin-based tension in the cardiac sarcomere: Molecular origin and physiological adaptations. *Prog Biophys Mol Biol*, 110:204–17. doi: 10.1016/j.pbiomolbio.2012.08.003.

Arata T, Mukohata Y, Tonomura Y (1979) Coupling of movement of crossbridges with ATP splitting studied in terms of the acceleration of the ATPase activity of glycerol-treated muscle fibers on applying various types of repetitive stretch-release cycles. *J Biochem*, 86: 525–542.

Balaban RS (2002) Cardiac energy metabolism homeostasis: Role of cytosolic calcium. *J Mol Cell Cardiol*, 34: 1259–1271.

Barclay CJ, Constable JK, Gibbs CL (1993) Energetics of fast- and slow-twitch muscles of the mouse. *J Physiol*, 472: 61–80.

Barclay CJ, Widén C (2010) Efficiency of cross-bridges and mitochondria in mouse cardiac muscle. *Adv Exp Med Biol*, 682: 267–278. doi: 10.1007/978-1-4419-6366-6_15.

Barsotti RJ, Ferenczi MA (1988) Kinetics of ATP hydrolysis and tension production in skinned cardiac muscle of the guinea pig. *J Biol Chem*, 263: 16750–16756.

Bottinelli R, Canepari M, Reggiani C, Stienen GJ (1994) Myofibrillar ATPase activity during isometric contraction and isomyosin composition in rat single skinned muscle fibres. *J Physiol*, 481(Pt 3): 663–75.

Brenner B (1988) Effect of Ca^{2+} on cross-bridge turnover kinetics in skinned single rabbit psoas fibers: Implications for regulation of muscle contraction. *Proc Natl Acad Sci U S A*, 85: 3265–3269.

Brenner B, Eisenberg E (1986) Rate of force generation in muscle: Correlation with actomyosin ATPase activity in solution. *Proc Natl Acad Sci U S A*, 83: 3542–3546.

Brunello E, Caremani M, Melli L, et al. (2014) The contributions of filaments and cross-bridges to sarcomere compliance in skeletal muscle. *J Physiol*, 592: 3881–3899. doi: 10.1113/jphysiol.2014.276196.

Brunello E, Reconditi M, Elangovan R, et al. (2007) Skeletal muscle resists stretch by rapid binding of the second motor domain of myosin to actin. *Proc Natl Acad Sci U S A*, 104: 20114–20119. doi: 10.1073/pnas.0707626104.

Buschman HP, van der Laarse WJ, Stienen GJ, Elzinga G (1996) Force-dependent and force-independent heat production in single slow- and fast-twitch muscle fibres from Xenopus laevis. *J Physiol*, 496(Pt 2): 503–519.

Caremani, M, Pinzauti F, Reconditi M et al. (2016) Size and speed of the working stroke of cardiac myosin in situ. *Proc Natl Acad Sci U S A*, 113: 3675–3680.

Chaen S, Shimada M, Sugi H (1988) Cooperative interactions of myosin two heads in muscle force generation. *Adv Exp Med Biol*, 226: 289–298.

Chandra M, Mamidi R, Ford S, et al. (2009) Nebulin alters cross-bridge cycling kinetics and increases thin filament activation: A novel mechanism for increasing tension and reducing tension cost. *J Biol Chem*, 284: 30889–30896. doi: 10.1074/jbc.M109.049718.

Cooke R, Franks K, Luciani GB, Pate E (1988) The inhibition of rabbit skeletal muscle contraction by hydrogen ions and phosphate. *J Physiol*, 395: 77–97.

De Haan A, Van Ingen Schenau GJ, Ettema GJ, et al. (1989) Efficiency of rat medial gastrocnemius muscle in contractions with and without an active prestretch. *J Exp Biol*, 141: 327–341.

De Tombe PP, Stienen GJ (1995) Protein kinase A does not alter economy of force maintenance in skinned rat cardiac trabeculae. *Circ Res*, 76: 734–741.

De Tombe PP, Stienen GJM (2007) Impact of temperature on cross-bridge cycling kinetics in rat myocardium. *J Physiol*, 584: 591–600. doi: 10.1113/jphysiol.2007.138693.

Ebus JP, Stienen GJ (1996a) Origin of concurrent ATPase activities in skinned cardiac trabeculae from rat. *J Physiol*, 492(Pt 3): 675–687.

Ebus JP, Stienen GJ (1996b) Effects of 2,3-butanedione monoxime on cross-bridge kinetics in rat cardiac muscle. *Pflugers Arch*, 432: 921–929.

Ebus JP, Stienen GJ, Elzinga G (1994) Influence of phosphate and pH on myofibrillar ATPase activity and force in skinned cardiac trabeculae from rat. *J Physiol*, 476: 501–516.

Fenn WO (1923) A quantitative comparison between the energy liberated and the work performed by the isolated sartorius muscle of the frog. *J Physiol*, 58: 175–203.

Finer JT, Simmons RM, Spudich JA (1994) Single myosin molecule mechanics: Piconewton forces and nanometre steps. *Nature*, 368: 113–119. doi: 10.1038/368113a0.

Glyn H, Sleep J (1985) Dependence of adenosine triphosphatase activity of rabbit psoas muscle fibres and myofibrils on substrate concentration. *J Physiol*, 365: 259–276.

Güth K, Wojciechowski R (1986) Perfusion cuvette for the simultaneous measurement of mechanical, optical and energetic parameters of skinned muscle fibres. *Pflugers Arch*, 407: 552–557.

He Z, Stienen GJ, Barends JP, Ferenczi MA (1998) Rate of phosphate release after photoliberation of adenosine 5'-triphosphate in slow and fast skeletal muscle fibers. *Biophys J*, 75: 2389–2401.

Hill AV (1970) *First and Last Experiments in Muscle Mechanics*. Cambridge University Press, New York, USA.

Huxley AF (1957) Muscle structure and theories of contraction. *Prog Biophys Biophys Chem*, 7: 255–318.

Huxley AF, Simmons RM (1971) Proposed mechanism of force generation in striated muscle. *Nature*, 233: 533–538.

Huxley HE, Stewart A, Sosa H, Irving T (1994) X-ray diffraction measurements of the extensibility of actin and myosin filaments in contracting muscle. *Biophys J*, 67: 2411–2421. doi: 10.1016/S0006-3495(94)80728-3.

Kawai M, Güth K, Winnikes K, et al. (1987) The effect of inorganic phosphate on the ATP hydrolysis rate and the tension transients in chemically skinned rabbit psoas fibers. *Pflugers Arch*, 408: 1–9.

Kentish JC, Stienen GJ (1994) Differential effects of length on maximum force production and myofibrillar ATPase activity in rat skinned cardiac muscle. *J Physiol*, 475: 175–184.

Kobayashi T, Saeki Y, Chaen S, et al. (2004) Effect of deuterium oxide on contraction characteristics and ATPase activity in glycerinated single rabbit skeletal muscle fibers. *Biochim Biophys Acta*, 1659: 46–51. doi: 10.1016/j.bbabio.2004.07.008.

Kuhn HJ, Bletz C, Rüegg JC (1990) Stretch-induced increase in the Ca^{2+} sensitivity of myofibrillar ATPase activity in skinned fibres from pig ventricles. *Pflugers Arch*, 415: 741–746.

Lännergren J, Hoh JF (1984) Myosin isoenzymes in single muscle fibres of Xenopus laevis: Analysis of five different functional types. *Proc R Soc London Ser B Biol Sci*, 222: 401–408.

Levy RM, Umazume Y, Kushmerick MJ (1976) Ca^{2+} dependence of tension and ADP production in segments of chemically skinned muscle fibers. *Biochim Biophys Acta*, 430: 352–365.

Linari M, Piazzesi G, Lombardi V (2009) The effect of myofilament compliance on kinetics of force generation by myosin motors in muscle. *Biophys J*, 96: 583–592. doi: 10.1016/j.bpj.2008.09.026.

Maloiy GM, Heglund NC, Prager LM, et al. (1986) Energetic cost of carrying loads: Have African women discovered an economic way? *Nature*, 319: 668–669. doi: 10.1038/319668a0.

Moss RL, Fitzsimons DP, Ralphe JC (2015) Cardiac MyBP-C Regulates the Rate and Force of Contraction in Mammalian Myocardium. *Circ Res*, 116: 183–192. doi: 10.1161/CIRCRESAHA.116.300561.

Narolska NA, Eiras S, van Loon RB, et al. (2005a) Myosin heavy chain composition and the economy of contraction in healthy and diseased human myocardium. *J Muscle Res Cell Motil*, 26: 39–48. doi: 10.1007/s10974-005-9005-x.

Narolska NA, van Loon RB, Boontje NM, et al. (2005b) Myocardial contraction is 5-fold more economical in ventricular than in atrial human tissue. *Cardiovasc Res*, 65: 221–229. doi: 10.1016/j.cardiores.2004.09.029.

Natori R (1954) The property and contraction process of isolated myofibrils. *Jikeikai nmd J*, 1: 119–126.

Papp Z, Szabó A, Barends JP, Stienen GJM (2002) The mechanism of the force enhancement by MgADP under simulated ischaemic conditions in rat cardiac myocytes. *J Physiol*, 543: 177–189.

Pate E, Cooke R (1989) A model of crossbridge action: The effects of ATP, ADP and Pi. *J Muscle Res Cell Motil*, 10: 181–196.

Potma EJ, Stienen GJ (1996) Increase in ATP consumption during shortening in skinned fibres from rabbit psoas muscle: Effects of inorganic phosphate. *J Physiol*, 496(Pt 1): 1–12.

Potma EJ, Stienen GJ, Barends JP, Elzinga G (1994a) Myofibrillar ATPase activity and mechanical performance of skinned fibres from rabbit psoas muscle. *J Physiol*, 474: 303–317.

Potma EJ, van Graas IA, Stienen GJ (1994b) Effects of pH on myofibrillar ATPase activity in fast and slow skeletal muscle fibers of the rabbit. *Biophys J*, 67: 2404–2410. doi: 10.1016/S0006-3495(94)80727-1.

Potma EJ, van Graas IA, Stienen GJ (1995) Influence of inorganic phosphate and pH on ATP utilization in fast and slow skeletal muscle fibers. *Biophys J*, 69: 2580–2589. doi: 10.1016/S0006-3495(95)80129-3.

Rall JA (1982) Sense and nonsense about the Fenn effect. *Am J Physiol*, 242: H1–H6.

Reconditi M, Linari M, Lucii L, et al. (2004) The myosin motor in muscle generates a smaller and slower working stroke at higher load. *Nature*, 428: 578–581. doi: 10.1038/nature02380.

Reggiani C, Potma EJ, Bottinelli R, et al. (1997) Chemo-mechanical energy transduction in relation to myosin isoform composition in skeletal muscle fibres of the rat. *J Physiol*, 502(Pt 2): 449–460.

Rome LC, Cook C, Syme DA, et al. (1999) Trading force for speed: Why superfast crossbridge kinetics leads to superlow forces. *Proc Natl Acad Sci U S A*, 96: 5826–5831.

Smith NP, Barclay CJ, Loiselle DS (2005) The efficiency of muscle contraction. *Prog Biophys Mol Biol*, 88: 1–58. doi: 10.1016/j.pbiomolbio.2003.11.014.

Stephenson DG, Stewart AW, Wilson GJ (1989) Dissociation of force from myofibrillar MgATPase and stiffness at short sarcomere lengths in rat and toad skeletal muscle. *J Physiol*, 410: 351–366.

Stienen GJ (2000) Chronicle of skinned muscle fibres. *J Physiol*, 527(Pt 1): 1.

Stienen GJM (2015) Pathomechanisms in heart failure: The contractile connection. *J Muscle Res Cell Motil*, 36: 47–60. doi: 10.1007/s10974-014-9395-8.

Stienen GJ, Kiers JL, Bottinelli R, Reggiani C (1996) Myofibrillar ATPase activity in skinned human skeletal muscle fibres: Fibre type and temperature dependence. *J Physiol*, 493(Pt 2): 299–307.

Stienen GJ, Papp Z, Elzinga G (1993) Calcium modulates the influence of length changes on the myofibrillar adenosine triphosphatase activity in rat skinned cardiac trabeculae. *Pflugers Arch*, 425: 199–207.

Stienen GJ, Roosemalen MC, Wilson MG, Elzinga G (1990) Depression of force by phosphate in skinned skeletal muscle fibers of the frog. *Am J Physiol*, 259: C349–C357.

Stienen GJ, Zaremba R, Elzinga G (1995) ATP utilization for calcium uptake and force production in skinned muscle fibres of Xenopus laevis. *J Physiol*, 482(Pt 1): 109–122.

Sugi H, Iwamoto H, Akimoto T, Kishi H (2003) High mechanical efficiency of the cross-bridge powerstroke in skeletal muscle. *J Exp Biol*, 206: 1201–1206.

Sugi H, Iwamoto H, Akimoto T, Ushitani H (1998) Evidence for the load-dependent mechanical efficiency of individual myosin heads in skeletal muscle fibers activated by laser flash photolysis of caged calcium in the presence of a limited amount of ATP. *Proc Natl Acad Sci U S A*, 95: 2273–2278.

Sugi H, Minoda H, Inayoshi Y, et al. (2008) Direct demonstration of the cross-bridge recovery stroke in muscle thick filaments in aqueous solution by using the hydration chamber. *Proc Natl Acad Sci U S A*, 105: 17396–17401. doi: 10.1073/pnas.0809581105.

Sun Y-B, Hilber K, Irving M (2001) Effect of active shortening on the rate of ATP utilisation by rabbit psoas muscle fibres. *J Physiol*, 531: 781–791. doi: 10.1111/j.1469-7793.2001.0781h.x.

Szentesi P, Bekedam MA, van Beek-Harmsen BJ, et al. (2005) Depression of force production and ATPase activity in different types of human skeletal muscle fibers from patients with chronic heart failure. *J Appl Physiol*, 99: 2189–2195. doi: 10.1152/japplphysiol.00542.2005.

Szentesi P, Zaremba R, van Mechelen W, Stienen GJ (2001) ATP utilization for calcium uptake and force production in different types of human skeletal muscle fibres *J Physiol*, 531: 393–403.

Takagi Y, Homsher EE, Goldman YE, Shuman H (2006) Force generation in single conventional actomyosin complexes under high dynamic load. *Biophys J*, 90: 1295–1307. doi: 10.1529/biophysj.105.068429.

Van der Velden J, Borgdorff P, Stienen GJ (1999) Minoxidil-induced cardiac hypertrophy in guinea pigs. *Cell Mol Life Sci*, 55: 788–798.

Van der Velden J, Moorman AF, Stienen GJ (1998) Age-dependent changes in myosin composition correlate with enhanced economy of contraction in guinea-pig hearts. *J Physiol*, 507(Pt 2): 497–510.

Witjas-Paalberends ER, Güçlü A, Germans T, et al. (2014) Gene-specific increase in energetic cost of contraction in hypertrophic cardiomyopathy caused by thick filament mutations. *Cardiovasc Res*, doi: 10.1093/cvr/cvu127.

Witjas-Paalberends ER, Piroddi N, Stam K, et al. (2013) Mutations in MYH7 reduce the force generating capacity of sarcomeres in human familial hypertrophic cardiomyopathy. *Cardiovasc Res*, 99: 432–441. doi: 10.1093/cvr/cvt119.

Woledge RC, Curtin NA, Homsher E (1985) Energetic aspects of muscle contraction. *Monogr Physiol Soc*, 41: 1–357.

Chapter 11

Essential Myosin Light Chains Regulate Myosin Function and Muscle Contraction

Ingo Morano

Max-Delbrück-Center for Molecular Medicine,
Department of Molecular Muscle Physiology, and University Medicine Charité,
Robert-Rössle-Str. 10, 13125 Berlin, Germany

imorano@mdc-berlin.de

The molecular motor myosin II consists of two myosin heavy chains (MyHC), two regulatory (RLC), and two essential (ELC) myosin light chains. In particular, crystal structures, functional analysis of single myosin molecules, and genetically engineered animal models with mutated myosin subunits shed new light on the functional roles of ELC. This review will concentrate on the physiological and pathophysiological impact of the ELC in vertebrate muscle types, i.e., skeletal muscle fibers, cardiomyocytes, and smooth muscle cells. First, structures, interaction interfaces, and phosphorylation of ELC will be considered. Second, functional roles of ELC interactions domains and ELC isoforms and, third, pathophysiological aspects of ELC will be discussed. Based

Muscle Contraction and Cell Motility: Fundamentals and Developments
Edited by Haruo Sugi
Copyright © 2017 Pan Stanford Publishing Pte. Ltd.
ISBN 978-981-4745-16-1 (Hardcover), 978-981-4745-17-8 (eBook)
www.panstanford.com

on its multiple protein–protein interaction interfaces, ELC is demonstrated here as a multitasking protein factor, modulating a variety of myosin motor and, hence muscle contraction features. These comprise the following: (i) ELC/MyHC interaction, which regulates stiffness, unitary force development, working stroke, duty cycle, and in vitro sliding velocity of actin filaments of myosin molecules but not actin-activated myosin ATPase activity. Furthermore, ELC/MyHC interaction could be associated with the regulation of interfilament space of the sarcomeres and elementary steps of the cross-bridge cycle, association constant of ATP binding and equilibrium constant of the cross-bridge detachment step, Ca^{2+} sensitivity of force generation, power output, relaxation of the heart, and the stretch activation response. Perturbation of ELC/MyHC interaction interfaces by a variety of missense mutations associates with familial hypertrophic cardiomyopathy, further documenting the important functional roles of ELC. (ii) ELC/actin interaction slows down maximal shortening velocity, cross-bridge detachment rate, and in vitro sliding velocity of actin filaments and increases fiber stiffness (rigor), but does not affect isometric force generation.

11.1 Structure and Interaction Interfaces of Essential Myosin Light Chains

11.1.1 Structure of Myosin II

The myosin II molecule, the first molecular motor purified from skeletal muscle and named as early as 1864 by Willy Kühne ("Über das Protoplasma und die Contractilität"; Verlag von Wilhelm Engelmann, Leipzig) drives contraction of all muscle types by cyclical interactions with the thin (actin) filament while consuming ATP. The first 3D structure from chicken skeletal muscle myosin derived from crystallographic studies (Rayment et al., 1993b) and combined 3D-modeling approaches (Aydt et al., 2007) (Fig. 11.1). Please note that in this review the amino acids involved in the interaction motifs are numbered according to the 3D structure and sequence of chicken skeletal muscle myosin II (Rayment et al., 1993b). Native myosin II consists of two ≈200 kDa heavy chains (MyHC), each associated with two different

≈16–28 kDa light chains (MLC). Hence, a native myosin II molecule consists of two MyHC and four MLC, i.e., two essential MLCs (ELC, MLC-1) and two regulatory MLCs (RLC, MLC-2) (Weeds and Lowey, 1971; Rayment et al., 1993a; Rayment et al., 1993b), which sum up to ≈500 kDa. MyHC can be separated into (i) an N-terminal pear-shaped "head" or subfragment 1 (S-1) consisting of a pear-shaped 16.5 nm motor domain with ATPase and actin binding sites, connected to an α-helical 8.5 nm lever arm which binds two different MLC (Rayment et al., 1993a; Rayment et al., 1993b); (ii) 150 nm α-helical rod, consisting of a 60 nm subfragment 2 (S-2), and the C-terminal 90 nm light meromyosin (LMM). S-2 and LMM of two MyHC form a coiled-coil, while their S-1 fragments remain mobile (Weeds and Lowey, 1971; Rayment et al., 1993a; Rayment et al., 1993b).

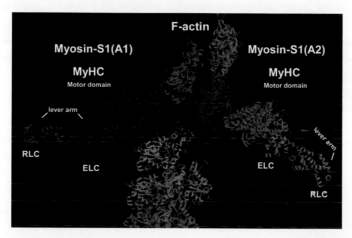

Figure 11.1 3D structure of myosin-S-1/actin interactions from chicken skeletal muscle myosin derived from crystallographic studies (Rayment et al., 1993b) and combined 3D-modeling approaches (Aydt et al., 2007) in the strongly bounded state. MyHC: myosin heavy chain; RLC: regulatory light chain; ELC: essential myosin light chain. A1, A2: alkali light chains.

The α-helical lever arm of the MyHC has a characteristic fold, with a ≈40° bend in the middle (the "elbow") and a sharp ≈60° turn (the "hook") downstream. The lever arm contains two IQ motifs in tandem, designated as IQ1 (consensus [FILV]Qxxx[RK]Gxxx [RK]xx[FILVWY], where x is any amino acid and brackets indicate an alternative), and downstream IQ2 (consensus [FILV]

Qxxx[RK]xxxxxxxx). IQ1 binds ELC and IQ2 binds RLC (Weeds and Lowey, 1971; Rayment et al., 1993a; Rayment et al., 1993b).

11.1.2 Structure of Essential Myosin Light Chains

Treatment of myosin II with alkali (Weeds and Lowey, 1971) leads to a dissociation of ELC from myosin. ELC, therefore, was named alkali light chains (A). SDS-PAGE showed the presence of two forms of alkali light chains in fast-twitch skeletal muscle a high molecular weight form with 25 kDa (A1) and a low-molecular-weight form with 16 kDa (A2) (Lowey and Risby, 1971; Weeds and Lowey, 1971). About 1/3 of adult fast skeletal muscle ELC is of the A2 type, while the predominant ELC represents the A1 isoform (Hoh, 1978; Lowey and Risby, 1971). A1 and A2 isoforms from fast-twitch skeletal muscle are the product of a single gene (MLY1), generated by alternative splicing, with A2 lacking the N-terminal aa 1–46 (Frank and Weeds, 1974; Nabeshima et al., 1984; Periasamy et al., 1984). In smooth muscle tissues, a distinct ELC gene (MYL6) is expressed, giving rise to a 17 kDa protein (MLC17), without N-terminal extension (Cavaille et al., 1986; Hasegawa and Morita, 1992; Lenz et al., 1989; Nabeshima et al., 1984; Nabeshima et al., 1987). ELCs expressed in cardiac and slow skeletal muscle are exclusively of the A1 isoform (Hoh, 1978; Lowey and Risby, 1971). In the heart, distinct A1 light chain genes are expressed, MYL4 and MYL3, coding for the atrial myosin light chain (ALC-1) and the ventricular myosin light chain (VLC-1), respectively. MYL3 is also expressed in slow-twitch skeletal muscle (Barton and Buckingham, 1985). Since MYL4 is expressed in all striated muscle types during embryonic development, it is also called MLC-1emb (Barton and Buckingham, 1985) (Table 11.1).

Table 11.1   ELC genes of human muscular tissues, number of amino acids (AA), and chromosomal (Chr) locations (Barton and Buckingham, 1985; Fodor et al., 1989)

Gene	Acc. No.	AA	Syn	Tissue	Chr
MYL1	CAG33150.1	194	MLC1f	Fast skeletal	2q33
MYL3	NP_000249.1	195	VLC-1	Ventricle, slow skeletal	3p21
MYL4	NP_001002841.1	197	ALC-1	Atrium	17q21
MYL6	CAG33295.1	151	MLC17	Smooth muscle	12q13.2

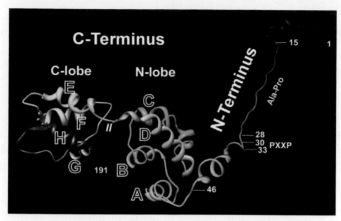

Figure 11.2A 3D structure of A1 from chicken skeletal muscle myosin derived from crystallographic studies (Rayment et al., 1993b) and combined 3D-modeling approaches (Aydt et al., 2007). A–H: helices of EF-hand domains. *II*: linker between N-lobe and C-lobe. Numbers indicate amino acid positions.

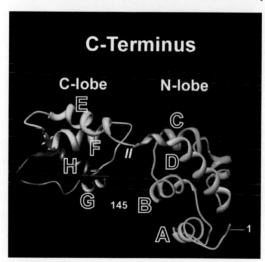

Figure 11.2B 3D structure of A2 from chicken skeletal muscle myosin derived from crystallographic studies (Rayment et al., 1993b). A–H: helices of EF-hand domains. *II*: linker between N-lobe and C-lobe. Numbers indicate amino acid positions.

Although ELCs belong to the superfamily of EF-hand Ca^{2+}-binding proteins sharing structural homologies with troponin C and calmodulin (Moncrief et al., 1990; Houdusse and Cohen,

1995), vertebrate ELCs lost their ability to bind Ca^{2+}. ELC isoforms of the A1 type can be separated into an antenna-like N-terminal (aa1–46) and a large dumbbell-like C-terminus (aa 47–191) composed of four EF hand domains (Aydt et al., 2007; Houdusse and Cohen, 1995; Milligan, 1996; Rayment et al., 1993b) (Fig. 11.2A). Hence, A2 (ELC without N-terminus) consists solely of the C-terminal four EF hand domains (Fig. 11.2B). The four C-terminal EF-hand domains form a N-lobe (EF-hand domains I–II with helices A–D) and a C-lobe (EF-hand domains III–IV with helices E–H), connected by the flexible linker *II* (Houdusse and Cohen, 1995; Rayment et al., 1993b; Xie et al., 1994) (Fig. 11.2A,B).

The proximal N-terminus of A1 (aa1-15) contains a "sticky" element of several charged amino acids, in particular lysines and downstream a repetitive stretched AP-rich segment (aa ≈15–28) (Fodor et al., 1989; Seharaseyon et al., 1990). Molecular modeling of the N-terminal A1 segment showed an antenna-like structure with a length of 91 Å (Aydt et al., 2007) (Fig. 11.1), i.e., long enough to bridge the gap between the C-terminal EF-hand domains and actin. However, FRET analysis of the N-terminus of A1 showed only a length of 41 Å (Lowey et al., 2007).

11.1.3   ELC Interaction Interfaces

The ELC forms multiple contacts with the motor domain of the MyHC, RLC, actin, and the IQ1-domain of the lever arm of the MyHC (Aydt et al., 2007; Milligan, 1996; Rayment et al., 1993b) (Fig. 11.3). Hence, ELC stabilizes myosin II structures and tunes a variety of myosin II functions.

The C-terminal EF-hand domains of ELC wrap around a broad binding interface of 23 aa of the α-helical lever arm (aa 783–806) (Houdusse and Cohen, 1995; Rayment et al., 1993b). Most of the amino acid residues of the hydrophobic α-helical lever arm make contacts with the ELC. Thus, ELC/lever arm binding is strong, with ≈10% hydrogen bonds, and ≈90% van der Waals contacts (Milligan, 1996; Xie et al., 1994). The C-lobe of ELC locates more upstream of the α-helical lever arm and interacts through EF-hand domains III and IV with the first half of the IQ1 recognition site. The N-lobe of ELC locates more downstream and binds to the second half of the IQ1 of the lever arm (Figs. 11.1 and 11.2) (Houdusse and Cohen, 1995; Rayment et al., 1993b).

Besides ELC/lever arm binding, the 3D structures of myosin-S-1 suggest close contacts between the C-lobe of ELC and the motor domain of the MyHC. Helix E of ELC is close to a short helix of the MyHC (aa 720–730) which is part of the converter domain (Fig. 11.3). Furthermore, helix F of the ELC contacts the N-terminal aa 21–31 of the MyHC (Fig. 11.3) (Milligan, 1996; Rayment et al., 1993b). ELC may also bind via its N-terminal antenna to the Src-homology 3 (SH3)-like subdomain (aa 35–79) of the MyHC (Lowey et al., 2007). A PXXP-consensus motif (Alexandropoulos et al., 1995) for SH3 interaction is present in the N-terminus of A1 (aa 30–33). An additional contact area of the ELC with the RLC involves the first loop of the ELC and the F-G linker of the RLC (Rayment et al., 1993b; Xie et al., 1994).

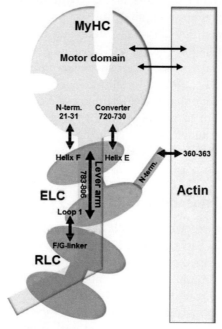

Figure 11.3 Scheme of A1-myosin-S-1/actin interactions. There are multiple interactions between the motor domain of the myosin heavy chain (MyHC) (black double arrows). Please note the interaction domains of the essential myosin light chain (ELC; red) with the MyHC (N-term., converter, lever arm), actin, and regulatory light chain (RLC) (red double arrows). Numbers correspond to amino acid position of the MyHC and actin.

Recent biochemical data indicated that the "sticky" lysine-rich N-terminus, but not the A-P-rich segment of A1 (aa 1–15) binds to a cluster of negatively charged amino acids of actin (aa 360–363) (Sutoh, 1982; Stepkowski, 1995; Timson et al., 1999; Trayer et al., 1987; Hayashibara and Miyanishi, 1994). This could be supported by structural data comparing actin decorated with S-1(A1) and S-1(A2), showing an extra-binding site in the C-terminal domain of actin with S-1(A1) (Milligan, 1996). However, it is not yet unequivocally clear if the N-terminus of ELC is long enough to bridge the gap to the actin filament of 78.9 Å, but there is a large structural and functional body of evidence that suggest this. Thus, the length of the N-terminus was 91 Å in a molecular modeling study (Aydt et al., 2007), i.e., long enough to tether the C-terminal EF-hand domains of ELC with the actin filament. In line, the distance between aa 16 of the N-terminus of A1 and aa 374 of actin was only 2.9 nm, as demonstrated by a recent FRET analysis (Guhathakurta et al., 2015). Furthermore, cardiac fibers of transgenic rats which overexpressed an N-terminally 43 aa truncated VLC-1 (TgM$^{\Delta 1-43}$) showed decreased rigor stiffness, providing evidence that the N-terminus of A1 tethers myosin with the actin filament (Wang et al., 2013). However, in a biochemical FRET study the length of the N-terminus was only 41 Å (Lowey et al., 2007), i.e., not sufficiently long to tether the C-terminal EF-hand domains with actin.

11.1.4 ELC Phosphorylation

Recent in vitro experiments with ELC and MLCK showed multiple phosphorylation sites, i.e., T127 (or T129 or Y130), S179, and Y185 by mass-spectroscopy (Cadete et al., 2012). Surprisingly, inhibition of MLCK by ML-7 abolished phosphorylation of S179 and Y186, but not of T127, T129, or Y130. This complex phosphorylation pattern of cardiac ELC was even raised upon ischemia-reperfusion experiments of rat hearts, leading to ELC phosphorylation of T69, T77 or Y78, T132 or T134 or Y135, T164, S184, and Y190 (Cadete et al., 2012). Furthermore, phosphorylation of ELC was reported upon activation of protein kinase C by adenosine-stimulation of rabbit cardiomyocytes at T64 und S195 (Arrell et al., 2001). However, recent works could not confirm the phosphorylation of purified cardiac ELC or ELC of skinned cardiac fibers (Morano et al., 1990; Bialojan et al., 1988).

11.2   Functional Roles of ELCs?

11.2.1   ELC/MyHC Interactions

11.2.1.1   ELC/lever arm interactions

Studies with single myosin II molecules lacking selectively ELCs revealed about one-third of the force generated by complete myosin as well as significantly reduced in vitro sliding velocity of actin filaments while actin-activated myosin ATPase activity remained unaffected (VanBuren et al., 1994). In these experiments, part of the reduced force generation of ELC-denuded myosin may be due to a reduced duty cycle (the fraction of the total cross-bridge cycle that myosin is attached to actin in a force-generating state). Interestingly, the RLC seems not to be involved in the modulation of unitary force production of myosin molecules. In fact, the force-generating properties of RLC denuded myosin remained almost normal (VanBuren et al., 1994).

The intense ELC/lever arm interaction suggests a structural stabilization of the compliant α-helical lever arm. This is an important functional aspect since the lever arm is considered to be the elastic element which amplifies the very small conformational changes in the motor domain to a large movement of around 5–10 nm (Piazzesi et al., 2007). Hence, flexural rigidity "κ" ("stiffness") and the movement distance "x" of the lever arm affect force generation "F" since $F = \kappa\, x$ (Howard and Spudich, 1996; Adamovic et al., 2008). Molecular simulations of the stiffness $\kappa = 3kTL_\mathrm{p}/L^3$ (where k is the Boltzmann constant, T is the temperature, L is the length of the lever arm, and L_p is the persistence length) of a 10 nm poly-alanine α-helix (mimicking the lever arm) revealed a stiffness of 1.2 pN/nm (Adamovic et al., 2008). Stiffness modeling of a coiled-coil, which ought to be similar to the lever arm occupied with two MLC, led to $\kappa \approx 3.1$ pN/nm (Adamovic et al., 2008; Howard and Spudich, 1996) calculated a $\kappa \approx 2$ pN/nm (based on $L_\mathrm{p} = 100$ nm, $L = 8$ nm, $kT = 4$ pN·nm). Fiber experiments (rabbit psoas) revealed myosin stiffness of 1.7 pN/nm (Linari et al., 2007) and 3.3 pN/nm (Piazzesi et al., 2007) while optical trap experiments with single myosin molecules (rabbit psoas) led to a stiffness of 1.79 pN/nm (Lewalle et al., 2008). However, stiffness values obtained with single myosin molecules are controversial: Maximally 1.4 pN/nm and 0.4 pN/nm for fast and slow rat skeletal muscle, respectively (Capitanio et al., 2006), and 0.7 pN/nm for rabbit skeletal muscle myosin (Veigel et al., 1998) were also reported. In fact, stiffness of single myosin molecules could be modulated by selectively replacing the mouse VLC-1 by the human VLC-1 isoform or a E56G mutated variant of the human VLC-1 in a transgenic mouse model (Lossie et al., 2014).

11.2.1.2 ELC/motor domain couplings

ELC appeared initially to be "essential" for ATPase activity of myosin (Stracher, 1969; Dreizen and Gershman, 1970). However, a decade later it could be shown that myosin heavy chains prepared from skeletal (Wagner and Giniger, 1981; Sivaramakrishnan and Burke, 1982) and cardiac muscle (Mathern and Burke, 1986) with selectively removed ELC revealed both normal Ca^{2+} or K^+/EDTA activated as well as actin-activated myosin ATPase activities. However, there are some experimental observations that suggest that ELC/motor domain interactions could well modulate some aspects of chemo-mechanical energy transformation of myosin.

The Helix F of the C-lobe of the ELC is in close contact to the N-terminal aa 21–31 of the motor domain of the MyHC while E-helix of the C-lobe of ELC interact with part of the converter (aa 720–730) of the motor domain (Rayment et al., 1993b; Milligan, 1996). The N-terminal aa 1–80 of the motor domain is an important functional structure since its deletion destroyed normal myosin-II functions (abnormal actin affinity and ADP release rate) and could not rescue *Dictyostelium* myosin-II null cells (Fujita-Becker et al., 2006). Therefore, the C-lobe of the ELC could well modulate properties of myosin II via interaction with the N-terminus of the MyHC. In fact, studies using ELC labeled with fluorescence probes could demonstrate a reciprocal coupling between the catalytic site of the MyHC and the C-lobe of the ELC upon ATP binding to the motor domain (Borejdo et al., 2001; Marsh and Lowey, 1980). The converter (aa 711–781) is a critical component during force generation of myosin II. Crystal structures showed large conformational change of the converter domain, which led to a rotation of the lever arm by 60°, i.e., the power stroke (Geeves and Holmes, 2005). Besides the lever arm (Uyeda et al., 1996), the converter domain was suggested to represent an alternative element of elastic distortion during force generation

(Dobbie et al., 1998). In fact, cardiomyopathy mutations within the converter domain could substantially modify myosin cross-bridge stiffness in fibers (Kohler et al., 2002; Seebohm et al., 2009). Hence, interaction with the ELC could well modulate rotation and stiffness of the converter and, therefore, force generation. The functional importance of the C-lobe of the ELC could recently be documented in an in vivo model. Truncation of the C-lobe of cardiac ELC in "*Lazy susan*," a mutant zebrafish, generated severely reduced cardiac functions (Meder et al., 2009).

One more interaction interface has been suggested between the N-terminal antenna of the ELC and the protruding SH3-subdomain (aa 35–79) of the motor domain of MyHC (Lowey et al., 2007). Selective deletion of the SH3 subdomain from the motor domain of MyHC deteriorated actin affinity and ADP release rate of myosin in vitro (Fujita-Becker et al., 2006), showing an important functional role of the SH3 domain for myosin function. It is not yet clear which part of the N-terminus of ELC binds to the SH3 domain of the MyHC, but a consensus PXXP motif in the N-Terminus of the ELC (aa 30–33; Fig. 11.2A) for SH3-binding exists. However, the distance between the PXXP-motiv of the ELC to the SH3 subdomain of the MyHC is probably too large to enable SH3/PXXP-interaction (c.f. Figs. 11.1 and 11.2).

11.2.2   ELC/Actin Interaction

Different sets of experimental approaches in vitro suggest that actin-binding of the N-terminal antenna of A1 slows down myosin motor function. Convincing experiments with muscle preparations containing distinct amounts of A1 and A2, i.e., ELC with or without the N-terminal antenna but the same C-Terminus were performed. Myosin with A2, i.e., ELC without N-terminal antenna, will not interact with actin. Hence, actin concentration required for half-maximal ATPase activity was lower for S-1(A1) than for S-1(A2), and the maximum turnover rate in the presence of actin for S-1(A2) was higher than for S-1(A1) (Weeds and Taylor, 1975; Chalovich et al., 1984). However, these differences could only be observed at low ionic strength conditions and disappear at higher (more physiologic) ionic strength (Chalovich et al., 1984; Wagner et al., 1979). In physiologic functional experiments maximal shortening velocity rose with increasing amounts of

A2, i.e., the alternatively spliced ELC isoform without the actin-binding N-terminus (Bottinelli et al., 1994). A linear correlation between the A1/A2 ratio and shortening velocity has also been reported for single muscle fibers of mammalian muscle fibers (Greaser et al., 1988; Sweeney et al., 1988). In vitro motility experiments demonstrated that myosin with A2 translocated actin filaments with a higher velocity than A1 myosin (Lowey et al., 1993a,b). Deletion of the sticky actin-binding N-terminus (aa 1–13) from ELC produced the same accelerating effect in the in vitro motility assay and actin-activated myosin ATPase activity than deletion of the whole N-terminal antenna (aa 1–43) (Lowey et al., 2007). Additional experimental evidence for the functional importance of the ELC/actin interaction could be obtained by weakening the ELC/actin binding by peptide competition. N-terminal ELC peptides which bind to actin increased shortening velocity of electrically driven primary adult cardiomyocytes (Haase et al., 2006; Morano et al., 1995; Tunnemann et al., 2007). Likewise, weakening the A1–actin interaction by N-terminal A1 peptides, which competitively bind to actin, increased myosin ATPase activity in vitro (Hayashibara and Miyanishi, 1994; Rarick et al., 1996; Timson et al., 1999). In line, recent transgenic overexpression of an N-terminally truncated ventricular A1 (TgM$^{\Delta 1-43}$) in the heart accelerated the ADP-dependent cross-bridge detachment step (Wang et al., 2013) which critically determines maximal shortening velocity (Siemankowski et al., 1985). In contrast to maximal shortening velocity, isometric tension generation of cardiac fibers prepared from TgM$^{\Delta 1-43}$ remained normal (Wang et al., 2013). These experiments suggest that binding of the N terminus of ELC to actin represents a mechanical load placed on the cross-bridges thereby slowing down the actin transport velocity (Lowey et al., 1993b).

Recent FRET-analysis showed that the amplitude of the myosin lever arm displacement was larger with A1 if compared with the movement of the lever arm associated with A2 (Guhathakurta et al., 2015). Since the movement distant of the lever arm "x" determines myosin force generation (c.f. above), force generation of muscle fibers with A2 should be lower than force generation of muscle fibers with A1. However, isometric tension generation of cardiac fibers prepared from transgenic rats overexpressing an N-terminally truncated VLC-1 (TgM$^{\Delta 1-43}$)

remained normal (Wang et al., 2013) (c.f. above). The reason for this contradiction is yet not understood.

Binding of the N-Terminus of A1 to actin seems to be controlled by the thin filament regulatory system. The presence of tropomyosin increased the affinity of N-terminal ELC peptides to F-actin, while binding of troponin I abolished A1 peptide binding to F-actin-tropomyosin (Patchell et al., 2002; Trayer and Trayer, 1985). This could explain the Ca^{2+} dependency of A1 binding to regulated actin (Trayer and Trayer, 1985) since Ca^{2+} binding to troponin C reduced actin affinity of troponin I (Stoker et al., 2012) and may, therefore, lose its inhibitory effect on A1 binding to actin during muscle activation. Furthermore, in the presence of Ca^{2+}, a more extended conformation of the N-terminus of A1 occurs (Stepkowski, 1995), and a much tighter binding to actin could be observed (Nieznanski et al., 2003). Furthermore, orientation and mobility of the N-Terminus of ELC changed in the presence of ATP or actin and differed in relaxed and rigor-contracted skeletal muscle fibers, where the N-Terminus of ELC was strongly immobilized (Borejdo et al., 2001).

11.2.3 ELC/RLC Interaction

Information of the functional role of ELC/RLC-interaction derived in particular from smooth muscle myosin-II and scallop myosin-II. Mutations introduced to disrupt the ELC/RLC binding interface abolished in vitro motility and reduced ATPase activity of myosin with phosphorylated RLC (Ni et al., 2012). Scallop myosin with mutated ELC/RLC interface showed a very weak RLC-binding to the ELC-MyHC complex and no Ca^{2+} sensitive ATPase or Ca^{2+} binding (Jancso and Szent-Gyorgyi, 1994).

11.2.4 Phosphorylation of ELC

The physiological function of cardiac ELC phosphorylation is still not well understood, but results of experiments available suggest a critical role. Expression of a C-terminally truncated cardiac ELC in the *laz*-zebrafish caused severely reduced heart functions, which could be compensated by overexpression of normal cardiac ELC (Meder et al., 2009). In contrast, overexpression of an alanine-substituted S195 cardiac ELC could not compensate the *laz* cardiac phenotype, while a cardiac ELC with a phosphor-mimetic amino acid at position S195 rescued the failing *laz* heart (Meder et al., 2009).

11.2.5 Functional Roles of ELC Isoforms

11.2.5.1 Striated muscle ELC isoforms

The modulatory role of fast-twitch skeletal muscle A1 and A2 isoforms are already discussed above (c.f. Section 11.2.2). Besides alternatively spliced A1 and A2 from fast-twitch skeletal muscle, two ELC isoforms generated by the expression of different genes in the slow twitch skeletal muscle and ventricle (Myl3) as well as in embryonic skeletal muscle and atrium (Myl4) coding VLC-1 and ALC-1, respectively, exist (Barton and Buckingham, 1985; Fodor et al., 1989). Expression of ELC isoforms is tissue specific and developmentally regulated. Human embryos express large amounts of ALC-1 both in the whole heart and in skeletal muscle (Barton and Buckingham, 1985). ALC-1 protein levels decrease in the skeletal muscle and ventricle to undetectable levels during early postnatal development but persists in the atrium throughout the whole life (Cummins et al., 1980; Price et al., 1980). The hypertrophied heart of children with Tetralogy of Fallot expresses large amounts of VLC-1 in the atrium (Auckland et al., 1986) and ALC-1 in the ventricle, up to adulthood (Shi et al., 1991). Similarly, the hypertrophied left ventricle of patients with ischemic, dilative, and hypertrophic cardiomyopathy expresses ALC-1 (Morano et al., 1997; Ritter et al., 1999; Schaub et al., 1984). Surgical intervention and subsequent normalization of the hemodynamic state decrease ALC-1 expression in these patients (Sutsch et al., 1992). In fact, activation of hypertrophic signaling pathways in cardiomyoblasts and subsequent activation of the Ca^{2+}-calmodulin-calcineurin-NFAT as well as the Ca^{2+}-calmodulin-CaMK (in particular, the CaMKIV) rose human MYL4 promoter activity (Woischwill et al., 2005).

The distinct primary sequences of ELC-isoforms suggest that ELC isoforms could differentially modulate myosin function and, therefore, muscle contraction. Evidence for this idea came from heart preparations expressing different amounts of VLC-1 and ALC-1. Skinned cardiac fibers of human and pig with partial replacements of VLC-1 by ALC-1 showed dose-dependent increased isometric force generation and maximal shortening velocity (Morano et al., 1996; Morano et al., 1997). A positive inotropic effect of ALC-1 could also been shown in transgenic mice overexpressing mouse ALC-1 as well as in transgenic rats overexpressing human ALC-1 (hALC-1) in the ventricle (Abdelaziz et al., 2004). In both transgenic animal models, ALC-1 expression in the ventricle rose maximal shortening velocity of skinned fibers, velocity of actin filament translocation, and the velocity of isovolumetric force generation of whole heart preparations. Expression of hALC-1 could significantly attenuate functional defects of experimentally failing hearts of transgenic rats (Abdelaziz et al., 2005).

We could recently show that actin affinity of the N-terminus of ALC-1 was significantly weaker than actin affinity of VLC-1 (Petzhold et al., 2014). Considering the above-mentioned results showing that ELC/actin interaction slows down maximal shortening velocity (c.f. Section 11.2.2) we hypothesized that the weak actin affinity of ALC-1 provides a lower mechanical load placed on the cross-bridges than the VLC-1, thus allowing a higher maximal shortening velocity. In line, the duration of the force generation step of myosin with atrial MLC was significantly smaller compared with myosin with ventricular MLC (Yamashita et al., 2003), allowing a higher detachment rate, i.e., the prerequisite for an increased maximal shortening velocity.

We observed a significantly (more than threefold) lower KD of the hALC-1/MyHC complex than of the VLC-1/MyHC complex (Petzhold et al., 2011). The distinct lever arm binding properties of hALC-1 and hVLC-1 isoforms are supported by different primary structures (84% identity). The different binding properties could be due to higher amounts of A1 interaction sites along the IQ1 motif than hVLC-1. We suggested that the strong binding of hALC-1 to myosin could increase the stiffness of the lever arm and therefore force generation of the individual myosin cross-bridge above the value obtained upon VLC-1 binding. This could explain the above-mentioned observation that heart fibers with ALC-1 could well generate more force than fibers with VLC-1. An increased force generation per single myosin cross-bridge in the presence of ALC-1 suggests improved muscle contraction without additional activating free Ca^{2+}. In fact, overexpression of ALC-1 using adenoviral vectors in primary adult rat cardiomyocytes increased significantly contraction amplitude while the systolic-free Ca^{2+}

levels and Ca^{2+} cycling kinetics remained unchanged (Petzhold et al., 2011). In contrast, maximal force development of skinned ventricular fibers of transgenic mice with mALC-1 was not altered compared to wild-type mice (Fewell et al., 1998). Likewise, unitary force generation of myosin with atrial MLC was not different from unitary force generation of myosin with ventricular MLC (Yamashita et al., 2003). In these studies, however, myosin preparations did not selectively differ by the essential MLC isoforms but by both essential and regulatory MLC forms.

11.2.5.2   Smooth muscle ELC isoforms

Two ELC isoforms with 17 kDa (LC17a, LC17b) are formed in smooth muscle by alternative splicing from a single gene (Cavaille et al., 1986; Nabeshima et al., 1987; Lash et al., 1990). LC17a/b belongs to the A2-type, i.e., without N-terminal antenna. LC17a/b consists of 151 amino acids that only differ in their most C-terminal nine amino acids (i.e., helix H), which is according to the 3D structures (Figs. 11.1 and 11.2) in close contact to the lever arm. Thus, LC17 isoforms could have distinct binding properties with the lever arm and could serve, therefore, as interesting natural variants to investigate the functional roles of ELC/lever arm interactions. However, studies analyzing the functional roles of LC17 isoforms are controversial. It has been reported that LC17b slows the actin activated ATPase (Hasegawa and Morita, 1992) and shortening velocity (Malmqvist and Arner, 1991; Sjuve et al., 1996). Specific extraction/reconstitution of the LC17a/b and overexpression in isolated smooth muscle cells have also shown a slowing effect of LC17b on contractile kinetics (Huang et al., 1999; Matthew et al., 1998). However, other studies on isolated cells have not found a correlation between LC17 isoforms and contraction (Sherwood and Eddinger, 2002). ELC exchange experiments on isolated myosin (Sherwood and Eddinger, 2002) and in vitro motility experiments using expressed myosin fragments have also failed to show effects of LC17 isoforms on ATPase and actin translocation velocity (Kelley et al., 1993).

11.3   Pathophysiology of ELC

Twelve different disease-related missense mutations of the hVLC-1 gene MYL3 (accession NC_000003.12) were yet reported, all of them associate with familial hypertrophic cardiomyopathy (FHC). Hypertrophic cardiomyopathy is an autosomal dominant genetic disease that affects 0.2% of the population (Maron and Maron, 2012). FHC-associated hVLC-1 mutations comprise M149V, R154H (Poetter et al., 1996), E143K (Olson et al., 2002), A57G (Lee et al., 2001), E56G (Richard et al., 2003), H155D, E152K (Kaski et al., 2009), R81H (Fokstuen et al., 2011), G128C (Garcia-Pavia et al., 2011), V79I (Andersen et al., 2012), R63C (Chiou et al., 2015) and E177G (Jay et al., 2013) (Fig. 11.4). Although the individual MYL3 mutations are rare, they may sum up to a considerable significance in the pathogenesis of FHC. Human MYL3 consists of seven exons, while six are expressed since the last exon is in the 3′-untranslated sequence (Fodor et al., 1989) (Fig. 11.4). No mutation has yet been detected in the hALC-1-gene. Almost all VLC-1 mutations are located in the myosin-binding C-Terminus, which is coded by exons II-VI (Fig. 11.4). VLC mutations mainly group in exon III and exon IV.

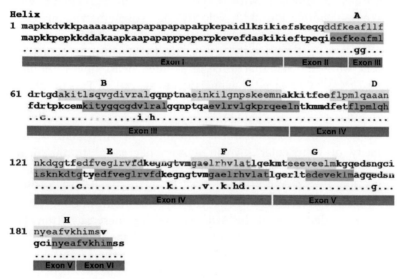

Figure 11.4 Comparisons of amino acid sequences (one letter code) and α-helix positions between ELC of chicken fast skeletal muscle (Sk; blue letters) and ELC of human ventricle (hVLC-1; red letters). Exon boundaries refer to the hVLC-1. Positions of the α-helices A-H are labeled in yellow (Sk) or green (hVLC-1). Positions of the FHC-causing hVLC-1 mutations are given in the third (dotted) lines.

Transgenic animal models overexpressing VLC-1 with missense mutations showed to be valuable to identify functional roles of ELC. Thus, ELC could be associated with the regulation of interfilament space of the sarcomeres and elementary steps of the cross-bridge cycle, association constant of ATP binding and equilibrium constant of the cross-bridge detachment step (Muthu et al., 2011), Ca^{2+} sensitivity of force generation, power output and relaxation of the heart (Sanbe et al., 2000), and the stretch activation response (Fewell et al., 1998). VLC-1 missense mutations associated with FHC decreased binding affinity to the myosin lever arm, which follow the following order: hVLC-1 > hVLC-1^{A57G} > hVLC-1^{E143K} > hVLC-1^{M149V} > hVLC-1^{R154H} > hVLC-1^{E56G} (Lossie et al., 2012; Muthu et al., 2011; Petzhold et al., 2011). Investigations of functional features of single myosin motor molecules associated with mutated ELC revealed distinct stiffness, working stroke, and force generating properties, which declined in the order: hVLC-1 > hVLC-1^{E56G} (Lossie et al., 2014). Reduced functions of myosin with hVLC-1^{E56G} is in agreement with decreased amplitude and velocity of isovolumetric pressure development of isolated perfused hearts decreased in the order TgM^{hVLC-1} > TgM^{E56G} (Lossie et al., 2014). In close agreement, a recent work showed that force of skinned heart fibers obtained from TgM^{A57G} decreased (Kazmierczak et al., 2013). This predicts a "loss of function" of myosin with hVLC-1^{A57G} similar to the myosin with hVLC-1^{E56G} investigated herein.

Similar to myosin stiffness and force, in vitro actin sliding velocity of ventricular myosin motors decreased in the order hVLC-1 > hVLC-1^{E56G}. In vitro actin sliding velocity (Vf) decreases if the duty time (ts) of XBs increases, $Vf = x/ts$ (Uyeda et al., 1990). Duty time is determined by the ADP release rate from the catalytic site, i.e., ts decreases if the ADP release rate increases (Homsher et al., 1992; Siemankowski et al., 1985). The 3D structures of myosin-S-1 suggest close contacts between helix F and the N-terminal antenna of the ELC to the N-Terminus of the myosin-MD (Milligan, 1996; Rayment et al., 1993a; Lowey et al., 2007). Hence, ELC may affect in vitro actin sliding velocity by modulation of the ADP release rate upon interaction with the N-terminus of the myosin-MD and/or by regulation of the extent of the working stroke x, i.e., the working stroke (c.f. above).

The deleterious effects of mutated hVLC-1 on myosin and cardiac functions may provide a valuable molecular pathomechanism provoking the development of FHC. Those mutations weaken the myosin-LA affinity, reduce stiffness and unitary force generation of the single myosin molecule, slow down actin filament sliding velocity, and depress cardiac performance. The resulting cardiac hypo-contractility, which may occur in the human ventricle, could then activate hypertrophic pathways leading to the FHC phenotype (Ashrafian et al., 2003).

Acknowledgment

I thank Shokoufeh Mahmoodzadeh for critically reading the manuscript.

References

Abdelaziz, A. I., Pagel, I., Schlegel, W. P., Kott, M., Monti, J., Haase, H., and Morano, I. (2005). Human atrial myosin light chain 1 expression attenuates heart failure. *Adv. Exp. Med. Biol.*, 565, 283–292; discussion 292, 405–215.

Abdelaziz, A. I., Segaric, J., Bartsch, H., Petzhold, D., Schlegel, W. P., Kott, M., Seefeldt, I., Klose, J., Bader, M., Haase, H., et al. (2004). Functional characterization of the human atrial essential myosin light chain (hALC-1) in a transgenic rat model. *J. Mol. Med. (Berl.)*, 82, 265–274.

Adamovic, I., Mijailovich, S. M., and Karplus, M. (2008). The elastic properties of the structurally characterized myosin II S2 subdomain: A molecular dynamics and normal mode analysis. *Biophys. J.*, 94, 3779–3789.

Alexandropoulos, K., Cheng, G., and Baltimore, D. (1995). Proline-rich sequences that bind to Src homology 3 domains with individual specificities. *Proc. Natl. Acad. Sci. U. S. A.*, 92, 3110–3114.

Andersen, P. S., Hedley, P. L., Page, S. P., Syrris, P., Moolman-Smook, J. C., McKenna, W. J., Elliott, P. M., and Christiansen, M. (2012). A novel Myosin essential light chain mutation causes hypertrophic cardiomyopathy with late onset and low expressivity. *Biochem. Res. Int.*, 2012, 685108.

Arrell, D. K., Neverova, I., Fraser, H., Marban, E., and Van Eyk, J. E. (2001). Proteomic analysis of pharmacologically preconditioned cardiomyocytes reveals novel phosphorylation of myosin light chain 1. *Circ. Res.*, 89, 480–487.

Ashrafian, H., Redwood, C., Blair, E., and Watkins, H. (2003). Hypertrophic cardiomyopathy: A paradigm for myocardial energy depletion. *Trends Genet*, 19, 263–268.

Auckland, L. M., Lambert, S. J., and Cummins, P. (1986). Cardiac myosin light and heavy chain isotypes in tetralogy of Fallot. *Cardiovasc. Res.*, 20, 828–836.

Aydt, E. M., Wolff, G., and Morano, I. (2007). Molecular modeling of the myosin-S1(A1) isoform. *J. Struct. Biol.*, 159, 158–163.

Barton, P. J., and Buckingham, M. E. (1985). The myosin alkali light chain proteins and their genes. *Biochem. J.*, 231, 249–261.

Bialojan, C., Morano, I., and Ruegg, J. C. (1988). Different phosphorylation patterns of cardiac myosin light chains using ATP and ATP gamma S as substrates. *J. Mol. Cell Cardiol.*, 20, 575–578.

Borejdo, J., Ushakov, D. S., Moreland, R., Akopova, I., Reshetnyak, Y., Saraswat, L. D., Kamm, K., and Lowey, S. (2001). The power stroke causes changes in the orientation and mobility of the termini of essential light chain 1 of myosin. *Biochemistry*, 40, 3796–3803.

Bottinelli, R., Betto, R., Schiaffino, S., and Reggiani, C. (1994). Unloaded shortening velocity and myosin heavy chain and alkali light chain isoform composition in rat skeletal muscle fibres. *J. Physiol.*, 478(Pt 2), 341–349.

Cadete, V. J., Sawicka, J., Jaswal, J. S., Lopaschuk, G. D., Schulz, R., Szczesna-Cordary, D., and Sawicki, G. (2012). Ischemia/reperfusion-induced myosin light chain 1 phosphorylation increases its degradation by matrix metalloproteinase 2. *FEBS J.*, 279, 2444–2454.

Capitanio, M., Canepari, M., Cacciafesta, P., Lombardi, V., Cicchi, R., Maffei, M., Pavone, F. S., and Bottinelli, R. (2006). Two independent mechanical events in the interaction cycle of skeletal muscle myosin with actin. *Proc. Natl. Acad. Sci. U. S. A.*, 103, 87–92.

Cavaille, F., Janmot, C., Ropert, S., and d'Albis, A. (1986). Isoforms of myosin and actin in human, monkey and rat myometrium. Comparison of pregnant and non-pregnant uterus proteins. *Eur. J. Biochem.*, 160, 507–513.

Chalovich, J. M., Stein, L. A., Greene, L. E., and Eisenberg, E. (1984). Interaction of isozymes of myosin subfragment 1 with actin: Effect of ionic strength and nucleotide. *Biochemistry*, 23, 4885–4889.

Chiou, K. R., Chu, C. T., and Charng, M. J. (2015). Detection of mutations in symptomatic patients with hypertrophic cardiomyopathy in Taiwan. *J. Cardiol.*, 65, 250–256.

Cummins, P., Price, K. M., and Littler, W. A. (1980). Foetal myosin light chain in human ventricle. *J. Muscle Res. Cell Motil.*, 1, 357–366.

Dobbie, I., Linari, M., Piazzesi, G., Reconditi, M., Koubassova, N., Ferenczi, M. A., Lombardi, V., and Irving, M. (1998). Elastic bending and active tilting of myosin heads during muscle contraction. *Nature*, 396, 383–387.

Dreizen, P., and Gershman, L. C. (1970). Relationship of structure to function in myosin. II. Salt denaturation and recombination experiments. *Biochemistry*, 9, 1688–1693.

Fewell, J. G., Hewett, T. E., Sanbe, A., Klevitsky, R., Hayes, E., Warshaw, D., Maughan, D., and Robbins, J. (1998). Functional significance of cardiac myosin essential light chain isoform switching in transgenic mice. *J. Clin. Invest.*, 101, 2630–2639.

Fodor, W. L., Darras, B., Seharaseyon, J., Falkenthal, S., Francke, U., and Vanin, E. F. (1989). Human ventricular/slow twitch myosin alkali light chain gene characterization, sequence, and chromosomal location. *J. Biol. Chem.*, 264, 2143–2149.

Fokstuen, S., Munoz, A., Melacini, P., Iliceto, S., Perrot, A., Ozcelik, C., Jeanrenaud, X., Rieubland, C., Farr, M., Faber, L., et al. (2011). Rapid detection of genetic variants in hypertrophic cardiomyopathy by custom DNA resequencing array in clinical practice. *J. Med. Genet.*, 40, 572 576.

Frank, G., and Weeds, A. G. (1974). The amino-acid sequence of the alkali light chains of rabbit skeletal-muscle myosin. *Eur. J. Biochem.*, 44, 317–334.

Fujita-Becker, S., Tsiavaliaris, G., Ohkura, R., Shimada, T., Manstein, D. J., and Sutoh, K. (2006). Functional characterization of the N-terminal region of myosin-2. *J. Biol. Chem.*, 281, 36102–36109.

Garcia-Pavia, P., Vazquez, M. E., Segovia, J., Salas, C., Avellana, P., Gomez-Bueno, M., Vilches, C., Gallardo, M. E., Garesse, R., Molano, J., et al. (2011). Genetic basis of end-stage hypertrophic cardiomyopathy. *Eur. J. Heart Fail.*, 13, 1193–1201.

Geeves, M. A., and Holmes, K. C. (2005). The molecular mechanism of muscle contraction. *Adv. Protein Chem.*, 71, 161–193.

Greaser, M. L., Moss, R. L., and Reiser, P. J. (1988). Variations in contractile properties of rabbit single muscle fibres in relation to troponin T isoforms and myosin light chains. *J. Physiol.*, 406, 85–98.

Guhathakurta, P., Prochniewicz, E., and Thomas, D. D. (2015). Amplitude of the actomyosin power stroke depends strongly on the isoform of the myosin essential light chain. *Proc. Natl. Acad. Sci. U. S. A.*, 112, 4660–4665.

Haase, H., Dobbernack, G., Tunnemann, G., Karczewski, P., Cardoso, C., Petzhold, D., Schlegel, W. P., Lutter, S., Pierschalek, P., Behlke, J., et al. (2006). Minigenes encoding N-terminal domains of human cardiac myosin light chain-1 improve heart function of transgenic rats. *FASEB J.*, 20, 865–873.

Hasegawa, Y., and Morita, F. (1992). Role of 17-kDa essential light chain isoforms of aorta smooth muscle myosin. *J. Biochem.*, 111, 804–809.

Hayashibara, T., and Miyanishi, T. (1994). Binding of the amino-terminal region of myosin alkali 1 light chain to actin and its effect on actin-myosin interaction. *Biochemistry*, 33, 12821–12827.

Hoh, J. F. (1978). Light chain distribution of chicken skeletal muscle myosin isoenzymes. *FEBS Lett.*, 90, 297–300.

Homsher, E., Wang, F., and Sellers, J. R. (1992). Factors affecting movement of F-actin filaments propelled by skeletal muscle heavy meromyosin. *Am. J. Physiol.*, 262, C714–C723.

Houdusse, A., and Cohen, C. (1995). Target sequence recognition by the calmodulin superfamily: Implications from light chain binding to the regulatory domain of scallop myosin. *Proc. Natl. Acad. Sci. U. S. A.*, 92, 10644–10647.

Howard, J., and Spudich, J. A. (1996). Is the lever arm of myosin a molecular elastic element? *Proc. Natl. Acad. Sci. U. S. A.*, 93, 4462–4464.

Huang, Q. Q., Fisher, S. A., and Brozovich, F. V. (1999). Forced expression of essential myosin light chain isoforms demonstrates their role in smooth muscle force production. *J. Biol. Chem.*, 274, 35095–35098.

Jancso, A., and Szent-Gyorgyi, A. G. (1994). Regulation of scallop myosin by the regulatory light chain depends on a single glycine residue. *Proc. Natl. Acad. Sci. U. S. A.*, 91, 8762–8766.

Jay, A., Chikarmane, R., Poulik, J., and Misra, V. K. (2013). Infantile hypertrophic cardiomyopathy associated with a novel MYL3 mutation. *Cardiology*, 124, 248–251.

Kaski, J. P., Syrris, P., Esteban, M. T., Jenkins, S., Pantazis, A., Deanfield, J. E., McKenna, W. J., and Elliott, P. M. (2009). Prevalence of sarcomere protein gene mutations in preadolescent children with hypertrophic cardiomyopathy. *Circ. Cardiovasc. Genet.*, 2, 436–441.

Kazmierczak, K., Paulino, E. C., Huang, W., Muthu, P., Liang, J., Yuan, C. C., Rojas, A. I., Hare, J. M., and Szczesna-Cordary, D. (2013). Discrete effects of A57G-myosin essential light chain mutation associated with familial hypertrophic cardiomyopathy. *Am. J. Physiol. Heart Circ. Physiol.*, 305, H575–H589.

Kelley, C. A., Takahashi, M., Yu, J. H., and Adelstein, R. S. (1993). An insert of seven amino acids confers functional differences between smooth muscle myosins from the intestines and vasculature. *J. Biol. Chem.*, 268, 12848–12854.

Kohler, J., Winkler, G., Schulte, I., Scholz, T., McKenna, W., Brenner, B., and Kraft, T. (2002). Mutation of the myosin converter domain alters cross-bridge elasticity. *Proc. Natl. Acad. Sci. U. S. A.*, 99, 3557–3562.

Lash, J. A., Helper, D. J., Klug, M., Nicolozakes, A. W., and Hathaway, D. R. (1990). Nucleotide and deduced amino acid sequence of cDNAs encoding two isoforms for the 17,000 dalton myosin light chain in bovine aortic smooth muscle. *Nucleic Acids Res.*, 18, 7176.

Lee, W., Hwang, T. H., Kimura, A., Park, S. W., Satoh, M., Nishi, H., Harada, H., Toyama, J., and Park, J. E. (2001). Different expressivity of a ventricular essential myosin light chain gene Ala57Gly mutation in familial hypertrophic cardiomyopathy. *Am. Heart J.*, 141, 184–189.

Lenz, S., Lohse, P., Seidel, U., and Arnold, H. H. (1989). The alkali light chains of human smooth and nonmuscle myosins are encoded by a single gene. Tissue-specific expression by alternative splicing pathways. *J. Biol. Chem.*, 264, 9009–9015.

Lewalle, A., Steffen, W., Stevenson, O., Ouyang, Z., and Sleep, J. (2008). Single-molecule measurement of the stiffness of the rigor myosin head. *Biophys. J.*, 94, 2160–2169.

Linari, M., Caremani, M., Piperio, C., Brandt, P., and Lombardi, V. (2007). Stiffness and fraction of Myosin motors responsible for active force in permeabilized muscle fibers from rabbit psoas. *Biophys. J.*, 92, 2476–2490.

Lossie, J., Kohncke, C., Mahmoodzadeh, S., Steffen, W., Canepari, M., Maffei, M., Taube, M., Larcheveque, O., Baumert, P., Haase, H., et al. (2014). Molecular mechanism regulating myosin and cardiac functions by ELC. *Biochem. Biophys. Res. Commun.*, 450, 464–469.

Lossie, J., Ushakov, D. S., Ferenczi, M. A., Werner, S., Keller, S., Haase, H., and Morano, I. (2012). Mutations of ventricular essential myosin light chain disturb myosin binding and sarcomeric sorting. *Cardiovasc. Res.*, 93, 390–396.

Lowey, S., and Risby, D. (1971). Light chains from fast and slow muscle myosins. *Nature*, 234, 81–85.

Lowey, S., Saraswat, L. D., Liu, H., Volkmann, N., and Hanein, D. (2007). Evidence for an interaction between the SH3 domain and the N-terminal extension of the essential light chain in class II myosins. *J. Mol. Biol.*, 371, 902–913.

Lowey, S., Waller, G. S., and Trybus, K. M. (1993a). Function of skeletal muscle myosin heavy and light chain isoforms by an in vitro motility assay. *J. Biol. Chem.*, 268, 20414–20418.

Lowey, S., Waller, G. S., and Trybus, K. M. (1993b). Skeletal muscle myosin light chains are essential for physiological speeds of shortening. *Nature*, 365, 454–456.

Malmqvist, U., and Arner, A. (1991). Correlation between isoform composition of the 17 kDa myosin light chain and maximal shortening velocity in smooth muscle. *Pflugers Archiv. Eur. J. Physiol.*, 418, 523–530.

Maron, B. J., and Maron, M. S. (2012). Hypertrophic cardiomyopathy. *Lancet*, 381, 242–255.

Marsh, D. J., and Lowey, S. (1980). Fluorescence energy transfer in myosin subfragment-1. *Biochemistry*, 19, 774–784.

Mathern, B. E., and Burke, M. (1986). Stability and substructure of cardiac myosin subfragment 1 and isolation and properties of its heavy-chain subunit. *Biochemistry*, 25, 884–889.

Matthew, J. D., Khromov, A. S., Trybus, K. M., Somlyo, A. P., and Somlyo, A. V. (1998). Myosin essential light chain isoforms modulate the velocity of shortening propelled by nonphosphorylated cross-bridges. *J. Biol. Chem.*, 273, 31289–31296.

Meder, B., Laufer, C., Hassel, D., Just, S., Marquart, S., Vogel, B., Hess, A., Fishman, M. C., Katus, H. A., and Rottbauer, W. (2009). A single serine in the carboxyl terminus of cardiac essential myosin light chain-1 controls cardiomyocyte contractility in vivo. *Circ. Res.*, 104, 650–659.

Milligan, R. A. (1996). Protein-protein interactions in the rigor actomyosin complex. *Proc. Natl. Acad. Sci. U. S. A.*, 93, 21–26.

Moncrief, N. D., Kretsinger, R. H., and Goodman, M. (1990). Evolution of EF-hand calcium-modulated proteins. I. Relationships based on amino acid sequences. *J. Mol. Evol.*, 30, 522–562.

Morano, I., Hadicke, K., Haase, H., Bohm, M., Erdmann, E., and Schaub, M. C. (1997). Changes in essential myosin light chain isoform expression provide a molecular basis for isometric force regulation in the failing human heart. *J. Mol. Cell Cardiol.*, 29, 1177–1187.

Morano, I., Ritter, O., Bonz, A., Timek, T., Vahl, C. F., and Michel, G. (1995). Myosin light chain-actin interaction regulates cardiac contractility. *Circ. Res.*, 76, 720–725.

Morano, I., Rosch, J., Arner, A., and Ruegg, J. C. (1990). Phosphorylation and thiophosphorylation by myosin light chain kinase: Different effects on mechanical properties of chemically skinned ventricular fibers from the pig. *J. Mol. Cell Cardiol.*, 22, 805–813.

Morano, M., Zacharzowski, U., Maier, M., Lange, P. E., Alexi-Meskishvili, V., Haase, H., and Morano, I. (1996). Regulation of human heart contractility by essential myosin light chain isoforms. *J. Clin. Invest.*, 98, 467–473.

Muthu, P., Wang, L., Yuan, C. C., Kazmierczak, K., Huang, W., Hernandez, O. M., Kawai, M., Irving, T. C., and Szczesna-Cordary, D. (2011). Structural and functional aspects of the myosin essential light chain in cardiac muscle contraction. *Faseb J.*, 25, 4394–4405.

Nabeshima, Y., Fujii-Kuriyama, Y., Muramatsu, M., and Ogata, K. (1984). Alternative transcription and two modes of splicing results in two myosin light chains from one gene. *Nature*, 308, 333–338.

Nabeshima, Y., Nonomura, Y., and Fujii-Kuriyama, Y. (1987). Nonmuscle and smooth muscle myosin light chain mRNAs are generated from a single gene by the tissue-specific alternative RNA splicing. *J. Biol. Chem.*, 262, 10608–10612.

Ni, S., Hong, F., Haldeman, B. D., Baker, J. E., Facemyer, K. C., and Cremo, C. R. (2012). Modification of interface between regulatory and essential light chains hampers phosphorylation-dependent activation of smooth muscle myosin. *J. Biol. Chem.*, 287, 22068–22079.

Nieznanski, K., Nieznanska, H., Skowronek, K., Kasprzak, A. A., and Stepkowski, D. (2003). Ca^{2+} binding to myosin regulatory light chain affects the conformation of the N-terminus of essential light chain and its binding to actin. *Arch. Biochem. Biophys.*, 417, 153–158.

Olson, T. M., Karst, M. L., Whitby, F. G., and Driscoll, D. J. (2002). Myosin light chain mutation causes autosomal recessive cardiomyopathy with mid-cavitary hypertrophy and restrictive physiology. *Circulation*, 105, 2337–2340.

Patchell, V. B., Gallon, C. E., Hodgkin, M. A., Fattoum, A., Perry, S. V., and Levine, B. A. (2002). The inhibitory region of troponin-I alters the ability of F-actin to interact with different segments of myosin. *Eur. J. Biochem.*, 269, 5088–5100.

Periasamy, M., Strehler, E. E., Garfinkel, L. I., Gubits, R. M., Ruiz-Opazo, N., and Nadal-Ginard, B. (1984). Fast skeletal muscle myosin light chains 1 and 3 are produced from a single gene by a combined process of differential RNA transcription and splicing. *J. Biol. Chem.*, 259, 13595–13604.

Petzhold, D., Lossie, J., Keller, S., Werner, S., Haase, H., and Morano, I. (2011). Human essential myosin light chain isoforms revealed distinct myosin binding, sarcomeric sorting, and inotropic activity. *Cardiovasc. Res.*, 90, 513–520.

Petzhold, D., Simsek, B., Meissner, R., Mahmoodzadeh, S., and Morano, I. (2014). Distinct interactions between actin and essential myosin light chain isoforms. *Biochem. Biophys. Res. Commun.*, 449, 284–288.

Piazzesi, G., Reconditi, M., Linari, M., Lucii, L., Bianco, P., Brunello, E., Decostre, V., Stewart, A., Gore, D. B., Irving, T. C., et al. (2007). Skeletal muscle performance determined by modulation of number of myosin motors rather than motor force or stroke size. *Cell*, 131, 784–795.

Poetter, K., Jiang, H., Hassanzadeh, S., Master, S. R., Chang, A., Dalakas, M. C., Rayment, I., Sellers, J. R., Fananapazir, L., and Epstein, N. D. (1996). Mutations in either the essential or regulatory light chains of myosin are associated with a rare myopathy in human heart and skeletal muscle. *Nat. Genet.*, 13, 63–69.

Price, K. M., Littler, W. A., and Cummins, P. (1980). Human atrial and ventricular myosin light-chains subunits in the adult and during development. *Biochem. J.*, 191, 571–580.

Rarick, H. M., Opgenorth, T. J., von Geldern, T. W., Wu-Wong, J. R., and Solaro, R. J. (1996). An essential myosin light chain peptide induces supramaximal stimulation of cardiac myofibrillar ATPase activity. *J. Biol. Chem.*, 271, 27039–27043.

Rayment, I., Holden, H. M., Whittaker, M., Yohn, C. B., Lorenz, M., Holmes, K. C., and Milligan, R. A. (1993a). Structure of the actin-myosin complex and its implications for muscle contraction. *Science*, 261, 58–65.

Rayment, I., Rypniewski, W. R., Schmidt-Base, K., Smith, R., Tomchick, D. R., Benning, M. M., Winkelmann, D. A., Wesenberg, G., and Holden, H. M. (1993b). Three-dimensional structure of myosin subfragment-1: A molecular motor. *Science*, 261, 50–58.

Richard, P., Charron, P., Carrier, L., Ledeuil, C., Cheav, T., Pichereau, C., Benaiche, A., Isnard, R., Dubourg, O., Burban, M., et al. (2003). Hypertrophic cardiomyopathy: Distribution of disease genes, spectrum of mutations, and implications for a molecular diagnosis strategy. *Circulation*, 107, 2227–2232.

Ritter, O., Haase, H., Schulte, H. D., Lange, P. E., and Morano, I. (1999). Remodeling of the hypertrophied human myocardium by cardiac bHLH transcription factors. *J. Cell. Biochem.*, 74, 551–561.

Sanbe, A., Nelson, D., Gulick, J., Setser, E., Osinska, H., Wang, X., Hewett, T. E., Klevitsky, R., Hayes, E., Warshaw, D. M., et al. (2000). In vivo analysis of an essential myosin light chain mutation linked to familial hypertrophic cardiomyopathy. *Circ. Res.*, 87, 296–302.

Schaub, M. C., Tuchschmid, C. R., Srihari, T., and Hirzel, H. O. (1984). Myosin isoenzymes in human hypertrophic hearts. Shift in atrial myosin heavy chains and in ventricular myosin light chains. *Eur. heart J.*, 5 Suppl F, 85–93.

Seebohm, B., Matinmehr, F., Kohler, J., Francino, A., Navarro-Lopez, F., Perrot, A., Ozcelik, C., McKenna, W. J., Brenner, B., and Kraft, T. (2009). Cardiomyopathy mutations reveal variable region of myosin converter as major element of cross-bridge compliance. *Biophys. J.*, 97, 806–824.

Seharaseyon, J., Bober, E., Hsieh, C. L., Fodor, W. L., Francke, U., Arnold, H. H., and Vanin, E. F. (1990). Human embryonic/atrial myosin alkali light chain gene: Characterization, sequence, and chromosomal location. *Genomics*, 7, 289–293.

Sherwood, J. J., and Eddinger, T. J. (2002). Shortening velocity and myosin heavy- and light-chain isoform mRNA in rabbit arterial smooth muscle cells. *Am. J. Physiol. Cell Physiol.*, 282, C1093–C1102.

Shi, Q. W., Danilczyk, U., Wang, J. X., See, Y. P., Williams, W. G., Trusler, G. A., Beaulieu, R., Rose, V., and Jackowski, G. (1991). Expression of ventricular myosin subunits in the atria of children with congenital heart malformations. *Circ. Res.*, 69, 1601–1607.

Siemankowski, R. F., Wiseman, M. O., and White, H. D. (1985). ADP dissociation from actomyosin subfragment 1 is sufficiently slow to limit the unloaded shortening velocity in vertebrate muscle. *Proc. Natl. Acad. Sci. U. S. A.*, 82, 658–662.

Sivaramakrishnan, M., and Burke, M. (1982). The free heavy chain of vertebrate skeletal myosin subfragment 1 shows full enzymatic activity. *J. Biol. Chem.*, 257, 1102–1105.

Sjuve, R., Haase, H., Morano, I., Uvelius, B., and Arner, A. (1996). Contraction kinetics and myosin isoform composition in smooth muscle from hypertrophied rat urinary bladder. *J. Cell. Biochem.*, 63, 86–93.

Stepkowski, D. (1995). The role of the skeletal muscle myosin light chains N-terminal fragments. *FEBS Lett.*, 374, 6–11.

Stoker, M. A., Forbes, J. A., Hanif, R., Cooper, C., Nian, H., Konrad, P. E., and Neimat, J. S. (2012). Decreased rate of CSF leakage associated with complete reconstruction of suboccipital cranial defects. *J. Neurol. Surg. Part B Skull Base*, 73, 281–286.

Stracher, A. (1969). Evidence for the involvement of light chains in the biological functioning of myosin. *Biochem. Biophys. Res. Commun.*, 35, 519–525.

Sutoh, K. (1982). Identification of myosin-binding sites on the actin sequence. *Biochemistry*, 21, 3654–3661.

Sutsch, G., Brunner, U. T., von Schulthess, C., Hirzel, H. O., Hess, O. M., Turina, M., Krayenbuehl, H. P., and Schaub, M. C. (1992). Hemodynamic performance and myosin light chain-1 expression of the hypertrophied left ventricle in aortic valve disease before and after valve replacement. *Circ. Res.*, 70, 1035–1043.

Sweeney, H. L., Kushmerick, M. J., Mabuchi, K., Sreter, F. A., and Gergely, J. (1988). Myosin alkali light chain and heavy chain variations correlate with altered shortening velocity of isolated skeletal muscle fibers. *J. Biol. Chem.*, 263, 9034–9039.

Timson, D. J., Trayer, H. R., Smith, K. J., and Trayer, I. P. (1999). Size and charge requirements for kinetic modulation and actin binding by alkali 1-type myosin essential light chains. *J. Biol. Chem.*, 274, 18271–18277.

Trayer, H. R., and Trayer, I. P. (1985). Differential binding of rabbit fast muscle myosin light chain isoenzymes to regulated actin. *FEBS Lett.*, 180, 170–173.

Trayer, I. P., Trayer, H. R., and Levine, B. A. (1987). Evidence that the N-terminal region of A1-light chain of myosin interacts directly with the C-terminal region of actin. A proton magnetic resonance study. *Eur. J. Biochem.*, 164, 259–266.

Tunnemann, G., Karczewski, P., Haase, H., Cardoso, M. C., and Morano, I. (2007). Modulation of muscle contraction by a cell-permeable peptide. *J. Mol. Med. (Berl.)*, 85, 1405–1412.

Uyeda, T. Q., Abramson, P. D., and Spudich, J. A. (1996). The neck region of the myosin motor domain acts as a lever arm to generate movement. *Proc. Natl. Acad. Sci. U. S. A.*, 93, 4459–4464.

Uyeda, T. Q., Kron, S. J., and Spudich, J. A. (1990). Myosin step size. Estimation from slow sliding movement of actin over low densities of heavy meromyosin. *J. Mol. Biol.*, 214, 699–710.

VanBuren, P., Waller, G. S., Harris, D. E., Trybus, K. M., Warshaw, D. M., and Lowey, S. (1994). The essential light chain is required for full force production by skeletal muscle myosin. *Proc. Natl. Acad. Sci. U. S. A.*, 91, 12403–12407.

Veigel, C., Bartoo, M. L., White, D. C., Sparrow, J. C., and Molloy, J. E. (1998). The stiffness of rabbit skeletal actomyosin cross-bridges determined with an optical tweezers transducer. *Biophys. J.*, 75, 1424–1438.

Wagner, P. D., and Giniger, E. (1981). Hydrolysis of ATP and reversible binding to F-actin by myosin heavy chains free of all light chains. *Nature*, 292, 560–562.

Wagner, P. D., Slater, C. S., Pope, B., and Weeds, A. G. (1979). Studies on the actin activation of myosin subfragment-1 isoezymes and the role of myosin light chains. *Eur. J. Biochem.*, 99, 385–394.

Wang, L., Muthu, P., Szczesna-Cordary, D., and Kawai, M. (2013). Characterizations of myosin essential light chain's N-terminal truncation mutant Delta43 in transgenic mouse papillary muscles by using tension transients in response to sinusoidal length alterations. *J. Muscle Res. Cell Motil.*, 34, 93–105.

Weeds, A. G., and Lowey, S. (1971). Substructure of the myosin molecule. II. The light chains of myosin. *J. Mol. Biol.*, 61, 701–725.

Weeds, A. G., and Taylor, R. S. (1975). Separation of subfragment-1 isoenzymes from rabbit skeletal muscle myosin. *Nature*, 257, 54–56.

Woischwill, C., Karczewski, P., Bartsch, H., Luther, H. P., Kott, M., Haase, H., and Morano, I. (2005). Regulation of the human atrial myosin light chain 1 promoter by Ca^{2+}-calmodulin-dependent signaling pathways. *FASEB J.*, 19, 503–511.

Xie, X., Harrison, D. H., Schlichting, I., Sweet, R. M., Kalabokis, V. N., Szent-Gyorgyi, A. G., and Cohen, C. (1994). Structure of the regulatory domain of scallop myosin at 2.8 A resolution. *Nature*, 368, 306–312.

Yamashita, H., Sugiura, S., Fujita, H., Yasuda, S., Nagai, R., Saeki, Y., Sunagawa, K., and Sugi, H. (2003). Myosin light chain isoforms modify force-generating ability of cardiac myosin by changing the kinetics of actin-myosin interaction. *Cardiovasc. Res.*, 60, 580–588.

Chapter 12

Regulation of Calcium Uptake into the Sarcoplasmic Reticulum in the Heart

Susumu Minamisawa

Department of Cell Physiology, The Jikei University School of Medicine,
3-25-8 Nishi-Shimbashi, Minatoku, 105-8461 Tokyo, Japan

sminamis@jikei.ac.jp

Muscles require Ca^{2+} to contract. In the heart, periodic changes in Ca^{2+} concentration in cardiomyocytes are essential for contraction and relaxation. The sarcoplasmic reticulum (SR), an extensive intracellular membrane system integrally regulates the intracellular Ca^{2+} concentration by releasing and uptaking Ca^{2+} through two fundamental proteins: cardiac ryanodine receptor (RyR2) and SR calcium ATPase 2a (SERCA2a), respectively. The Ca^{2+} concentration is also fine-tuned by their intrinsic regulatory domains and associated SR proteins. A growing body of evidence, including studies using genetically engineered mouse models, has shown that Ca^{2+} cycling and Ca^{2+}-dependent signaling pathways play a critical role in maintaining cardiac function. Dysregulation of Ca^{2+} cycling causes cardiac dysfunction such as heart failure and cardiomyopathy. Improvement of the SR function may ameliorate the effects on cardiac pump function, which is a potential therapeutic target for

Muscle Contraction and Cell Motility: Fundamentals and Developments
Edited by Haruo Sugi
Copyright © 2017 Pan Stanford Publishing Pte. Ltd.
ISBN 978-981-4745-16-1 (Hardcover), 978-981-4745-17-8 (eBook)
www.panstanford.com

heart failure. This chapter focuses on the advances in knowledge concerning Ca^{2+} re-uptake into the SR.

12.1 Introduction

Calcium cycling in the heart is integrally regulated by cardiac sarcoplasmic reticulum (SR) in which Ca^{2+} is released through a calcium channel (the ryanodine receptor type 2: RyR2) and is restored from cytosol through SR calcium ATPase 2a (SERCA2a). A major protein controlling Ca^{2+} re-uptake is SERCA2a that is located in the longitudinal region of the SR membrane. The activity of SERCA2a protein is mainly regulated by its inhibitory proteins: phospholamban (PLN), a 52-amino acid phosphoprotein and sarcolipin (SLN), a 31-amino acid transmembrane protein. The regulatory mechanism of PLN on SERCA2a has been extensively investigated. The physiological relevance of SLN in the heart, however, remains under investigation. Phosphorylation of PLN by protein kinase A (PKA) and Ca^{2+}/calmodulin-dependent protein kinase II (CAMKII) releases its inhibitory effect on SERCA2a through direct molecular interaction and augments the systolic and diastolic parameters. A growing body of evidence has indicated that SERCA2a pump activity is a major determinant of cardiac muscle contractility and relaxation (Frank et al., 2003; Periasamy and Janssen, 2008; Kranias and Hajjar, 2012). In addition to pump function, impaired Ca^{2+} cycling may be involved in tissue viability, because Ca^{2+} is also an integral signaling molecule for numerous other cellular processes including survival and cell death. Recent studies have demonstrated that enhancement of the SR function by disrupting the interaction between SERCA2a and PLN or by simply increasing the expression of SERCA2a improves cardiac function and structure in animal models of heart failure. In addition to PLN and SLN, other regulators of SR Ca^{2+}-transport were identified. One such regulator is SUMO (small ubiquitin-like modifier)-1 that causes SUMOylation on SERCA2a protein, which improves protein stability and activity. Another regulator is sarcalumenin, which is located in the SR lumen and interacts with SERCA2a to stabilize SERCA2a protein. In this chapter, I will review the advances in knowledge concerning Ca^{2+} re-uptake into the SR (Fig. 12.1).

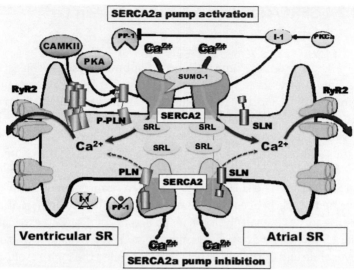

Figure 12.1 A fine-tuning regulatory system of SERCA2a Ca^{2+} uptake
function. Ca^{2+} is released through RyR2 and is restored from
cytosol through SERCA2a. The upper- or lower-half part of
the figure illustrates an increase or a decrease in SERCA2
pump activity, respectively. The left- or right-half part
of the figure shows the SERCA2 pump regulation in
the ventricle or in the atrium, respectively. The activity
of SERCA2a protein is mainly regulated by PLN in the
ventricle and SLN in the atrium. When the monomer form of
PLN and SLN physically interacts with SERCA2a, SERCA2a
pump activity is strongly inhibited. PLN also forms a pentamer,
the inhibitory effect of which on SERCA2a is weaker than that
of the monomeric form. Phosphorylation of PLN by PKA and
CAMKII releases its inhibitory effect on SERCA2a, whereas
SLN lacks such a phosphorylation site. In addition to PLN
and SLN, SUMO (small ubiquitin-like modifier)-1 causes
SUMOylation that improves the protein stability and activity
of SERCA2a. SAR is located in the SR lumen and interacts
with SERCA2a to stabilize SERCA2a protein. When PP-1
dephosphorylates PLN, PLN-interacts with SERCA2a to
inhibit its function. I-1 activated by PKA or PKCα inhibits PP-1
activity and then increases the phosphorylated PLN. **SR:**
sarcoplasmic reticulum; **SERCA2a:** SR calcium ATPase 2a;
RyR2: the ryanodine receptor type 2; **PLN:** phospholamban;
P-PLN: phosphorylated phospholamban; **SLN**; sarcolipin;
SRL: sarcalumenin; **PKA**: protein kinase A; **CAMKII**: $Ca^{2+}/$
calmodulin-dependent protein kinase II; **SUMO-1**: small
ubiquitin-like modifier-1; **PP-1**: protein phosphatase-1; **I-1**:
phosphatase inhibitor-1; **PKCα**: protein kinase Cα.

12.2   SERCA2a Plays a Central Role in Ca^{2+} Uptake

The SR, an extensive intracellular membrane system consisting of a lipid bilayer that surrounds each myofibril, is a fundamental organelle that coordinates the movement of cytosolic Ca^{2+} during each cycle of cardiac contraction and relaxation.

SERCA2a, a cardiac and slow-twitch skeletal muscle isoform of the SERCA family that is a P-type ATPases, is the primary regulator of the rate of Ca^{2+} re-uptake during relaxation in the heart. SERCA2a is a 110 kD integral membrane protein with ten transmembrane helices that pumps Ca^{2+} into the SR lumen at the expense of ATP hydrolysis. The structure of SERCA, especially SERCA1 has been extensively investigated by Toyoshima's group (Toyoshima and Cornelius, 2013). They unveiled the high-resolution structure of SERCA1a, a fast-twitch skeletal muscle isoform of the SERCA family, providing in-depth information to understand the fundamental SERCA function for Ca^{2+} transport (Toyoshima and Mizutani, 2004). Since SERCA1a has high homology with SERCA 2a and is known to interact similarly with PLN and SLN, the high-resolution structure of SERCA1a should also provide new insights into the SERCA 2a pump function.

Numerous studies have demonstrated that the expression levels and enzymatic activity of SERCA2a are significantly decreased in human and experimental heart failure (Gwathmey et al., 1987; Lehnart et al., 2009). This can be simply confirmed by gene-targeting experiments in which heterozygous SERCA2a gene knockout mice have depressed cardiac function and myocyte contractility (Ji et al., 2000; Periasamy et al., 1999). Although the reduction in the expression levels of SERCA2a protein is critical to maintaining normal cardiac function, compensatory mechanisms could come into play. An inducible model with cardiac-specific deletion of SERCA2a was generated in order to investigate the mechanisms of SERCA2a deficiency (Andersson et al., 2009). Cardiac function was only moderately impaired 4 weeks after SERCA2a deletion in adult mice (Andersson et al., 2009). Heinis et al. also demonstrated that cardiac function was practically normal with SERCA2a protein levels at 32% of control hearts one week after initiating the down-regulation of SERCA2a (Heinis et al., 2013). Therefore, compensatory mechanisms help the heart with the down-regulation of SERCA2a to maintain its function for a limited time before going into a deteriorated state. The increases in the expression and activity of the Na^+/Ca^{2+} exchanger and the plasma membrane Ca^{2+}-ATPase and the decrease in phospholamban protein may collectively compensate for SERCA2a-deficient cardiac function for a limited time.

The endogenous inhibitor PLN tightly regulates the activity of SERCA2a, which will be discussed later. Although SERCA2a is known to be phosphorylated at serine-38 by Ca^2/calmodulin-dependent protein kinase, the biophysiological significance of phosphorylated SERCA2a for Ca^{2+} transport is controversial (Rodriguez et al., 2004; Frank et al., 2003).

Instead of phosphorylation, it is now evident that SERCA2 activity is regulated by post-translational modification such as SUMOylation. SUMOylation is a ubiquitin-like reversible post-translational modifications where SUMO (small ubiquitin-like modifier) covalently attaches at two lysine residues of SERCA2a. SERCA2a activity is specifically modified by SUMO-1 that binds SERCA2a to improve protein stability and activity (Kho et al., 2011).

12.3 Phospholamban: A Critical Regulator of SERCA2a

PLN, a 52-amino acid SR transmembrane phosphoprotein, principally controls the activity of SERCA2a in cardiac myocytes. The structure and function of PLN are extensively reviewed elsewhere (Shaikh et al., 2016; Haghighi et al., 2014; Traaseth et al., 2008). SERCA2a and PLN interact at their cytoplasmic and transmembrane domains. PLN inhibits SERCA2a activity through direct interaction in its unphosphorylated form, whereas the phosphorylated form of PLN dissociates from SERCA2a. Phosphorylation of PLN relieves its inhibition on SERCA2a activity following phosphorylation at serine-16 and threonine-17 by PKA and CAMKII, respectively (Kranias and Hajjar, 2012; Frank et al., 2003). Under physiological conditions, phosphorylation at serine-16 by PKA contributes to an increase in the rate of Ca^{2+} uptake into the SR. PLN is mostly responsible for the lucitropic effect on the heart by β-adrenergic stimulation, because PLN-deficient mice display maximal cardiac contractility and relaxation in the absence of any catecholamine stimulation (Luo et al., 1996).

In addition to phosphorylation, PLN activity is tightly regulated in a yin-yang mode: myocardial SR possesses phosphatase activity capable of dephosphorylating PLN by phosphatase type 1 and type 2A (Nicolaou and Kranias, 2009; Ikeda et al., 2008). The activity of phosphatase type 1 (PP-1) is decreased by phosphatase inhibitor-1 (I-1). I-1 can be activated through phosphorylation by PKA (Gupta et al., 1996; Huang et al., 1999; El-Armouche et al., 2003) or PKCα (Braz et al., 2004). Successful PP-1 isoform knockdown was achieved for each isoform without affecting the expression of the other isoforms. PP-1β knockdown most significantly enhanced the Ca^{2+} transient and cell shortening by augmenting PLN phosphorylation at baseline and with low-dose isoproterenol stimulation (10 nM). Interestingly, PP-1β was preferentially associated with SERCA and PLN in GFP-PP-1-transfected cardiomyocytes, as well as in canine longitudinal SR preparations (Aoyama et al., 2011).

12.4 PLN Mutations Related to Human Cardiomyopathy

Cardiomyopathy is defined as a disease of the myocardium associated with structural and/or functional disorder. Gene mutations, including the PLN gene, have been intensively identified in cardiomyopathy (Haghighi et al., 2014; Fatkin et al., 2014). Most of the mutations in the PLN gene cause dilated cardiomyopathy, a distinct form of myocardium disorder characterized by heart chamber dilation with severe contractile dysfunction and a frequent association with heart failure. Among PLN mutations, R9C and R14del mutations in the PLN gene have the highest incidence and are associated with an increase in the inhibitory effect on the Ca^{2+} affinity of SERCA2 (Schmitt et al., 2003; van Rijsingen et al., 2014; Fish et al., 2016; Ablorh and Thomas, 2015). These findings were also observed in mouse models harboring the R9C or R14del-PLN mutation (Schmitt et al., 2003; Haghighi et al., 2012). Heterozygous R14del carriers develop left ventricular dilation, episodic ventricular arrhythmias, and death by middle age (Haghighi et al., 2006). R14del-PLN transgenic mice exhibited super-inhibition of SERCA2a, which resulted in early death (Haghighi et al., 2006). Interestingly, this PLN mutation also causes arrhythmiogenic right ventricular cardiomyopathy in humans (van Rijsingen et al., 2014). The third human PLN mutation (Leu 39 stop) was also identified to cause dilated cardiomyopathy and premature death in the homozygous state, although heterozygous carriers showed asymptomatic cardiac hypertrophy without alterations in cardiac function (Haghighi et al., 2003). In addition, a recent study has identified a novel R25C mutation that is associated with super-inhibition of Ca^{2+} cycling and ventricular arrhythmia (Liu et al., 2015). Moreover, mutations in the promoter region of the PLN gene have also been identified in patients with cardiomyopathy by several researchers, including us (Minamisawa et al., 2003; Haghighi et al., 2008; Medin et al., 2007).

It should be noted that SERCA2 mutations have been found in patients with Darier disease, which is characterized by keratotic papules of seborrheic areas of the skin (keratosis follicularis) (Dhitavat et al., 2004; Savignac et al., 2011), but not reported in patients with cardiomyopathy. Intriguingly, patients with Darier disease do not usually display any cardiac dysfunction (Tavadia et al., 2001). So far there is no clear explanation why SERCA2 mutations do not cause cardiomyopathy even though several mutations in Darier disease have been found in the PLN-interacting region.

12.5 Enhancement of SR Function Is a Novel Therapeutic Target for Heart Failure

Impaired Ca^{2+} cycling and a decrease in Ca^{2+} uptake are central physiological hallmarks of a number of animal models of heart failure, as well as in human failing hearts (Periasamy and Janssen, 2008; Gwathmey et al., 1987). Our prime goal is to increase or restore SERCA2a activity in failing hearts. Genetic approaches and pharmacological interventions, designed to increase SERCA2a activity, may prove valuable in preventing or reversing impaired Ca^{2+} cycling in heart failure. There are several potential options to achieve this goal: (1) by increasing SERCA2a protein in the heart, (2) by modulating SERCA2a itself to increase the Ca^{2+} transport, (3) by decreasing PLN protein in the heart, and (4) by increasing the phosphorylated PLN.

12.5.1 Strategies to Increase SERCA2a Protein in Heart Failure

An initial approach to increasing SERCA2a activity could be simply increasing the expression level of SERCA2a, because impaired Ca^{2+} cycling is associated with a decrease in SERCA2a expression (Periasamy and Janssen, 2008; Gwathmey et al., 1987). Encouraging experimental data have come from Dr. Hajjar's group, demonstrating that muscle contractility and relaxation were restored by adenovirus-mediated SERCA2a gene transfer in an animal model of heart failure (Miyamoto et al., 2000; del Monte et al., 2001) as well as isolated failing human cardiac myocytes (del Monte et al., 1999). In addition, using adeno-associated virus (AAV) vectors, long-term overexpression of SERCA2a preserved cardiac function and improved ventricular remodeling in a large-animal, volume-overload model of heart failure (Kawase et al., 2008). Furthermore, the first clinical trial of SERCA2a gene therapy in humans, the CUPID trial (Calcium Upregulation by Percutaneous administration of gene therapy In cardiac Disease) showed a promising result, demonstrating improvement in clinical data with no increases in adverse effects (Jessup et al., 2011). Moreover, after a single SERCA2a gene delivery by an AAV vector, beneficial effects persisted in patients with advanced heart failure for a long period (>3 years) (Zsebo et al., 2014). Thus, AAV-mediated SERCA2a gene transfer seems to be a promising strategy for treating heart failure. However, SERCA2a lacks the tissue specificity and SERCA2b, an alternative spliced isoform of SERCA2a, is expressed in all types of non-muscle cells. Therefore, it remains unclear whether unexpected biological outcomes may result from the broad-spread gene delivery of SERCA2a. Further studies are needed to establish its efficiency and safety.

To achieve an increase in SERCA2a protein, a better understanding of its transcriptional regulation is required. Thyroid hormone is known to increase SERCA2 expression at the transcriptional level (Moriscot et al., 1997; Iordanidou et al., 2010). However, in addition to the chronotropic effect, lack of tissue specificity in thyroid hormone may be problematic for its use as a therapeutic agent for heart failure. Other hormones such as norepinephrine, angiotensin II, endothelin-1, parathyroid hormone, and prostaglandin-F2α, as well as cytokines tumor necrosis factor-α and interleukin-6 also downregulate SERCA2 expression (Zarain-Herzberg et al., 2012). In term of transcription factors, Brady et al. demonstrated that Sp1 and Sp3 transcription factors activated human SERCA2 gene expression in cardiomyocytes (Brady et al., 2003). Zarain-Herzberg et al. suggested that overexpression of Sp1 factor in cardiac hypertrophy downregulated SERCA2 gene expression (Zarain-Herzberg et al., 2012). Arai et al. demonstrated that Egr-1 is a transcriptional inhibitor of the SERCA2 gene through p44/42 MAPK in doxorubicin-induced cardiomyopathy and that Egr-1 antisense oligonucleotides blocked the doxorubicin-induced reduction in SERCA2 mRNA (Arai et al., 2000). Brady et al. demonstrated that Egr-1 did not significantly alter the promoter activity of the human SERCA2 gene (Brady et al., 2003). Recently, the calcineurin-NFAT (nuclear factor of activated T cells) pathway has been suggested to regulate SERCA2 transcription (Prasad and Inesi, 2012; Zarain-Herzberg et al., 2012). Thus, a key transcriptional factor for SERCA2 gene remains under investigation.

12.5.2 Strategies to Modulate SERCA2a to Increase Ca^{2+} Transport

Instead of gene transfer, pharmacological interventions designed to modulate SERCA2a itself to increase SERCA2a activity and Ca^{2+} uptake should be considered. As described previously, SERCA2 activity is regulated by SUMOylation. Previous studies have demonstrated that the expression levels of SUMO-1 and SUMOylation of SERCA2a are decreased in both experimental animal and human heart failure (Kho et al., 2011). Gene transfer of SUMO-1 showed ameliorated effects on cardiac function and survival rate in animal models of heart failure (Kho et al., 2011; Tilemann et al., 2013). Furthermore, Hajjar's group has recently identified a small molecule, N106 (N-(4-methoxybenzo[d]thiazol-2-yl)-5-(4-methoxyphenyl)-1,3,4-oxadiazol-2-amine, that enhanced SERCA2a SUMOylation and increased contractility both in in vitro and in vivo experiments. In addition, they characterized the putative site on E1 ligase where N106 docks and confers its activity (Kho et al., 2015).

Although several compounds have been known to directly increase SERCA activity and Ca^{2+} uptake in in vitro experiments, no commercially available medicine targeting SERCA activity has been developed so far. Nelson et al. recently found that the muscle-specific endogenous peptide DWORF, which was encoded by a transcript annotated as long noncoding RNA, localized to the SR membrane and enhanced SERCA activity by physical interaction (Nelson et al., 2016).

12.5.3 Strategies to Decrease PLN Protein in Heart Failure

There are several rationales to consider PLN as a prime target to increase SERCA activity: (1) PLN is the strongest endogenous inhibitor of SERCA2a. (2) PLN is a terminal effector of β-adrenergic signaling pathways. The inhibitory effect of PLN is almost abolished when PLN is phosphorylated by PKA whereas PKA has various biological effects and some of them may have adverse effects on failing hearts. Therefore, targeting PLN may restrict the adverse effects caused by PKA. (3) PLN is predominantly expressed in the heart, especially ventricles. This tissue specificity is a great advantage to design interventions. (4) PLN inhibition does not affect chronotropic responses, which is beneficial for patients with heart failure. (5) PLN is a small protein consisting of 52 amino acids. It should be easy to manipulate genetic modification on the PLN gene. (6) PLN-deficient mice have not displayed any adverse events so far and can survive a normal life span. In heart failure, ameliorated effects of PLN gene ablation last at least a year without increased mortality in mice (Minamisawa et al., 1999). (7) PLN is remarkably conserved among the species. Recent technologies enable us to design siRNAs or microRNA for RNA silencing (Grobetal et al., 2014; Fechner et al., 2007; Watanabe et al., 2004; Bish et al., 2011) or antisenses (He et al., 1999) for the PLN gene to decrease the PLN transcripts and then reduce SERCA2a activity. Although AAV serotype 6 vector-mediated cardiac gene transfer of short hairpin RNA effectively decreased the expression levels of PLN expression, long-term and high-level short hairpin RNA expression might cause oversaturation of the endogenous cellular microRNA pathways, resulting in cardiac cytotoxicity (Bish et al., 2011; Grimm et al., 2006). These strategies have not been tested for in vivo hearts or myocytes from failing hearts. Further investigation is required to validate its efficiency and safety for clinical introduction.

12.5.4 Strategies to Disrupt the Interaction between SERCA2a and PLN

To disrupt the interaction between SERCA2a and PLN, interventions could be designed to increase the phosphorylated state of PLN, or to interrupt PLN to associate with SERCA2a. Increasing PLN phosphorylation can be achieved by enhancing PKA and CAMKII activity or decreasing phosphatase activity. As described earlier, the activity of PP-1, especially PP-1β is decreased by I-1 (Gupta et al., 1996; Huang et al., 1999; El-Armouche et al., 2003; Aoyama et al., 2011). Thus, kinases, phosphatases, and their regulators are orchestrated to fine-tune the activity of SERCA2a. Indeed, modulation of this fine-tuned system could be a target for restoring the impairment of SERCA activity in failing hearts (Haghighi et al., 2014; Miyazaki et al., 2012; Weber et al., 2015).

Several nucleotide mimetics such as quercetin and its derivatives (McKenna et al., 1996; Blaskovic et al., 2013), gingerol (Antipenko et al., 1999; Namekata et al., 2013), and ellagic acid (Antipenko et al., 1999; Namekata et al., 2013) have been reported to increase SERCA2a activity through inhibiting the interaction between SERCA and PLN. These compounds interact with the nucleotide-binding site of SERCA, not directly with PLN, suggesting that conformational change of this site may affect the binding affinity of SERCA and PLN. However, to date, none have been tested for in vivo animals and the validity for clinical application remains to be elucidated.

Previous studies have identified that the PLN residues affect SERCA2a activity and Ca^{2+} uptake. Several in vitro and in vivo studies have demonstrated that certain PLN mutants such as S16E pseudo-phosphorylated PLN increased the contractility and relaxation of myocytes and cardiac muscles (Minamisawa et al., 1999; Hoshijima et al., 2002; Iwanaga et al., 2004) (Fig. 12.2). The molecular mechanism of how the PLN mutants disrupt the interaction between SERCA2a and endogenous PLN is now under investigation (Lockamy et al., 2011; Hou et al., 2008; Ablorh and Thomas, 2015). Although further investigation is required, selective disruption of the interaction between SERCA2a and PLN is a novel promising strategy for genetic and pharmacological interventions to prevent or reverse cardiac performance in heart failure.

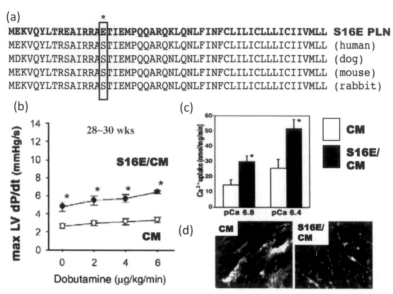

(a)

```
                       *
MEKVQYLTREAIRRAETIEMPQQARQKLQNLFINFCLILICLLLICIIVMLL  S16E PLN
MEKVQYLTRSAIRRASTIEMPQQARQKLQNLFINFCLILICLLLICIIVMLL  (human)
MDKVQYLTRSAIRRASTIEMPQQARQNLQNLFINFCLILICLLLICIIVMLL  (dog)
MEKVQYLTRSAIRRASTIEMPQQARQNLQNLFINFCLILICLLLICIIVMLL  (mouse)
MEKVQYLTRSAIRRASTIEMPQQARQNLQNLFINFCLILICLLLICIIVMLL  (rabbit)
```

(b) **(c)** **(d)**

Figure 12.2 (a) The phosphorylation site of serine (E) 16 is highly conserved among species. When serine 16 is changed to glutamate, a negative charge residue, S16E mutant PLN mimics phosphorylated PLN. Pseudophosphorylated S16E PLN gene transfer dramatically increased cardiac contractility and Ca^{2+} uptake, and restored histological abnormality in BIO14.6 cardiomyopathic (CM) Syrian hamster. (b) LV contractility (max LV dP/dt) was assessed at baseline and in response to increasing doses of dobutamine at 28 to 30 weeks after rAAV/S16EPLN delivery ($n = 4$ for both groups). *, $p < 0.05$ versus rAAV/lacZ. Error bars are the means ± standard error of the mean (SEM) (c) Ca^{2+} uptake activity at the submaximal free Ca^{2+} concentration in 12-week rAAV/S16E PLN-treated CM hamster hearts (black bar, $n = 4$) compared with rAAV/lacZ-treated hearts (white bar, $n = 3$). *, $p < 0.05$ versus rAAV/lacZ. (d) Trichrome (Masson) staining of sections from CM hamster ventricles 5 weeks after rAAV/lacZ (left) and rAAV/S16E PLN (right) gene deliveries. Fibrotic aria is significantly restricted in the ventricle after S16E PLN gene delivery. (Modified from Hoshijima et al., *Nat Med*, 2002). **PLN**: phospholamban; **CM**: cardiomyopathy; **S16E**: serine 16 to glutamate mutant PLN.

12.6 Sarcolipin, a Homologue of PLN, Is an Atrium-Specific Inhibitor of SERCA2a

Sarcolipin (SLN) is a 31-amino acid proteolipid that inhibits the activity of SERCA1 and SERCA2a (Wawrzynow et al., 1992; Odermatt et al., 1998; Toyoshima et al., 2013), which is a homolog of PLN that is abundant in ventricular muscle. SLN is expressed specifically in atrial muscle of the heart from early developmental stages around embryonic day 10.5 and throughout life (Minamisawa et al., 2006; Nakano et al., 2011) (Fig. 12.3). SLN is also expressed most abundantly in fast-twitch skeletal muscle and less abundantly in slow-twitch skeletal muscle (Minamisawa et al., 2006; Babu et al., 2007a; Bhupathy et al., 2007; Vangheluwe et al., 2005). The regulatory mechanism of the atrium-specific expression of SLN remains unclear, although transcription factors play an important role in the chamber-specific expression of a certain gene. Interestingly, SLN is ectopically expressed in the ventricle of NKX2.5 knockout mice (Pashmforoush et al., 2004; Terada et al., 2011), suggesting that NKX2.5 is a critical transcription factor

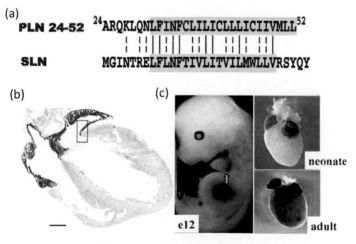

Figure 12.3 (a) SLN is an atrial homologue of PLN. (b) Endogenous SLN expression is restricted in the atrium. (c) LacZ-knockin-SLN knockout mice demonstrates that LacZ stain is observed only in the atria throughout life. A white arrow indicates the atria at embryonic day 12 of a mouse. (Modified from Minamisawa et al. *J Biol Chem*, 2006; Nakano et al. *J Mol Cell Cardiol*, 2011). **SLN:** sarcolipin; **PLN:** phospholamban.

to prevent the induction of SLN expression from the ventricle. Tanwar et al. recently reported that Gremlin 2, a secreted BMP antagonist, increased the expression of SLN in embryonic stem cells by activation of the JNK signaling pathway (Tanwar et al., 2014). The expression of SLN is also known to be up-regulated by corticosteroid (Gayan-Ramirez et al., 2000) and certain types of cardiac diseases such as mitral regurgitation (Zheng et al., 2014) and Tako-Tsubo cardiomyopathy (Nef et al., 2009) and down-regulated by thyroid hormone (Minamisawa et al., 2006) and stresses such as pressure overload (Shimura et al., 2005) and arrhythmias (Uemura et al., 2004; Shanmugam et al., 2011).

Previous studies, including our unpublished data using SLN-deficient and SLN-overexpressed mice, have confirmed the inhibitory effects of SLN on SERCA2a activity in the heart (Morimoto et al., 2014; Gramolini et al., 2006; Asahi et al., 2004; Babu et al., 2007b) (Fig. 12.4). Its SERCA inhibition is characterized as follows: SLN primarily lowers the V_{max} of SERCA-mediated Ca^{2+} uptake, but not pump affinity for Ca^{2+} (Sahoo et al., 2013; Maurya and Periasamy, 2015), which is different from PLN in cardiac muscle. Recent studies have revealed that the exact binding site of SLN is located in a groove close to helices M2, M6, and M9 of SERCA1a in two similar three-dimensional structures of the SLN-SERCA1a complex (Winther et al., 2013; Toyoshima et al., 2013). It should be noted that SLN lacks a distinct PKA- or CAMKII-mediated phosphorylation site in its cytosolic region. Therefore, unlike PLN, which possesses a fine-tuning activation/inhibition system involving kinases, phosphatases, and their modulators, physiological regulation of SLN activity remains unclear. Interestingly, a recent study demonstrated that, in addition to phosphorylation, s-acylation/deacylation and s-oleoylation also regulates SLN activity (Montigny et al., 2014). The physiological relevance of this finding remains to be elucidated in future.

A recent important finding is that SLN in skeletal muscles plays a critical role in the regulation of muscle-based thermogenesis in mammals, which has not been found in PLN (Mall et al., 2006; Bal et al., 2012; Maurya and Periasamy, 2015). Periasamy's group discovered that SLN-deficient mice were not able to maintain their core body temperature and developed hypothermia when exposed to acute cold stress (4°C) (Bal et al., 2012), suggesting that SLN plays a fundamental role in SERCA-mediated nonshivering thermogenesis and energy metabolism in skeletal muscles. The physiological significance of this finding remains controversial (Butler et al., 2015). The role of SLN in energy metabolism in cardiac muscles has not yet been investigated.

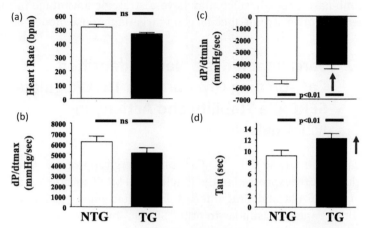

Figure 12.4 Hemodynamic data were assessed at baseline at 10–12 week from the heart-specific overexpression of SLN mice ($n = 4 \sim 6$ for both groups) (unpublished data). (a) Heart rate; (b) LV contractility (max LV dP/dt); (c) LV relaxation (min LV dP/dt); (d) LV relaxation (Tau). LV relaxation was impaired in SLN transgenic mice. Error bars are means ±SEM. **NTG**: non-transgenic mice; **TG**: SLN transgenic mice; **ns**: not significant.

The characteristic expression pattern of SLN can be technologically utilized for the atrium-specific gene targeting using the Cre-loxP system. The SLN-deficient, Cre recombinase-knock-in mouse has already been generated (Nakano et al., 2011; Shimura et al., 2016). Efficient atrium-specific gene deletion has been validated in several studies (Groenke et al., 2013; Nakashima et al., 2014). However, because SLN in this knock-in mouse is deleted because of the Cre recombinase insertion, the effects of heterozygous SLN deletion in the atrium must be evaluated since complete SLN deletion is reported to enhance SERCA2a pump activity and to cause atrial remodeling such as augmented fibrosis in SLN-null mice (Xie et al., 2012; Babu et al., 2007a). Although both SLN mRNA and protein levels were decreased by approximately ~50% in SLN heterozygous mouse atria, there were no morphological, physiological, or molecular biological abnormalities (Shimura et al., 2016). The properties of contractility and Ca^{2+} handling were also similar to wild-type mice (Shimura et al., 2016), suggesting that SLN heterozygous deletion does not significantly affect the character of the mouse atrium. Therefore, SLN-deficient, Cre recombinase-knock-in mice may have a strategic advantage in the generation of an atrium-specific gene-targeting mouse model.

12.7 Sarcalumenin Is a Newly Identified Ca^{2+}-Binding Glycoprotein That Regulates SERCA2a Stability and Activity in Mammals

Sarcalumenin (SAR) is a Ca^{2+}-binding glycoprotein that is predominantly expressed in the longitudinal SR of striated muscles, where SERCA2a and SAR interact in cardiac muscles. The Ca^{2+}-binding property is similar to other SR Ca^{2+}-binding proteins such as calsequestrin that is mainly localized at the terminal cisternae of the SR (Leberer et al., 1990; Rossi and Dirksen, 2006). SAR consists of 436 acidic amino acids (160 KDa) and includes a Ca^{2+}-binding insert that is sandwiched between a negatively charged N-terminal residues and the C-terminal region where several putative nucleotide-binding motifs for P-loop–containing ATPase/GTPase are located (Yoshida et al., 2005; Leberer et al., 1989) (Fig. 12.5a).

SAR exhibits high capacity (~35 mol Ca^{2+}/mol) and moderate affinity (kD of ~300 μM) of Ca^{2+}-binding properties (Leberer et al., 1990; Rossi and Dirksen, 2006). A 53 kDa alternative splice variant of SAR has also been reported, although its function remains unclear. The expression levels of SAR isoforms increase markedly during muscle development (Froemming and Ohlendieck, 1998) and decrease in heart failure (Shimura et al., 2008; Jiao et al., 2012).

To implicate the role of SAR in the heart, SAR-deficient mice are now available (Yoshida et al., 2005). We found that SAR-deficient mice exhibited an altered longitudinal SR ultrastructural morphology and mildly weakened Ca^{2+} uptake activity, although the mutant mice were apparently normal in growth, health, and reproduction (Yoshida et al., 2005). We have demonstrated that SAR-deficient mice showed mild cardiac dysfunction at rest (Yoshida et al., 2005) and that SAR deficiency-induced progressive heart failure in response to pressure overload (Shimura et al., 2008) (Fig. 12.5b). SAR plays a critical role in maintaining cardiac function under physiological stresses, such as endurance exercise (Jiao et al., 2009) and aging (Jiao et al., 2012), by regulating the Ca^{2+} transport activity into the SR. These results indicate that SAR plays an essential role in maintaining the Ca^{2+} cycling of the SR in the heart.

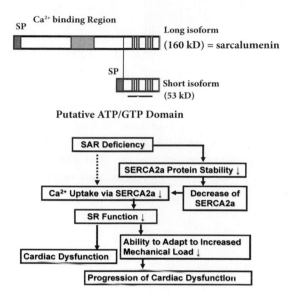

Figure 12.5 (a) Structure of sarcalumenin (SAR). SAR consists of 436 acidic amino acids (160 KDa) and includes a Ca^{2+}-binding insert that is sandwiched between negatively charged N-terminal residues and C-terminal region where several putative nucleotide-binding motifs for P-loop–containing ATPase/GTPase are located. A 53 kDa alternative splice variant of SAR has also been reported, although its function remains unclear. (b) SAR deficient mice exhibit cardiac dysfunction. **SP**: signal peptide sequence.

Interestingly, the expression levels of SERCA protein were decreased in both the fast-skeletal muscle (expressing SERCA1) and slow-skeletal or cardiac muscle (expressing SERCA2a) of SAR deficient mice whereas the mRNA levels were maintained. Different animal models of cardiac hypertrophy showed that the down-regulation of SAR protein preceded the down-regulation of SERCA2 protein (Shimura et al., 2008; Jiao et al., 2012). Our studies suggest that SAR plays a role in stabilizing the SERCA structure and increasing its apparent Ca^{2+} affinity with no change in its turnover rate (Shimura et al., 2008). The age-related decrease in SAR expression is also found in human skeletal muscle (Gueugneau et al., 2014) and rat skeletal muscle (O'Connell et al., 2008).

12.8 Conclusion

Ca^{2+} uptake into the cardiac SR is regulated by the fine-tuning activation of SERCA2a. Impairment of SERCA2a pump activity plays a critical role in cardiac function and human heart diseases, especially heart failure. Although there is an accumulating evidence of proof-of-concept in this field, further exploration is required to effectively achieve enhanced SERCA2a activity, or to find a specific compound or gene that disrupts the interaction between SERCA2a and PLN. Furthermore, whether or not augmented SERCA2a activity has potential adverse effects on morbidity and mortality for clinical applications also needs to be examined.

References

Ablorh, N. A., and Thomas, D. D. 2015. Phospholamban phosphorylation, mutation, and structural dynamics: A biophysical approach to understanding and treating cardiomyopathy. *Biophys. Rev., 7*: 63–76.

Andersson, K. B., Birkeland, J. A., Finsen, A. V., Louch, W. E., Sjaastad, I., Wang, Y., Chen, J., Molkentin, J. D., Chien, K. R., Sejersted, O. M., and Christensen, G. 2009. Moderate heart dysfunction in mice with inducible cardiomyocyte-specific excision of the Serca2 gene. *J. Mol. Cell. Cardiol., 47*: 180–187.

Antipenko, A. Y., Spielman, A. I., and Kirchberger, M. A. 1999. Interactions of 6-gingerol and ellagic acid with the cardiac sarcoplasmic reticulum Ca^{2+}-ATPase. *J. Pharmacol. Exp. Ther., 290*: 227–234.

Aoyama, H., Ikeda, Y., Miyazaki, Y., Yoshimura, K., Nishino, S., Yamamoto, T., Yano, M., Inui, M., Aoki, H., and Matsuzaki, M. 2011. Isoform-specific roles of protein phosphatase 1 catalytic subunits in sarcoplasmic reticulum-mediated Ca^{2+} cycling. *Cardiovasc. Res., 89*: 79–88.

Arai, M., Yoguchi, A., Takizawa, T., Yokoyama, T., Kanda, T., Kurabayashi, M., and Nagai, R. 2000. Mechanism of doxorubicin-induced inhibition of sarcoplasmic reticulum Ca^{2+}-ATPase gene transcription. *Circ. Res., 86*: 8–14.

Asahi, M., Otsu, K., Nakayama, H., Hikoso, S., Takeda, T., Gramolini, A. O., Trivieri, M. G., Oudit, G. Y., Morita, T., Kusakari, Y., Hirano, S., Hongo, K., Hirotani, S., Yamaguchi, O., Peterson, A., Backx, P. H., Kurihara, S., Hori, M., and MacLennan, D. H. 2004. Cardiac-specific overexpression of sarcolipin inhibits sarco(endo)plasmic reticulum Ca^{2+} ATPase (SERCA2a) activity and impairs cardiac function in mice. *Proc. Natl. Acad. Sci. U. S. A.,* **101**: 9199–9204.

Babu, G. J., Bhupathy, P., Carnes, C. A., Billman, G. E., and Periasamy, M. 2007a. Differential expression of sarcolipin protein during muscle development and cardiac pathophysiology. *J. Mol. Cell. Cardiol.,* **43**: 215–222.

Babu, G. J., Bhupathy, P., Timofeyev, V., Petrashevskaya, N. N., Reiser, P. J., Chiamvimonvat, N., and Periasamy, M. 2007b. Ablation of sarcolipin enhances sarcoplasmic reticulum calcium transport and atrial contractility. *Proc. Natl. Acad. Sci. U. S. A.,* **104**: 17867–17872.

Bal, N. C., Maurya, S. K., Sopariwala, D. H., Sahoo, S. K., Gupta, S. C., Shaikh, S. A., Pant, M., Rowland, L. A., Bombardier, E., Goonasekera, S. A., Tupling, A. R., Molkentin, J. D., and Periasamy, M. 2012. Sarcolipin is a newly identified regulator of muscle-based thermogenesis in mammals. *Nat. Med.,* **18**: 1575–1579.

Bhupathy, P., Babu, G. J., and Periasamy, M. 2007. Sarcolipin and phospholamban as regulators of cardiac sarcoplasmic reticulum Ca^{2+} ATPase. *J. Mol. Cell. Cardiol.,* **42**: 903–911.

Bish, L. T., Sleeper, M. M., Reynolds, C., Gazzara, J., Withnall, E., Singletary, G. E., Duchlis, C., Hui, D., High, K. A., Gao, G., Wilson, J. M., and Sweeney, H. L. 2011. Cardiac gene transfer of short hairpin RNA directed against phospholamban effectively knocks down gene expression but causes cellular toxicity in canines. *Human Gene Ther.,* **22**: 969–977.

Blaskovic, D., Zizkova, P., Drzik, F., Viskupicova, J., Veverka, M., and Horakova, L. 2013. Modulation of rabbit muscle sarcoplasmic reticulum Ca^{2+}-ATPase activity by novel quercetin derivatives. *Interdisciplinary Toxicol.,* **6**: 3–8.

Brady, M., Koban, M. U., Dellow, K. A., Yacoub, M., Boheler, K. R., and Fuller, S. J. 2003. Sp1 and Sp3 transcription factors are required for trans-activation of the human SERCA2 promoter in cardiomyocytes. *Cardiovasc. Res.,* **60**: 347–354.

Braz, J. C., Gregory, K., Pathak, A., Zhao, W., Sahin, B., Klevitsky, R., Kimball, T. F., Lorenz, J. N., Nairn, A. C., Liggett, S. B., Bodi, I., Wang, S., Schwartz, A., Lakatta, E. G., DePaoli-Roach, A. A., Robbins, J., Hewett, T. E., Bibb,

J. A., Westfall, M. V., Kranias, E. G., and Molkentin, J. D. 2004. PKC-alpha regulates cardiac contractility and propensity toward heart failure. *Nat. Med.,* **10**: 248–254.

Butler, J., Smyth, N., Broadbridge, R., Council, C. E., Lee, A. G., Stocker, C. J., Hislop, D. C., Arch, J. R., Cawthorne, M. A., and Malcolm East, J. 2015. The effects of sarcolipin over-expression in mouse skeletal muscle on metabolic activity. *Arch. Biochem. Biophys.,* **569**: 26–31.

del Monte, F., Harding, S. E., Schmidt, U., Matsui, T., Kang, Z. B., Dec, G. W., Gwathmey, J. K., Rosenzweig, A., and Hajjar, R. J. 1999. Restoration of contractile function in isolated cardiomyocytes from failing human hearts by gene transfer of SERCA2a. *Circulation,* **100**: 2308–2311.

del Monte, F., Williams, E., Lebeche, D., Schmidt, U., Rosenzweig, A., Gwathmey, J. K., Lewandowski, E. D., and Hajjar, R. J. 2001. Improvement in survival and cardiac metabolism after gene transfer of sarcoplasmic reticulum Ca^{2+}-ATPase in a rat model of heart failure. *Circulation,* **104**: 1424–1429.

Dhitavat, J., Fairclough, R. J., Hovnanian, A., and Burge, S. M. 2004. Calcium pumps and keratinocytes: Lessons from Darier's disease and Hailey-Hailey disease. *Br. J. Dermatol.,* **150**: 821–828.

El-Armouche, A., Rau, T., Zolk, O., Ditz, D., Pamminger, T., Zimmermann, W. H., Jackel, E., Harding, S. E., Boknik, P., Neumann, J., and Eschenhagen, T. 2003. Evidence for protein phosphatase inhibitor-1 playing an amplifier role in beta-adrenergic signaling in cardiac myocytes. *FASEB J.,* **17**: 437–439.

Fatkin, D., Seidman, C. E., and Seidman, J. G. 2014. Genetics and disease of ventricular muscle. *Cold Spring Harb. Perspect. Med.,* **4**: a021063.

Fechner, H., Suckau, L., Kurreck, J., Sipo, I., Wang, X., Pinkert, S., Loschen, S., Rekittke, J., Weger, S., Dekkers, D., Vetter, R., Erdmann, V. A., Schultheiss, H. P., Paul, M., Lamers, J., and Poller, W. 2007. Highly efficient and specific modulation of cardiac calcium homeostasis by adenovector-derived short hairpin RNA targeting phospholamban. *Gene Ther.,* **14**: 211–218.

Fish, M., Shaboodien, G., Kraus, S., Sliwa, K., Seidman, C. E., Burke, M. A., Crotti, L., Schwartz, P. J., and Mayosi, B. M. 2016. Mutation analysis of the phospholamban gene in 315 South Africans with dilated, hypertrophic, peripartum and arrhythmogenic right ventricular cardiomyopathies. *Sci. Rep.,* **6**: 22235.

Frank, K. F., Bolck, B., Erdmann, E., and Schwinger, R. H. 2003. Sarcoplasmic reticulum Ca^{2+}-ATPase modulates cardiac contraction and relaxation. *Cardiovasc. Res.,* **57**: 20–27.

Froemming, G. R., and Ohlendieck, K. 1998. Oligomerisation of Ca^{2+}-regulatory membrane components involved in the excitation-contraction-relaxation cycle during postnatal development of rabbit skeletal muscle. *Biochim. Biophys. Acta,* **1387**: 226–238.

Gayan-Ramirez, G., Vanzeir, L., Wuytack, F., and Decramer, M. 2000. Corticosteroids decrease mRNA levels of SERCA pumps, whereas they increase sarcolipin mRNA in the rat diaphragm. *J. Physiol.,* **524** (Pt 2): 387–397.

Gramolini, A. O., Trivieri, M. G., Oudit, G. Y., Kislinger, T., Li, W., Patel, M. M., Emili, A., Kranias, E. G., Backx, P. H., and Maclennan, D. H. 2006. Cardiac-specific overexpression of sarcolipin in phospholamban null mice impairs myocyte function that is restored by phosphorylation. *Proc. Natl. Acad. Sci. U. S. A.,* **103**: 2446–2451.

Grimm, D., Streetz, K. L., Jopling, C. L., Storm, T. A., Pandey, K., Davis, C. R., Marion, P., Salazar, F., and Kay, M. A. 2006. Fatality in mice due to oversaturation of cellular microRNA/short hairpin RNA pathways. *Nature,* **441**: 537–541.

Grobetal, T., Hammer, E., Bien-Moller, S., Geisler, A., Pinkert, S., Roger, C., Poller, W., Kurreck, J., Volker, U., Vetter, R., and Fechner, H. 2014. A novel artificial microRNA expressing AAV vector for phospholamban silencing in cardiomyocytes improves Ca^{2+} uptake into the sarcoplasmic reticulum. *PloS One,* **9**: e92188.

Groenke, S., Larson, E. D., Alber, S., Zhang, R., Lamp, S. T., Ren, X., Nakano, H., Jordan, M. C., Karagueuzian, H. S., Roos, K. P., Nakano, A., Proenza, C., Philipson, K. D., and Goldhaber, J. I. 2013. Complete atrial-specific knockout of sodium-calcium exchange eliminates sinoatrial node pacemaker activity. *PloS One,* **8**: e81633.

Gueugneau, M., Coudy-Gandilhon, C., Gourbeyre, O., Chambon, C., Combaret, L., Polge, C., Taillandier, D., Attaix, D., Friguet, B., Maier, A. B., Butler-Browne, G., and Bechet, D. 2014. Proteomics of muscle chronological ageing in post-menopausal women. *BMC Genomics,* **15**: 1165.

Gupta, R. C., Neumann, J., Watanabe, A. M., Lesch, M., and Sabbah, H. N. 1996. Evidence for presence and hormonal regulation of protein phosphatase inhibitor-1 in ventricular cardiomyocyte. *Am. J. Physiol.,* **270**: H1159–H1164.

Gwathmey, J. K., Copelas, L., MacKinnon, R., Schoen, F. J., Feldman, M. D., Grossman, W., and Morgan, J. P. 1987. Abnormal intracellular calcium handling in myocardium from patients with end-stage heart failure. *Circ. Res.,* **61**: 70–76.

Haghighi, K., Bidwell, P., and Kranias, E. G. 2014. Phospholamban interactome in cardiac contractility and survival: A new vision of an old friend. *J. Mol. Cell. Cardiol.,* **77**: 160–167.

Haghighi, K., Chen, G., Sato, Y., Fan, G. C., He, S., Kolokathis, F., Pater, L., Paraskevaidis, I., Jones, W. K., Dorn, G. W., 2nd, Kremastinos, D. T., and Kranias, E. G. 2008. A human phospholamban promoter polymorphism in dilated cardiomyopathy alters transcriptional regulation by glucocorticoids. *Hum. Mutat.* **29**: 640–647.

Haghighi, K., Kolokathis, F., Gramolini, A. O., Waggoner, J. R., Pater, L., Lynch, R. A., Fan, G. C., Tsiapras, D., Parekh, R. R., Dorn, G. W., 2nd, MacLennan, D. H., Kremastinos, D. T., and Kranias, E. G. 2006. A mutation in the human phospholamban gene, deleting arginine 14, results in lethal, hereditary cardiomyopathy. *Proc. Natl. Acad. Sci. U. S. A.,* **103**: 1388–1393.

Haghighi, K., Kolokathis, F., Pater, L., Lynch, R. A., Asahi, M., Gramolini, A. O., Fan, G. C., Tsiapras, D., Hahn, H. S., Adamopoulos, S., Liggett, S. B., Dorn, G. W., 2nd, MacLennan, D. H., Kremastinos, D. T., and Kranias, E. G. 2003. Human phospholamban null results in lethal dilated cardiomyopathy revealing a critical difference between mouse and human. *J. Clin. Invest.,* **111**: 869–876.

Haghighi, K., Pritchard, T., Bossuyt, J., Waggoner, J. R., Yuan, Q., Fan, G. C., Osinska, H., Anjak, A., Rubinstein, J., Robbins, J., Bers, D. M., and Kranias, E. G. 2012. The human phospholamban Arg14-deletion mutant localizes to plasma membrane and interacts with the Na/K-ATPase. *J. Mol. Cell. Cardiol.,* **52**: 773–782.

He, H., Meyer, M., Martin, J. L., McDonough, P. M., Ho, P., Lou, X., Lew, W. Y., Hilal-Dandan, R., and Dillmann, W. H. 1999. Effects of mutant and antisense RNA of phospholamban on SR Ca^{2+}-ATPase activity and cardiac myocyte contractility. *Circulation,* **100**: 974–980.

Heinis, F. I., Andersson, K. B., Christensen, G., and Metzger, J. M. 2013. Prominent heart organ-level performance deficits in a genetic model of targeted severe and progressive SERCA2 deficiency. *PloS One,* **8**: e79609.

Hoshijima, M., Ikeda, Y., Iwanaga, Y., Minamisawa, S., Date, M. O., Gu, Y., Iwatate, M., Li, M., Wang, L., Wilson, J. M., Wang, Y., Ross, J., Jr., and Chien, K. R. 2002. Chronic suppression of heart-failure progression by a pseudophosphorylated mutant of phospholamban via in vivo cardiac rAAV gene delivery. *Nat. Med.,* **8**: 864–871.

Hou, Z., Kelly, E. M., and Robia, S. L. 2008. Phosphomimetic mutations increase phospholamban oligomerization and alter the structure of its regulatory complex. *J. Biol. Chem.,* **283**: 28996–29003.

Huang, B., Wang, S., Qin, D., Boutjdir, M., and El-Sherif, N. 1999. Diminished basal phosphorylation level of phospholamban in the postinfarction remodeled rat ventricle: Role of beta-adrenergic pathway, G(i) protein, phosphodiesterase, and phosphatases. *Circ. Res.,* **85**: 848–855.

Ikeda, Y., Hoshijima, M., and Chien, K. R. 2008. Toward biologically targeted therapy of calcium cycling defects in heart failure. *Physiology (Bethesda),* **23**: 6–16.

Iordanidou, A., Hadzopoulou-Cladaras, M., and Lazou, A. 2010. Non-genomic effects of thyroid hormone in adult cardiac myocytes: Relevance to gene expression and cell growth. *Mol. Cell. Biochem.,* **340**: 291–300.

Iwanaga, Y., Hoshijima, M., Gu, Y., Iwatate, M., Dieterle, T., Ikeda, Y., Date, M. O., Chrast, J., Matsuzaki, M., Peterson, K. L., Chien, K. R., and Ross, J., Jr. 2004. Chronic phospholamban inhibition prevents progressive cardiac dysfunction and pathological remodeling after infarction in rats. *J. Clin. Invest.,* **113**: 727–736.

Jessup, M., Greenberg, B., Mancini, D., Cappola, T., Pauly, D. F., Jaski, B., Yaroshinsky, A., Zsebo, K. M., Dittrich, H., Hajjar, R. J., and Calcium Upregulation by Percutaneous Administration of Gene Therapy in Cardiac Disease, I. 2011. Calcium Upregulation by Percutaneous Administration of Gene Therapy in Cardiac Disease (CUPID): A phase 2 trial of intracoronary gene therapy of sarcoplasmic reticulum Ca^{2+}-ATPase in patients with advanced heart failure. *Circulation,* **124**: 304–313.

Ji, Y., Lalli, M. J., Babu, G. J., Xu, Y., Kirkpatrick, D. L., Liu, L. H., Chiamvimonvat, N., Walsh, R. A., Shull, G. E., and Periasamy, M. 2000. Disruption of a single copy of the SERCA2 gene results in altered Ca^{2+} homeostasis and cardiomyocyte function. *J. Biol. Chem.,* **275**: 38073–38080.

Jiao, Q., Bai, Y., Akaike, T., Takeshima, H., Ishikawa, Y., and Minamisawa, S. 2009. Sarcalumenin is essential for maintaining cardiac function during endurance exercise training. *Am. J. Physiol. Heart Circ. Physiol.,* **297**: H576–H582.

Jiao, Q., Takeshima, H., Ishikawa, Y., and Minamisawa, S. 2012. Sarcalumenin plays a critical role in age-related cardiac dysfunction due to decreases in SERCA2a expression and activity. *Cell Calcium,* **51**: 31–39.

Kawase, Y., Ly, H. Q., Prunier, F., Lebeche, D., Shi, Y., Jin, H., Hadri, L., Yoneyama, R., Hoshino, K., Takewa, Y., Sakata, S., Peluso, R., Zsebo, K., Gwathmey, J. K., Tardif, J. C., Tanguay, J. F., and Hajjar, R. J. 2008. Reversal of cardiac dysfunction after long-term expression of SERCA2a by gene transfer in a pre-clinical model of heart failure. *J. Am. Coll. Cardiol.,* **51**: 1112–1119.

Kho, C., Lee, A., Jeong, D., Oh, J. G., Chaanine, A. H., Kizana, E., Park, W. J., and Hajjar, R. J. 2011. SUMO1-dependent modulation of SERCA2a in heart failure. *Nature,* **477**: 601–605.

Kho, C., Lee, A., Jeong, D., Oh, J. G., Gorski, P. A., Fish, K., Sanchez, R., DeVita, R. J., Christensen, G., Dahl, R., and Hajjar, R. J. 2015. Small-molecule activation of SERCA2a SUMOylation for the treatment of heart failure. *Nat. Commun.,* **6**: 7229.

Kranias, E. G., and Hajjar, R. J. 2012. Modulation of cardiac contractility by the phospholamban/SERCA2a regulatome. *Circ. Res.,* **110**: 1646–1660.

Leberer, E., Charuk, J. H., Green, N. M., and MacLennan, D. H. 1989. Molecular cloning and expression of cDNA encoding a lumenal calcium binding glycoprotein from sarcoplasmic reticulum. *Proc. Natl. Acad. Sci. U. S. A.,* **86**: 6047–6051.

Leberer, E., Timms, B. G., Campbell, K. P., and MacLennan, D. H. 1990. Purification, calcium binding properties, and ultrastructural localization of the 53,000- and 160,000 (sarcalumenin)-dalton glycoproteins of the sarcoplasmic reticulum. *J. Biol. Chem.,* **265**: 10118–10124.

Lehnart, S. E., Maier, L. S., and Hasenfuss, G. 2009. Abnormalities of calcium metabolism and myocardial contractility depression in the failing heart. *Heart Fail. Rev.,* **14**: 213–224.

Liu, G. S., Morales, A., Vafiadaki, E., Lam, C. K., Cai, W. F., Haghighi, K., Adly, G., Hershberger, R. E., and Kranias, E. G. 2015. A novel human R25C-phospholamban mutation is associated with super-inhibition of calcium cycling and ventricular arrhythmia. *Cardiovasc. Res.,* **107**: 164–174.

Lockamy, E. L., Cornea, R. L., Karim, C. B., and Thomas, D. D. 2011. Functional and physical competition between phospholamban and its mutants provides insight into the molecular mechanism of gene therapy for heart failure. *Biochem. Biophys. Res. Commun.,* **408**: 388–392.

Luo, W., Wolska, B. M., Grupp, I. L., Harrer, J. M., Haghighi, K., Ferguson, D. G., Slack, J. P., Grupp, G., Doetschman, T., Solaro, R. J., and Kranias, E. G. 1996. Phospholamban gene dosage effects in the mammalian heart. *Circ. Res.,* **78**: 839–847.

Mall, S., Broadbridge, R., Harrison, S. L., Gore, M. G., Lee, A. G., and East, J. M. 2006. The presence of sarcolipin results in increased heat production by Ca^{2+}-ATPase. *J. Biol. Chem.,* **281**: 36597–36602.

Maurya, S. K., and Periasamy, M. 2015. Sarcolipin is a novel regulator of muscle metabolism and obesity. *Pharmacol. Res.* **102**: 270–275.

McKenna, E., Smith, J. S., Coll, K. E., Mazack, E. K., Mayer, E. J., Antanavage, J., Wiedmann, R. T., and Johnson, R. G., Jr. 1996. Dissociation of phospholamban regulation of cardiac sarcoplasmic reticulum Ca^{2+}ATPase by quercetin. *J. Biol. Chem.,* **271**: 24517–24525.

Medin, M., Hermida-Prieto, M., Monserrat, L., Laredo, R., Rodriguez-Rey, J. C., Fernandez, X., and Castro-Beiras, A. 2007. Mutational screening of phospholamban gene in hypertrophic and idiopathic dilated cardiomyopathy and functional study of the PLN-42 C>G mutation. *Eur. J. Heart Fail.,* **9**: 37–43.

Minamisawa, S., Hoshijima, M., Chu, G., Ward, C. A., Frank, K., Gu, Y., Martone, M. E., Wang, Y., Ross, J., Jr., Kranias, E. G., Giles, W. R., and Chien, K. R. 1999. Chronic phospholamban-sarcoplasmic reticulum calcium ATPase interaction is the critical calcium cycling defect in dilated cardiomyopathy. *Cell,* **99**: 313–322.

Minamisawa, S., Sato, Y., Tatsuguchi, Y., Fujino, T., Imamura, S., Uetsuka, Y., Nakazawa, M., and Matsuoka, R. 2003. Mutation of the phospholamban promoter associated with hypertrophic cardiomyopathy. *Biochem. Biophys. Res. Commun.,* **304**: 1–4.

Minamisawa, S., Uemura, N., Sato, Y., Yokoyama, U., Yamaguchi, T., Inoue, K., Nakagome, M., Bai, Y., Hori, H., Shimizu, M., Mochizuki, S., and Ishikawa, Y. 2006. Post-transcriptional downregulation of sarcolipin mRNA by triiodothyronine in the atrial myocardium. *FEBS Lett.,* **580**: 2247–2252.

Miyamoto, M. I., del Monte, F., Schmidt, U., DiSalvo, T. S., Kang, Z. B., Matsui, T., Guerrero, J. L., Gwathmey, J. K., Rosenzweig, A., and Hajjar, R. J. 2000. Adenoviral gene transfer of SERCA2a improves left-ventricular function in aortic-banded rats in transition to heart failure. *Proc. Natl. Acad. Sci. U. S. A.,* **97**: 793–798.

Miyazaki, Y., Ikeda, Y., Shiraishi, K., Fujimoto, S. N., Aoyama, H., Yoshimura, K., Inui, M., Hoshijima, M., Kasahara, H., Aoki, H., and Matsuzaki, M. 2012. Heart failure-inducible gene therapy targeting protein phosphatase 1 prevents progressive left ventricular remodeling. *PloS One,* **7**: e35875.

Montigny, C., Decottignies, P., Le Marechal, P., Capy, P., Bublitz, M., Olesen, C., Moller, J. V., Nissen, P., and le Maire, M. 2014. S-palmitoylation and s-oleoylation of rabbit and pig sarcolipin. *J. Biol. Chem.,* **289**: 33850–33861.

Morimoto, S., Hongo, K., Kusakari, Y., Komukai, K., Kawai, M., J, O. U., Nakayama, H., Asahi, M., Otsu, K., Yoshimura, M., and Kurihara, S. 2014. Genetic modulation of the SERCA activity does not affect the Ca^{2+} leak from the cardiac sarcoplasmic reticulum. *Cell Calcium,* **55**: 17–23.

Moriscot, A. S., Sayen, M. R., Hartong, R., Wu, P., and Dillmann, W. H. 1997. Transcription of the rat sarcoplasmic reticulum Ca^{2+} adenosine triphosphatase gene is increased by 3,5,3′-triiodothyronine receptor isoform-specific interactions with the myocyte-specific enhancer factor-2a. *Endocrinology,* **138**: 26–32.

Nakano, H., Williams, E., Hoshijima, M., Sasaki, M., Minamisawa, S., Chien, K. R., and Nakano, A. 2011. Cardiac origin of smooth muscle cells in the inflow tract. *J. Mol. Cell. Cardiol.,* **50**: 337–345.

Nakashima, Y., Yanez, D. A., Touma, M., Nakano, H., Jaroszewicz, A., Jordan, M. C., Pellegrini, M., Roos, K. P., and Nakano, A. 2014. Nkx2-5 suppresses the proliferation of atrial myocytes and conduction system. *Circ. Res.,* **114**: 1103–1113.

Namekata, I., Hamaguchi, S., Wakasugi, Y., Ohhara, M., Hirota, Y., and Tanaka, H. 2013. Ellagic acid and gingerol, activators of the sarco-endoplasmic reticulum Ca^{2+}-ATPase, ameliorate diabetes mellitus-induced diastolic dysfunction in isolated murine ventricular myocardia. *Eur. J. Pharmacol.,* **706**: 48–55.

Nef, H. M., Mollmann, H., Troidl, C., Kostin, S., Voss, S., Hilpert, P., Behrens, C. B., Rolf, A., Rixe, J., Weber, M., Hamm, C. W., and Elsasser, A. 2009. Abnormalities in intracellular Ca^{2+} regulation contribute to the pathomechanism of Tako-Tsubo cardiomyopathy. *Eur. Heart J.,* **30**: 2155–2164.

Nelson, B. R., Makarewich, C. A., Anderson, D. M., Winders, B. R., Troupes, C. D., Wu, F., Reese, A. L., McAnally, J. R., Chen, X., Kavalali, E. T., Cannon, S. C., Houser, S. R., Bassel-Duby, R., and Olson, E. N. 2016. A peptide encoded by a transcript annotated as long noncoding RNA enhances SERCA activity in muscle. *Science,* **351**: 271–275.

Nicolaou, P., and Kranias, E. G. 2009. Role of PP1 in the regulation of Ca cycling in cardiac physiology and pathophysiology. *Front. Biosci. (Landmark Ed),* **14**: 3571–3585.

O'Connell, K., Gannon, J., Doran, P., and Ohlendieck, K. 2008. Reduced expression of sarcalumenin and related Ca^{2+}-regulatory proteins in aged rat skeletal muscle. *Exp. Gerontol.,* **43**: 958–961.

Odermatt, A., Becker, S., Khanna, V. K., Kurzydlowski, K., Leisner, E., Pette, D., and MacLennan, D. H. 1998. Sarcolipin regulates the activity of SERCA1, the fast-twitch skeletal muscle sarcoplasmic reticulum Ca^{2+}-ATPase. *J. Biol. Chem.,* **273**: 12360–12369.

Pashmforoush, M., Lu, J. T., Chen, H., Amand, T. S., Kondo, R., Pradervand, S., Evans, S. M., Clark, B., Feramisco, J. R., Giles, W., Ho, S. Y., Benson,

D. W., Silberbach, M., Shou, W., and Chien, K. R. 2004. Nkx2-5 pathways and congenital heart disease; loss of ventricular myocyte lineage specification leads to progressive cardiomyopathy and complete heart block. *Cell,* **117**: 373–386.

Periasamy, M., and Janssen, P. M. 2008. Molecular basis of diastolic dysfunction. *Heart Fail. Clin.,* **4**: 13–21.

Periasamy, M., Reed, T. D., Liu, L. H., Ji, Y., Loukianov, E., Paul, R. J., Nieman, M. L., Riddle, T., Duffy, J. J., Doetschman, T., Lorenz, J. N., and Shull, G. E. 1999. Impaired cardiac performance in heterozygous mice with a null mutation in the sarco(endo)plasmic reticulum Ca^{2+}-ATPase isoform 2 (SERCA2) gene. *J. Biol. Chem.,* **274**: 2556–2562.

Prasad, A. M., and Inesi, G. 2012. Regulation and rate limiting mechanisms of Ca^{2+} ATPase (SERCA2) expression in cardiac myocytes. *Mol. Cell. Biochem.,* **361**: 85–96.

Rodriguez, P., Jackson, W. A., and Colyer, J. 2004. Critical evaluation of cardiac Ca^{2+}-ATPase phosphorylation on serine 38 using a phosphorylation site-specific antibody. *J. Biol. Chem.,* **279**: 17111–17119.

Rossi, A. E., and Dirksen, R. T. 2006. Sarcoplasmic reticulum: The dynamic calcium governor of muscle. *Muscle Nerve,* **33**: 715–731.

Sahoo, S. K., Shaikh, S. A., Sopariwala, D. H., Bal, N. C., and Periasamy, M. 2013. Sarcolipin protein interaction with sarco(endo)plasmic reticulum Ca^{2+} ATPase (SERCA) is distinct from phospholamban protein, and only sarcolipin can promote uncoupling of the SERCA pump. *J. Biol. Chem.,* **288**: 6881–6889.

Savignac, M., Edir, A., Simon, M., and Hovnanian, A. 2011. Darier disease: A disease model of impaired calcium homeostasis in the skin. *Biochim. Biophys. Acta,* **1813**: 1111–1117.

Schmitt, J. P., Kamisago, M., Asahi, M., Li, G. H., Ahmad, F., Mende, U., Kranias, E. G., MacLennan, D. H., Seidman, J. G., and Seidman, C. E. 2003. Dilated cardiomyopathy and heart failure caused by a mutation in phospholamban. *Science,* **299**: 1410–1413.

Shaikh, S. A., Sahoo, S. K., and Periasamy, M. 2016. Phospholamban and sarcolipin: Are they functionally redundant or distinct regulators of the sarco(endo)plasmic reticulum calcium ATPase? *J. Mol. Cell. Cardiol.,* **91**: 81–91.

Shanmugam, M., Molina, C. E., Gao, S., Severac-Bastide, R., Fischmeister, R., and Babu, G. J. 2011. Decreased sarcolipin protein expression and enhanced sarco(endo)plasmic reticulum Ca^{2+} uptake in human atrial fibrillation. *Biochem. Biophys. Res. Commun.,* **410**: 97–101.

Shimura, D., Kusakari, Y., Sasano, T., Nakashima, Y., Nakai, G., Jiao, Q., Jin, M., Yokota, T., Ishikawa, Y., Nakano, A., Goda, N., and Minamisawa, S. 2016. Heterozygous deletion of sarcolipin maintains normal cardiac function. *Am. J. Physiol. Heart Circ. Physiol.,* **310**: H92–H103.

Shimura, M., Minamisawa, S., Takeshima, H., Jiao, Q., Bai, Y., Umemura, S., and Ishikawa, Y. 2008. Sarcalumenin alleviates stress-induced cardiac dysfunction by improving Ca^{2+} handling of the sarcoplasmic reticulum. *Cardiovasc. Res.,* **77**: 362–370.

Shimura, M., Minamisawa, S., Yokoyama, U., Umemura, S., and Ishikawa, Y. 2005. Mechanical stress-dependent transcriptional regulation of sarcolipin gene in the rodent atrium. *Biochem. Biophys. Res. Commun.,* **334**: 861–866.

Tanwar, V., Bylund, J. B., Hu, J., Yan, J., Walthall, J. M., Mukherjee, A., Heaton, W. H., Wang, W. D., Potet, F., Rai, M., Kupershmidt, S., Knapik, E. W., and Hatzopoulos, A. K. 2014. Gremlin 2 promotes differentiation of embryonic stem cells to atrial fate by activation of the JNK signaling pathway. *Stem Cells,* **32**: 1774–1788.

Tavadia, S., Tait, R. C., McDonagh, T. A., and Munro, C. S. 2001. Platelet and cardiac function in Darier's disease. *Clin. Exp. Dermatol.,* **26**: 696–699.

Terada, R., Warren, S., Lu, J. T., Chien, K. R., Wessels, A., and Kasahara, H. 2011. Ablation of Nkx2-5 at mid-embryonic stage results in premature lethality and cardiac malformation. *Cardiovasc. Res.,* **91**: 289–299.

Tilemann, L., Lee, A., Ishikawa, K., Aguero, J., Rapti, K., Santos-Gallego, C., Kohlbrenner, E., Fish, K. M., Kho, C., and Hajjar, R. J. 2013. SUMO-1 gene transfer improves cardiac function in a large-animal model of heart failure. *Sci. Trans. Med.,* **5**: 211ra159.

Toyoshima, C., and Cornelius, F. 2013. New crystal structures of PII-type ATPases: Excitement continues. *Curr. Opin. Struct. Biol.,* **23**: 507–514.

Toyoshima, C., Iwasawa, S., Ogawa, H., Hirata, A., Tsueda, J., and Inesi, G. 2013. Crystal structures of the calcium pump and sarcolipin in the Mg^{2+}-bound E1 state. *Nature,* **495**: 260–264.

Toyoshima, C., and Mizutani, T. 2004. Crystal structure of the calcium pump with a bound ATP analogue. *Nature,* **430**: 529–535.

Traaseth, N. J., Ha, K. N., Verardi, R., Shi, L., Buffy, J. J., Masterson, L. R., and Veglia, G. 2008. Structural and dynamic basis of phospholamban and sarcolipin inhibition of Ca^{2+}-ATPase. *Biochemistry,* **47**: 3–13.

Uemura, N., Ohkusa, T., Hamano, K., Nakagome, M., Hori, H., Shimizu, M., Matsuzaki, M., Mochizuki, S., Minamisawa, S., and Ishikawa, Y. 2004. Down-regulation of sarcolipin mRNA expression in chronic atrial fibrillation. *Eur. J. Clin. Invest.,* **34**: 723–730.

van Rijsingen, I. A., van der Zwaag, P. A., Groeneweg, J. A., Nannenberg, E. A., Jongbloed, J. D., Zwinderman, A. H., Pinto, Y. M., Dit Deprez, R. H., Post, J. G., Tan, H. L., de Boer, R. A., Hauer, R. N., Christiaans, I., van den Berg, M. P., van Tintelen, J. P., and Wilde, A. A. 2014. Outcome in phospholamban R14del carriers: Results of a large multicentre cohort study. *Circ. Cardiovasc. Genet.,* **7**: 455–465.

Vangheluwe, P., Schuermans, M., Zador, E., Waelkens, E., Raeymaekers, L., and Wuytack, F. 2005. Sarcolipin and phospholamban mRNA and protein expression in cardiac and skeletal muscle of different species. *Biochem. J.,* **389**: 151–159.

Watanabe, A., Arai, M., Yamazaki, M., Koitabashi, N., Wuytack, F., and Kurabayashi, M. 2004. Phospholamban ablation by RNA interference increases Ca^{2+} uptake into rat cardiac myocyte sarcoplasmic reticulum. *J. Mol. Cell. Cardiol.,* **37**: 691–698.

Wawrzynow, A., Theibert, J. L., Murphy, C., Jona, I., Martonosi, A., and Collins, J. H. 1992. Sarcolipin, the "proteolipid" of skeletal muscle sarcoplasmic reticulum, is a unique, amphipathic, 31-residue peptide. *Arch. Biochem. Biophys.,* **298**: 620–623.

Weber, S., Meyer-Roxlau, S., Wagner, M., Dobrev, D., and El-Armouche, A. 2015. Counteracting protein kinase activity in the heart: The multiple roles of protein phosphatases. *Front. Pharmacol.,* **6**: 270.

Winther, A. M., Bublitz, M., Karlsen, J. L., Moller, J. V., Hansen, J. B., Nissen, P., and Buch-Pedersen, M. J. 2013. The sarcolipin-bound calcium pump stabilizes calcium sites exposed to the cytoplasm. *Nature,* **495**: 265–269.

Xie, L. H., Shanmugam, M., Park, J. Y., Zhao, Z., Wen, H., Tian, B., Periasamy, M., and Babu, G. J. 2012. Ablation of sarcolipin results in atrial remodeling. *Am. J. Physiol. Cell Physiol.,* **302**: C1762–C1771.

Yoshida, M., Minamisawa, S., Shimura, M., Komazaki, S., Kume, H., Zhang, M., Matsumura, K., Nishi, M., Saito, M., Saeki, Y., Ishikawa, Y., Yanagisawa, T., and Takeshima, H. 2005. Impaired Ca^{2+} store functions in skeletal and cardiac muscle cells from sarcalumenin-deficient mice. *J. Biol. Chem.,* **280**: 3500–3506.

Zarain-Herzberg, A., Estrada-Aviles, R., and Fragoso-Medina, J. 2012. Regulation of sarco(endo)plasmic reticulum Ca^{2+}-ATPase and calsequestrin gene expression in the heart. *Can. J. Physiol. Pharmacol.,* **90**: 1017–1028.

Zheng, J., Yancey, D. M., Ahmed, M. I., Wei, C. C., Powell, P. C., Shanmugam, M., Gupta, H., Lloyd, S. G., McGiffin, D. C., Schiros, C. G., Denney, T. S., Jr.,

Babu, G. J., and Dell'Italia, L. J. 2014. Increased sarcolipin expression and adrenergic drive in humans with preserved left ventricular ejection fraction and chronic isolated mitral regurgitation. *Circ. Heart Fail.,* **7**: 194–202.

Zsebo, K., Yaroshinsky, A., Rudy, J. J., Wagner, K., Greenberg, B., Jessup, M., and Hajjar, R. J. 2014. Long-term effects of AAV1/SERCA2a gene transfer in patients with severe heart failure: Analysis of recurrent cardiovascular events and mortality. *Circ. Res.,* **114**: 101–108.

Chapter 13

The Pivotal Role of Cholesterol and Membrane Lipid Rafts in the Ca²⁺-Sensitization of Vascular Smooth Muscle Contraction Leading to Vasospasm

Ying Zhang, Hiroko Kishi, Katsuko Kajiya, Tomoka Morita, and Sei Kobayashi

Department of Molecular and Cellular Physiology
Yamaguchi University, Graduate School of Medicine,
1-1-1 Minami-Kogushi, Ube, Yamaguchi 755-8505, Japan

seikoba@yamaguchi-u.ac.jp

The Rho-kinase (ROK)-mediated Ca²⁺-sensitization of vascular smooth muscles (VSM) contraction contributes to vasospasm, a major cause of acute-onset sudden death induced by vascular diseases such as ischemic heart and brain diseases. As an upstream signaling molecule of such abnormal vascular contraction leading to vasospasm, we previously identified sphingosylphosphorylcholine (SPC), which induced the Ca²⁺-sensitization of VSM contraction in vitro and in vivo through the sequential activation of Fyn (a Src family tyrosine kinase) and ROK. The marked elevation

Muscle Contraction and Cell Motility: Fundamentals and Developments
Edited by Haruo Sugi
Copyright © 2017 Pan Stanford Publishing Pte. Ltd.
ISBN 978-981-4745-16-1 (Hardcover), 978-981-4745-17-8 (eBook)
www.panstanford.com

of SPC level was also observed in the vasospastic patients, indicating that SPC is a causal factor of the Ca^{2+}-sensitization leading to vasospasm. Surprisingly we found the significant linkage between the extent of SPC-induced contraction and serum total cholesterol level in both human and rabbits. This review covers the current knowledge of the mechanisms by which SPC induces Ca^{2+}-sensitization leading to vasospasm and the roles of cholesterol and its enriched membrane microdomains, membrane lipid rafts.

13.1 Introduction

Vasospasm, such as coronary artery vasospasm and cerebral vasospasm, is one of causes for sudden death or a major lethal complication for patients after subarachnoid hemorrhage (SAH). This vasospasm means that a sustained abnormal contraction from coronary artery or cerebral vascular artery. Vascular smooth muscles (VSM) have two types of contraction, Ca^{2+}-dependent contraction and Ca^{2+}-independent contraction, i.e., Ca^{2+}-sensitization (Ganitkevich et al., 2002; Gao et al., 2013; Kureishi et al., 1997; Mizuno et al., 2008; Moreland et al., 1991; Somlyo and Somlyo, 2003). Ca^{2+}-dependent contraction plays an important role in the maintenance of physiologic blood pressure. This process is regulated by the Ca^{2+}/calmodulin-myosin light chain kinase (MLCK)-mediated pathway. When concentration of cytoplasmic Ca^{2+} increases, this increased Ca^{2+} binds to calmodulin and Ca^{2+}/calmodulin complex activates MLCK. The activated MLCK phosphorylates myosin light chain (MLC) and then produces the contraction (Gao et al., 2013; Mizuno et al., 2008; Moreland et al., 1991). On the other hand, Ca^{2+}-independent contraction, i.e., Ca^{2+}-sensitization induces abnormal contraction of VSM through a mechanism that is independent of Ca^{2+}/calmodulin-MLCK-mediated pathway (Gailly et al., 1997; Jensen et al., 1996; Kureishi et al., 1997; Kobayashi et al., 1991). It has been proposed a major cause of cardio-cerebral-vascular diseases, such as coronary artery vasospasm and cerebral vasospasm. In this review, we focused on and elaborated the mechanisms of Ca^{2+}-sensitization leading to vasospasm and the roles of cholesterol and its enriched membrane microdomains, membrane lipid rafts.

13.2 SPC Is a Causal Factor of Ca^{2+}-Sensitization Leading to Vasospasm

Accumulating evidence suggests that Rho-kinase (ROK) mediated Ca^{2+}-sensitization of VSM is associated with coronary artery and cerebral vasospasm (Chrissobolis and Sobey, 2006; Naraoka et al., 2013; Sato et al., 2000). The molecular mechanism underlying this Ca^{2+}-sensitization-induced vasospasm has not been completely elucidated. We previously demonstrated that sphingosyl-phosphorylcholine (SPC) as an upstream signaling molecule induced Ca^{2+}-sensitization contraction of VSM in porcine coronary arterial strips by activating ROK (Shirao et al., 2002; Todoroki-Ikeda et al., 2000). SPC, one of bioactive lipids, is involved in many functional effects as a signaling molecule including cell migration, invasion, and contractility (Kleger et al., 2011; Nixon et al., 2008). Here, we focus on its contractility in VSM and provide some evidence to demonstrate that SPC is a causal factor of the Ca^{2+}-sensitization leading to vasospasm.

In vitro study, SPC induced the Ca^{2+}-sensitization contraction of VSM in isolated porcine, rabbit, and human arterial strips (Morikage et al., 2006; Nakao et al., 2002). Simultaneous measurement of $[Ca^{2+}]_i$ and contraction of VSM showed that SPC induced the contraction of VSM without changing the concentration of Ca^{2+}, suggesting that SPC-induced contraction is Ca^{2+}-independent. In vivo study, Shirao et al. injected SPC into the cisterna magna and observed the changes in basilar artery diameter. They found that injection of SPC to the cisterna magna induced marked and long-lasting vasoconstriction of the basilar artery in canine (Shirao et al., 2008). This vasoconstriction was dependent on the concentration of SPC.

Another evidence indicating SPC is a causal factor of the Ca^{2+}-sensitization is the elevated concentration of SPC in the cerebrospinal fluid (CSF) of SAH patients. The concentration of SPC has been found to be significantly increased compared to that in patients with meningioma or normal pressure hydrocephalus. Under normal conditions, the concentration of SPC in plasma and serum is very low. The reported concentration of SPC was about 50 and 130 nM in plasma and in serum, respectively (Liliom et al., 2001). Although the exact concentration of SPC in normal human CSF is unknown, the concentration of SPC in CSF in patients after SAH was significantly higher (28 nM) than that in patients with meningioma (3.4 nM) or normal pressure hydrocephalus (1.6 nM) (Kurokawa et al., 2009). Taken together, these findings demonstrate that SPC is a causal factor of the Ca^{2+}-sensitization leading to vasospasm.

13.3 The Signaling Pathway of SPC-Induced Ca^{2+}-Sensitization Leading to Vasospasm

ROK is regarded as a molecule to mediate Ca^{2+}-sensitization. The active ROK phosphorylated myosin light chain phosphatase target subunit 1 (MYPT1) and inactivated myosin light chain phosphatase (Kimura et al., 1998). In addition, ROK can directly phosphorylate and activate MLC (Kureishi et al., 1997). Therefore, ROK mediates Ca^{2+}-sensitization through the inactivation of myosin light chain phosphatase and activation of MLC. However, the mechanism by which ROK mediated Ca^{2+}-sensitization has not been fully elucidated. We identified SPC as an upstream signaling molecule for the ROK-mediated Ca^{2+}-sensitization (Todoroki-Ikeda et al., 2000). First, SPC-induced Ca^{2+}-sensitization of VSM contraction was inhibited by the specific ROK inhibitor Y27632, suggesting that ROK involved in SPC-induced contraction. Second, both dominant negative ROK (dn-ROK) and ROK inhibiter Y27632 abolished SPC-induced Ca^{2+}-sensitization of VSM contraction in VSM strips permeabilized with β-escin. Conversely, dn-ROK showed no effect on Ca^{2+}-sensitization of VSM contraction induced by phorbol ester, a protein kinase C activator. In addition, SPC induced the translocation of ROK from the cytosol to the cell membrane of VSM cells (Shirao et al., 2002). This translocation was demonstrated to play an important role in the activation of ROK. These results indicated that SPC-ROK pathway mediates Ca^{2+}-sensitization of VSM contraction.

We then found that a member of the Src family protein tyrosine kinases (Src-PTKs), Fyn, was involved in the SPC-ROK pathway mediated Ca^{2+}-sensitization of contraction as an upstream signaling molecule of ROK (Nakao et al., 2002). PP1, a specific inhibitor of Src-PTKs, abolished SPC-induced Ca^{2+}-sensitization of contraction. However, PP3, an inactive analogue of PP1, could not inhibit SPC-induced contraction. In VSM cells stimulated by SPC, Fyn, not Src, translocated from the cytosol to the cell membrane (Nakao et al., 2002). This translocation is related to S-palmitoylation playing an important role in protein activation and signal transduction. 2-Bromopalmitate, a specific inhibitor for S-palmitoylation of Fyn, inhibited SPC-induced Ca^{2+}-sensitization of contraction. Eicosapentaenoic acid (EPA), an inhibitor for Fyn S-palmitoylation, inhibited both SPC-induced contraction and translocation of Fyn (Nakao et al., 2002). Moreover, PP1 and EPA inhibited SPC-induced Ca^{2+}-sensitization of VSM contraction, translocation, and activation of ROK. These results suggested that Fyn as an upstream signaling molecule of ROK is involved in SPC-ROK-mediated Ca^{2+}-sensitization leading to vasospasm.

13.4 Role of Cholesterol and Membrane Lipid Rafts in SPC-Induced Ca^{2+}-Sensitization Leading to Vasospasm

In arterial strips of patients with normal serum cholesterol levels, SPC-induced Ca^{2+}-sensitization of VSM contraction was not observed. In contrast, SPC significantly enhanced contraction of VSM in hypercholesterolemic patients (Morikage et al., 2006). This result indicated that cholesterol may be involved in SPC-induced Ca^{2+}-sensitization of VSM contraction. Cholesterol, a lipid molecule, is an essential structural component in the cell membrane. Much evidence demonstrated that serum cholesterol levels are strongly related to the occurrence of the cardiovascular events (Nelson, 2013; Peters et al., 2016; Unni et al., 2016). SPC-induced contraction was well correlated with the amount of total and LDL-cholesterol. This contraction showed a negative correlation with the amount of HDL-cholesterol (Morikage et al., 2006). In arterial strips of rabbits, the similar results that SPC-induced contraction was correlated with the levels of cholesterol in the serum were obtained. In rabbits, SPC-induced VSM contraction decreased upon oral administration of EPA or pravastatin, cholesterol-lowering agent. In addition, SPC-induced contraction was significantly decreased in the vascular strips which were deleted cholesterol with β-cyclodextrin (β-CD) (Morikage et al., 2006). Recently, we found that SPC-induced contraction was also correlated with the levels of cholesterol in the vascular smooth muscle tissue. In the range of 13.72–40.66 µg cholesterol/10 mg tissue, SPC-induced contraction was correlated with the amount of cholesterol in the tissue (unpublished data).

Although the relationship between cholesterol levels and Ca^{2+}-sensitization of VSM contraction has been clarified, the mechanism by which cholesterol enhances Ca^{2+}-sensitization of VSM contraction are not fully elucidated. Membrane lipid rafts are microdomains enriched in certain glycosphingolipids, gangliosides, and cholesterol (Róg and Vattulainen, 2014; Simons and Ikonen, 1997). They contain more cholesterol than other non-raft membranes in smooth muscle. The treatment with β-CD decreased not only cholesterol but also caveolin-1, a marker of lipid rafts, indicating that β-CD selectively disrupted the structure of cholesterol-enriched membrane microdomains, i.e., membrane lipid rafts. Membrane lipid rafts contain more cholesterol and recruits many signaling molecules (Head et al., 2014). In smooth muscle cells, we found that SPC induced the translocation of Fyn and ROK to the membrane lipid rafts using immunofluorescence staining and confocal microscopy (Morikage et al., 2006; Nakao et al., 2002). The treatment with β-CD inhibited this translocation of Fyn and ROK (unpublished data). These results suggested that membrane lipid rafts serving as a scaffold for recruiting Fyn and ROK play a vital role in SPC-induced Ca^{2+}-sensitization of VSM contraction (Fig. 13.1).

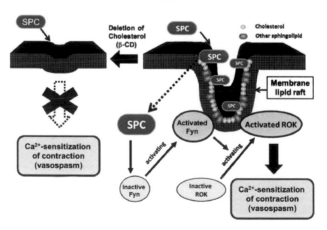

Figure 13.1 Signaling pathway of SPC-induced Ca^{2+}-sensitization and the roles of cholesterol and membrane lipid rafts in SPC-induced Ca^{2+}-sensitization leading to vasospasm.

13.5 Summary

In conclusion, SPC is a causal factor of the Ca^{2+}-sensitization leading to vasospasm. This Ca^{2+}-sensitization of VSM contraction is mediated by SPC/Fyn/ROK pathway and is dependent on the level of cholesterol. In addition, cholesterol-enriched membrane lipid rafts may be involved in SPC/Fyn/ROK pathway which mediates Ca^{2+}-sensitization leading to vasospasm.

Acknowledgments

This work was supported by Grants in Aid for Scientific Research from the Ministry of Education, Science, Technology, Sports and Culture of Japan.

References

Chrissobolis, S., and Sobey, C. G. (2006). Recent evidence for an involvement of rho-kinase in cerebral vascular disease. *Stroke*, 37, 2174–2180.

Gailly, P., Gong, M. C., Somlyo, A. V., and Somlyo, A. P. (1997). Possible role of atypical protein kinase C activated by arachidonic acid in Ca^{2+} sensitization of rabbit smooth muscle. *J Physiol.*, 500, 95–109.

Ganitkevich, V., Hasse, V., and Pfitzer, G. (2002). Ca^{2+}-dependent and Ca^{2+}-independent regulation of smooth muscle contraction. *J. Muscle Res. Cell Motil.*, 23, 47–52.

Gao, N., Huang, J., He, W., Zhu, M., Kamm, K. E., and Stull, J. T. (2013). Signaling through myosin light chain kinase in smooth muscles. *Biol. Chem.*, 288, 7596–7605.

Head, B. P., Patel, H. H., and Insel, P. A. (2014). Interaction of membrane/lipid rafts with the cytoskeleton: Impact on signaling and function: Membrane/lipid rafts, mediators of cytoskeletal arrangement and cell signaling. *Biochim. Biophys. Acta*, 1838, 532–545.

Jensen, P. E., Gong, M. C., Somlyo, A. V., and Somlyo, A. P. (1996). Separate upstream and convergent downstream pathways of G-protein- and phorbol ester-mediated Ca^{2+} sensitization of myosin light chain phosphorylation in smooth muscle. *Biochem. J.*, 318, 469–475.

Kimura, K., Fukata, Y., Matsuoka, Y., Bennett, V., Matsuura, Y., Okawa, K., Iwamatsu, A., and Kaibuchi, K. (1998). Regulation of the association of adducin with actin filaments by Rho-associated kinase (Rho-kinase) and myosin phosphatase. *J. Biol. Chem.*, 273, 5542–5548.

Kleger, A., Liebau, S., Lin, Q., von Wichert, G., and Seufferlein, T. (2011). The impact of bioactive lipids on cardiovascular development. *Stem Cells Int.*, 2011, 916180.

Kobayashi, S., Gong, M. C., Somlyo, A. V., and Somlyo, A. P. (1991). Ca²⁺ channel blockers distinguish between G protein-coupled pharmacomechanical Ca²⁺ release and Ca²⁺ sensitization. *Am. J. Physiol.*, 260, C364–C370.

Kureishi, Y., Kobayashi, S., Amano, M., Kimura, K., Kanaide, H., Nakano, T., Kaibuchi, K., and Ito, M. (1997). Rho-associated kinase directly induces smooth muscle contraction through myosin light chain phosphorylation. *J. Biol. Chem.*, 272, 12257–12260.

Kurokawa, T., Yumiya, Y., Fujisawa, H., Shirao, S., Kashiwagi, S., Sato, M., Kishi, H., Miwa, S., Mogami, K., Kato, S., Akimura, T., Soma, M., Ogasawara, K., Ogawa, A., Kobayashi, S., and Suzuki, M. (2009). Elevated concentrations of sphingosylphosphorylcholine in cerebrospinal fluid after subarachnoid hemorrhage: a possible role as a spasmogen. *J. Clin. Neurosci.*, 16, 1064–1068.

Liliom, K., Sun, G., Bünemann, M., Virág, T., Nusser, N., Baker, D. L., Wang, D. A., Fabian, M. J., Brandts, B., Bender, K., Eickel, A., Malik, K. U., Miller, D. D., Desiderio, D. M., Tigyi, G., and Pott, L. (2001). Sphingosylphosphocholine is a naturally occurring lipid mediator in blood plasma: A possible role in regulating cardiac function via sphingolipid receptors. *Biochem. J.*, 355, 189–197.

Mizuno, Y., Isotani, E., Huang, J., Ding, H., Stull, J. T., and Kamm, K. E. (2008). Myosin light chain kinase activation and calcium sensitization in smooth muscle in vivo. *Am. J. Physiol. Cell Physiol.*, 295, C358–C364.

Moreland, R. S., Cilea, J., and Moreland, S. (1991). Calcium dependent regulation of vascular smooth muscle contraction. *Adv. Exp. Med. Biol.*, 308, 81–94.

Morikage, N., Kishi, H., Sato, M., Guo, F., Shirao, S., Yano, T., Soma, M., Hamano, K., Esato, K., and Kobayashi, S. (2006). Cholesterol primes vascular smooth muscle to induce Ca²⁺ sensitization mediated by a sphingosylphosphorylcholine-Rho-kinase pathway: Possible role for membrane raft. *Circ. Res.*, 99, 299–306.

Nakao, F., Kobayashi, S., Mogami, K., Mizukami, Y., Shirao, S., Miwa, S., Todoroki-Ikeda, N., Ito, M., and Matsuzaki, M. (2002). Involvement of Src family protein tyrosine kinases in Ca⁽²⁺⁾ sensitization of coronary artery contraction mediated by a sphingosylphosphorylcholine-Rho-kinase pathway. *Circ. Res.*, 91, 953–960.

Naraoka, M., Munakata, A., Matsuda, N., Shimamura, N., and Ohkuma, H. (2013). Suppression of the Rho/Rho-kinase pathway and prevention of cerebral vasospasm by combination treatment with statin and fasudil after subarachnoid hemorrhage in rabbit. *Transl. Stroke Res.*, 4, 368–374.

Nelson, R. H. (2013). Hyperlipidemia as a risk factor for cardiovascular disease. *Prim Care* 40, 195–211.

Nixon, G. F., Mathieson, F. A., and Hunter, I. (2008). The multi-functional role of sphingosylphosphorylcholine. *Prog. Lipid Res.*, 47, 62–75.

Peters, S. A., Singhateh, Y., Mackay, D., Huxley, R. R., and Woodward, M. (2016). Total cholesterol as a risk factor for coronary heart disease and stroke in women compared with men: A systematic review and meta-analysis. *Atherosclerosis*, 248, 123–131.

Róg, T., and Vattulainen, I. (2014). Cholesterol, sphingolipids, and glycolipids: What do we know about their role in raft-like membranes? *Chem. Phys. Lipids,* 184, 82–104.

Sato, M., Tani, E., Fujikawa, H., and Kaibuchi, K. (2000). Involvement of Rho-kinase-mediated phosphorylation of myosin light chain in enhancement of cerebral vasospasm. *Circ. Res.*, 87, 195–200.

Shirao, S., Fujisawa, H., Kudo, A., Kurokawa, T., Yoneda, H., Kunitsugu, I., Ogasawara, K., Soma, M., Kobayashi, S., Ogawa, A., and Suzuki, M. (2008). Inhibitory effects of eicosapentaenoic acid on chronic cerebral vasospasm after subarachnoid hemorrhage: Possible involvement of a sphingosylphosphorylcholine-rho-kinase pathway. *Cerebrovasc. Dis.*, 26, 30–37.

Shirao, S., Kashiwagi, S., Sato, M., Miwa, S., Nakao, F., Kurokawa, T., Todoroki-Ikeda, N., Mogami, K., Mizukami, Y., Kuriyama, S., Haze, K., Suzuki, M., and Kobayashi, S. (2002). Sphingosylphosphorylcholine is a novel messenger for Rho-kinase-mediated Ca^{2+} sensitization in the bovine cerebral artery: Unimportant role for protein kinase C. *Circ. Res.*, 91, 112–119.

Simons, K., and Ikonen, E. (1997). Functional rafts in cell membranes. *Nature,* 387, 569–572.

Somlyo, A. P., and Somlyo, A. V. (2003). Ca^{2+} sensitivity of smooth muscle and nonmuscle myosin II: modulated by G proteins, kinases, and myosin phosphatase. *Physiol. Rev.*, 83, 1325–1358.

Todoroki-Ikeda, N., Mizukami, Y., Mogami, K., Kusuda, T., Yamamoto, K., Miyak, T., Sato, M., Suzuki, S., Yamagata, H., Hokazono, Y., and Kobayashi, S. (2000). Sphingosylphosphorylcholine induces Ca^{2+}-sensitization of vascular smooth muscle contraction: Possible involvement of rho-kinase. *FEBS Lett.*, 482, 85–90.

Unni, S. K., Quek, R. G., Biskupiak, J., Lee, V. C., Ye, X., and Gandra, S. R. (2016). Assessment of statin therapy, LDL-C levels, and cardiovascular events among high-risk patients in the United States. *J. Clin. Lipidol.*, 10, 63–71.

Chapter 14

The Catch State of Molluscan Smooth Muscle

Stefan Galler

Department of Cell Biology, University of Salzburg,
Hellbrunnerstraße 34, 5020 Salzburg, Austria

Stefan.Galler@sbg.ac.at

In terms of physics, holding a load without displacement should not require energy consumption. However, our muscles usually use energy for holding functions as this allows them to remain ready for movement. In contrast, certain specialized muscles are able to reduce energy consumption greatly when only holding but not movement is required. The saving of energy is most distinct in the so-called "catch" muscles of mollusks. For instance, when the shells of mussels are held closed, the responsible muscles enter in the "catch" state with very high holding efficiency. In experiments with isolated catch muscles of the mussel *Mytilus*, the neurotransmitter acetylcholine causes contraction; removal of acetylcholine is followed by only a slow or even no relaxation. This state of slowly decaying force in the absence of stimulation represents the catch state. Rapid relaxation and, thus, the termination of the catch state are attained by serotonin. Force and

Muscle Contraction and Cell Motility: Fundamentals and Developments
Edited by Haruo Sugi
Copyright © 2017 Pan Stanford Publishing Pte. Ltd.
ISBN 978-981-4745-16-1 (Hardcover), 978-981-4745-17-8 (eBook)
www.panstanford.com

even more the resistance to stretch are high in the catch state, but the consumption of ATP and the cytosolic free Ca^{2+} concentration are low as at resting condition. The catch state resembles the rigor state; however, unlike the rigor state the ATP concentration is high. Despite of more than one century of research the mechanism of catch is still unclear. The knowledge about molecular aspects has increased during the last decades; however, a convincing theory which satisfies the findings of various experimental approaches is still missing. This chapter gives a basic overview about the current state of research.

14.1 Background

Catch muscles have already attracted attention more than a century ago when the closing mechanism of the shells of mussels were investigated. The muscles responsible for permanent closing exert a great deal of pressure on a piece of wood inserted between the shells, but they are unable to shorten and close the shells when the piece of wood is removed (Grützner, 1904; von Uexküll, 1912). Later, the *anterior byssus retractor* muscle (ABRM) of the mussel *Mytilus* became a popular preparation for studying the phenomenon of catch.

Measurements of pulling force of isolated ABRMs showed that force decays only slowly or not at all after removing of the excitatory neurotransmitter acetylcholine from the bath solution. The remaining force (catch force) is rapidly relaxed after application of serotonin (Twarog, 1954; Fig. 14.1a). Neuronal activity is almost absent during the catch state (Flechter, 1937) and oxygen consumption is very low (Brecht et al., 1955; Baguet and Gillis, 1968). The catch state resembles the rigor state; however, unlike the rigor state the ATP concentration is high. Thus, the ATP level and the ATP/ADP ratio do not differ in catch and in other contractile states (Rüegg and Strassner, 1963, Rüegg, 1965). Direct measurements of intracellular Ca^{2+} showed that the free cytosolic Ca^{2+} concentration is at resting levels during catch (Ishii et al., 1989).

A catch-like state was also observed in skinned preparations of catch muscles (permeable cell membranes; Fig. 14.1b). Removal of Ca^{2+} after Ca^{2+} induced contractions is followed by incomplete relaxation. The remaining force decays only very slowly; addition of cAMP or the catalytic unit of protein kinase A (PKA) induces fast relaxation (Achazi et al., 1974; Cornelius, 1982; Pfitzer and Rüegg, 1982).

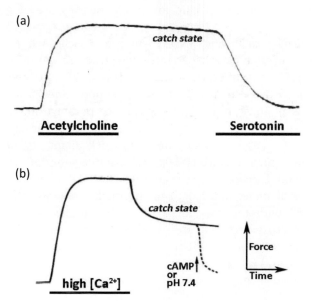

Figure 14.1 Time course of force during catch experiments on intact (a) and skinned ABRM preparations (b). The intact muscle is bathed in artificial seawater, the skinned preparations in solutions mimicking the intracellular ionic milieu at pH 6.7. (a) In the intact muscle, the catch state is established after washing out acetylcholine and terminated after addition of serotonin. In this example, the catch state is very distinct, since washing out acetylcholine has no appreciable influence on force. Such examples are not rare in fresh preparations. (b) In the skinned preparation the catch state is established after removal of Ca^{2+} down to basal levels and terminated after addition of cAMP (1–100 μM) or by alkalization (pH 7.4–7.7; for methodological details see Galler et al., 2005).

This chapter gives a basic overview about the current state of research on the catch phenomenon. More comprehensive reports were published by Rüegg (1965, 1971), Twarog (1976, 1979), Watabe and Hartshorne (1990), Hooper et al. (2008), Galler (2008), and Butler and Siegman (2010).

14.2   Structures of Catch Muscles

Peculiarities in function, such as the catch state, should have peculiarities in structure. Therefore, molecular and supramolecular structures which are specific for catch muscles are of greatest interest.

Molluscan catch muscles are smooth muscles which do not show any striation pattern of contractile structures. At the electron microscopic level, the large diameter (up to 75 nm) and length (25 μm) of thick filaments attract attention (Sobieszek, 1973). The basis of these filaments is a core of paramyosin. However, paramyosin is not specific for catch muscles. Paramyosin is observed in large amounts also in thick filaments of other invertebrate muscles such as *Limulus* telson levator, *Homarus* claw muscle, and *Lethocerus* air tube retractor (Elfvin et al., 1976). The large diameter and length of the thick filaments is probably associated with the high force of catch muscles. The maximal force per cross sectional area is three to four times higher in catch muscles than in other muscles of the animal kingdom (Twarog, 1967).

The myosin heavy chain, and in particular the myosin head, has no obvious peculiarity. The differences between myosin heavy chain isoforms of molluscan catch and striated muscle are not larger than those between fast and slow isoforms of mammalian skeletal muscle (Weiss et al., 1999) which are primarily known to differ in their kinetics of cross-bridge cycles (e.g. Pellegrino et al., 2003; Andruchov et al., 2004). Therefore, it is unlikely that the catch property is located in the myosin heads per se.

Catch muscles contain two different regulatory myosin light chain isoforms, one of which is also present in molluscan striated muscle; the other one is specific for catch muscle and contains a phosphorylation site (Morita and Kondo, 1982). The role of this specific isoform is not known.

Catch muscles contain appreciably high concentrations of twitchin. Twitchin is a short version of titin, the largest protein of vertebrate striated muscle. Twitchin is a molecular chain mainly consisting of 24 immunoglobulin-like and 15 fibronectin type III-like domains and has a molecular mass of about 530 kDa (Funabara et al., 2001, 2003, 2005). A kinase domain is located near the C-terminus. This kinase resembles the myosin light chain kinase (MLCK) of vertebrate smooth muscle. Twitchin can incorporate 3–4 phosphate groups. Catch muscles contain about three times more twitchin than other molluscan muscles. In ABRM of *Mytilus*, the molar ratio twitchin to myosin was reported to be about 1 to 15 (i.e., 7%) (Siegman et al., 1997).

A protein which seems to be specific for catch muscles is myorod (or catchin; Shelud'ko et al., 1999). Myorod is an alternative splicing product of the myosin heavy chain gene (Yamada et al., 2000). The main component of myorod is identical with about 75% of the rod part of myosin (light meromyosin and more than one third of subfragment 2). However, the two alpha-helices forming this rod are connected to a specific N-terminal domain of 156 amino acids forming a flexible globular structure. This domain has similarities with the regulatory myosin light chain and has several phosphorylation sites (Sobieszek et al., 2006). The molar concentration of myorod equals that of myosin.

Based on the abundance and specificity, myorod and twitchin should have important functions in catch muscles. As subsequently discussed, the role of twitchin was fairly explored; however, the role of myorod is still obscure.

14.3 Regulation of Catch

The catch state can be terminated by three main factors: (1) Serotonin with cAMP as intracellular messenger, (2) Increase in cytosolic Ca^{2+}, (3) Increase in cytosolic pH. Only the action of serotonin/cAMP was studied in considerable detail.

Siegman et al. (1997, 1998) showed that serotonin/cAMP induces relaxation of catch force by phosphorylation of twitchin. The phosphorylation is catalyzed by an enzyme similar to vertebrate protein kinase A (PKA) (Yamada et al., 2004; Bejar and Villamarin, 2006). Dephosphorylation of twitchin seems to induce the catch state. There is evidence that this dephosphorylation is catalyzed by a Ca^{2+} stimulated phosphatase similar to calcineurin (Castellani and Cohen, 1987, 1992).

In conclusion, the following scenario may take place when a catch state is developed (Fig. 14.2): The increase of cytosolic Ca^{2+} concentration due to excitatory stimulation induces two processes: (1) Ca^{2+} binding to the myosin necks which stimulates the myosin heads to execute force generating attachment/detachment cycles and (2) activation of the calcineurin-like phosphatase which dephosphorylates twitchin. The dephosphorylation of twitchin causes the formation of catch specific linkages between myofilaments ("catch linkages") which are ready to maintain catch force. When cytosolic Ca^{2+} is decreased to basal levels after the end of the excitatory stimulation, the stimulation of myosin heads and phosphatase is terminated. During the following phase, the catch phase, twitchin remains dephosphorylated and catch linkages persist sustaining the catch force. An increase in cAMP concentration after application of serotonin activates the PKA-like kinase which terminates the catch state by phosphorylating twitchin.

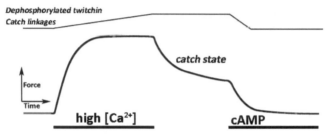

Figure 14.2 Model for the regulation of the catch state. Lower panel: Time course of the force response. Upper panel: Level of dephosphorylated twitchin and catch linkages. An increase in the cytosolic free Ca^{2+} concentration induces force generation and dephosphorylation of twitchin by a Ca^{2+}-stimulated calcineurin-like phosphatase. The formation of catch linkages is assumed to correspond to the dephosphorylation of twitchin. When Ca^{2+} is decreased, catch force maintained by catch linkages is observed. Catch force and catch linkages are decreased by cAMP which phosphorylates twitchin by a protein kinase A-like enzyme.

An increase of free Ca^{2+} concentration during the catch state transfers the muscle immediately back in the active state. The mechanism of this observation is unclear. Probably, Ca^{2+} activates the myosin heads in a way that the stretch-resistant catch linkages are outplayed in their action.

Cytosolic alkalization (e.g. an increase in pH from 6.7 to 7.4 or 7.7) prohibits and terminates the catch state. The effect was shown both on intact and on skinned muscle preparations. In intact preparation, acidification was induced by CO_2 and alkalization by ammonium chloride. The great pH sensitivity of catch is completely unexplained. Similar pH sensitivity was found in artificial paramyosin threads after stretch. Therefore, paramyosin was discussed as a possible candidate for explaining the pH sensitivity of catch some decades ago (Rüegg, 1964, 1965).

14.4 Catch Theories

After identification of myosin heads as the molecular force generators (Huxley, 1957; Huxley, 1969), it became plausible to assume that catch is due to an extreme slowing of the myosin head cycles (Lowy and Millman, 1963; Lowy et al., 1964). Myosin heads are thought to attach to and detach from the actin filament in a crawling-like manner. Under isometric conditions the cycles continue in a way that the myosin heads always bind to the same binding place on the actin filament. Each cycle is driven by the breakdown of one ATP molecule in the ATPase pocket of the myosin head. Therefore, a slowing of the cycle would result in higher energy efficiency for holding.

This theory was not compatible with Rüegg's finding that the catch state is unaffected by inhibiting the myosin function with thiourea (Rüegg, 1963). In these experiments, thiourea, which can penetrate the cell membranes, was applied to intact ABRM preparations. The results suggested that the force and resistance during catch are due to other structures than the myosin heads. Apparently, these structures are sensitive to serotonin/cAMP.

Also some ultrastructural studies suggested that catch is not simply a slowing of myosin head cycles. Thus, it was found that thick filaments aggregate during the catch state and they are dissociated again when catch is terminated by serotonin (Heumann and Zebe, 1968; Gilloteaux and Baguet, 1977; Hauck and Achazi, 1987). Since these findings were not confirmed in a later study (Bennet and Elliott, 1989), they were dismissed as artifacts. This later study appeared to be particularly reliable, because the preparations were fixed by fast freezing. As a consequence, it became more and more popular to assume that catch is due to a slowing of myosin heads. However, not all scientists agreed with this conclusion (Elliott, 1982, Kobayashi et al., 1985, Sugi et al., 1999).

In analogy to the vertebrate smooth muscle, the reason for the slowing of the myosin head cycle was thought to be a retardation of the release of ADP from the ATPase pocket of the myosin head after ATP hydrolysis ("locked ADP" hypothesis). The retardation was assumed to be caused by the decrease of cytosolic free Ca^{2+} concentration during catch leading to Ca^{2+}-free myosin heads with changed properties (Takahashi et al., 1988; Butler et al., 2001).

14.5   Challenges of the Traditional Myosin Head Theory

At the end of the 1990s, it was found that skinned ABRM preparations relax quickly from high-force rigor (halftime 0.28 s), when ATP is liberated by flash photolysis of caged-ATP in the absence of Ca^{2+} (Fig. 14.3; Galler et al., 1999). The relaxation is five times slower in the presence of 0.5 mM MgADP. These experiments indicate a fast detachment of Ca^{2+}-free myosin heads from actin filaments after binding of ATP even in the presence of ADP. If catch would be due to a hindered release of ADP (locked ADP hypothesis), then a much slower detachment is expected at least in the presence of ADP. Thus, the fast kinetics observed in these experiments is not favoring the assumption that catch is due to a retardation of myosin head detachment.

In this type of experiments, the ATP-induced relaxation was not complete; the remaining force decreased only very slowly, but dropped rapidly after addition of cAMP (Butler et al., 2001) indicating the presence of a catch state. Since >96% of myosin heads appeared to be nucleotide-free in the rigor state, these heads should readily bind ATP after flash photolysis of caged-ATP and detach from actin filaments resulting in <4% remaining force. However, the remaining force was considerably higher. This could mean that many myosin heads do not detach. However, usually all heads of various myosin classes detach from the actin filament after ATP binding. Therefore, the assumption is more likely that the high remaining force is sustained by other linkages (or by myosin heads which remain attached mediated by another protein; "tied myosin heads"; see below).

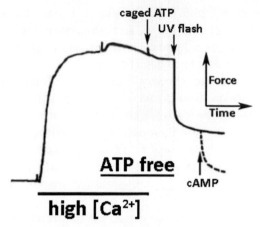

Figure 14.3   Time course of force during a caged ATP experiment on a skinned ABRM preparation. Caged ATP is applied after Ca^{2+} activation and subsequent ATP removal (high Ca^{2+} rigor state). ATP is suddenly photoliberated from caged ATP by a UV flash. The ATP increase is followed by a quick relaxation which indicates rapid myosin head detachment. The relaxation is not complete. The remaining force can be relaxed by cAMP which suggests the presence of a catch state. For methodological details see Galler et al. (1999).

A few years later, the effects of vanadate and other inhibitors of myosin function were investigated during the catch state of skinned ABRM preparations (Galler et al., 2005). Catch force was not influenced by these inhibitors, and therefore, the conclusion was made, that catch cannot be due to a simple retardation of myosin head detachment as assumed in the preceding decades.

This conclusion was further supported by the finding that the catch releasing factor cAMP does not increase the myosin head detachment rate (Andruchova et al., 2005). In this study, the detachment rate was determined in two types of experiments: (1) Sudden appearance of ATP in Ca^{2+}-free high force rigor via flash photolysis of caged-ATP (Fig. 14.3); (2) stretch experiments at maximal Ca^{2+} activation. If catch is due to a retardation of myosin head detachment, cAMP should accelerate the myosin head detachment.

Like cAMP, also alkalization does not accelerate the myosin head detachment measured in the same types of experiments

(Höpflinger et al., 2006). As mentioned above, alkalization is a strong catch releasing factor and should therefore accelerate the myosin head detachment if catch would really be based on a retardation of myosin head detachments.

Further evidence showing that catch cannot be a simple slowing of myosin head detachments came from experiments in which the myosin heads were blocked not only during the catch state but already before the onset of Ca^{2+} activation (Galler et al., 2010). In this study, the new inhibitor blebbistatin (Kovacs et al., 2004) was used in intact and skinned ABRM preparations. In experiments with skinned preparations, blebbistatin inhibited force generation after addition of Ca^{2+}. Interestingly, during ongoing presence of Ca^{2+} the resistance to stretch increased. This increase persisted after removal of Ca^{2+} indicating a catch-like state. The increased resistance decreased to initial levels after application of cAMP. In experiments with intact ABRMs, blebbistatin inhibited force generation after addition of acetylcholine. During ongoing presence of acetylcholine the resistance to stretch increased. Serotonin reduced the elevated stretch resistance back to initial values.

The acetylcholine/Ca^{2+} stimulated increase of stretch resistance which is sensitive to serotonin/cAMP suggests that a state of catch can develop still after prohibiting the myosin heads to produce force. In conclusion, these experiments provided final evidence for the assumption that catch cannot be due to a slowing of myosin head cycles by retardation of myosin head detachment.

14.6   Twitchin Bridges

Since catch cannot be due to a slowing of myosin head cycles, another mechanism has to be considered. One possibility is the existence of other linkages between thick and thin filaments different from myosin heads. An important candidate for these linkages is the catch regulator protein twitchin. Twitchin could form bridges between thick and thin filaments which are mechanically involved in maintaining the catch force. This assumption is mainly supported by the following findings:

(1) Biochemical experiments showed that twitchin connects myosin and actin filaments, if twitchin is not phosphorylated. Phosphorylation of twitchin by protein kinase A dissociates actin and myosin (Shelud'ko et al., 2004).

(2) In experiments with reconstituted thick and thin filaments investigated by light microscopy, it was found that thin filaments strongly bind to thick filaments under catch conditions (presence of ATP, basal free Ca^{2+} concentration) if the thick filaments contain twitchin in the unphosphorylated form (Yamada et al., 2001, 2004).

(3) A detailed study of myosin head kinetics showed only limited interactions between the catch linkages and myosin heads (Franke et al., 2007).

(4) In physiological experiments, it was found that diffusion-driven incorporation of twitchin in skinned fiber preparations of mammalian muscle induces an increase in stretch resistance (Avrova et al., 2009). This increase is reduced by addition of the catalytic unit of protein kinase A.

14.7 Myosin Heads Tied by Twitchin

The binding of twitchin to thick and thin filaments was investigated in further detail by using small segments of this large chain-like molecule (Funabara et al., 2007, 2009). A peptide was constructed which contained the phosphorylation site D2 and the flanking IG-like domains. This peptide exhibited high affinity for the thin filament, if it was not phosphorylated. The affinity considerably decreased after phosphorylation. The unphosphorylated D2 peptide has also the ability to bind to myosin. If both actin and myosin are present in the biochemical assay, the unphosphorylated peptide connects these proteins to form a stable complex.

More detailed experiments suggested that the binding of the D2 peptide to actin occurs in the region L10-P29 of actin. The binding to myosin seems to occur in the loop 2 region which is a segment of the myosin head. This loop binds to region D26-S27 of actin during usual attachment/detachment cycles of the myosin head. Thus, the D2 segment of twitchin and the loop 2 of the myosin head seem to have the same binding site on actin (which would allow competition). Taken together, it is plausible to assume that the D2 segment of twitchin lodges in the place between the myosin head and actin where these proteins normally interact among each other. In other words, the unphosphorylated D2 segment of twitchin seems to tie the myosin head to the actin filament.

The affinity of the binding of the D2 peptide to actin and myosin is only slightly decreased by increasing the pH from 6.7 to 7.4. Therefore, this small effect may not account for the large pH sensitivity of catch known from intact and skinned preparations (Funabara et al., 2010). Consequently, the large pH sensitivity of catch must have other reasons.

In a later study (Butler et al., 2010), further peptides corresponding to twitchin segments comprising the phosphorylation sites D1 and DX were constructed and used in biochemical and physiological assays. Like the D2 peptide, also these peptides bound to actin filaments (and myosin) in a phosphorylation dependent manner. Some modifications of these peptides were even able to relax catch force when applied to skinned fiber preparations.

Therefore, it seems that twitchin may have not only one but even two regions which are able to tie the myosin head to the actin filament in a phosphorylation depending manner: the D2 region near the C-terminus and the D1/DX region near the N-terminus.

Avrova et al. (2012) found on skinned fiber preparations that the phosphorylation state of twitchin modulates the orientation of the myosin heads during the ATPase cycle. This finding suggests that unphosphorylated twitchin, in fact, interferes with the myosin head cycle as predicted from the results of the biochemical assays.

In conclusion, twitchin may connect the myosin heads with the actin filament to form catch linkages. These "tied myosin heads" would represent force-bearing structures during the catch state. Phosphorylation of twitchin induced by cAMP would dissociate the tied myosin heads from the actin filament and cause relaxation of catch force.

Based on the assumptions that each twitchin molecule can connect two myosin heads to the actin filament and the molar ratio between twitchin and myosin is about 1: 15 (Siegman et al., 1997), every 7th or 8th myosin head could be tied to the actin filaments during catch. Considering the high stretch resistance during catch, it may appear doubtful, if this relatively small number of catch linkages is sufficient. Thus, it is possible that additional types of catch linkages exist.

14.8 Myorod

In theory, myorod could be an additional catch force bearing structure in addition of twitchin. As already mentioned, myorod is specific for catch muscles and highly concentrated (Shelud'ko et al., 1999; Yamada et al., 2000). During the past years, the knowledge about phosphorylation and interaction of myorod with other proteins has increased. Nevertheless, the function of myorod in the catch state is still unclear.

As already mentioned, the N-terminal domain of myorod exhibits similarities with the regulatory myosin light chains, and can be phosphorylated by vertebrate MLCK (Sobieszek et al., 2006). Thus, it would be plausible to assume that this structure is phosphorylated by the MLCK-like twitchin kinase. However, there is no consistent evidence for this assumption (Matusovsky et al., 2010; Sobieszek et al., 2010). In this context it is important to note that twitchin kinase seems to be stimulated by stretch (Butler and Siegman, 2011), a property which cannot be mimicked consistently in biochemical assays.

Unphosphorylated myorod is able to bind to actin filaments, whereas phosphorylated myorod is unable to bind (Matusovsky et al., 2011). The binding requires micromolar Ca^{2+} concentrations, and therefore, it may not account for a rigid actomyosin network during catch. Furthermore, phosphorylation of the N-terminal domain of myorod potentiates the actin-activated ATPase activity of the myosin head (Matusovsky et al., 2015).

In conclusion, myorod seems to have abilities to interact with other contractile structures in various biochemical assays. For elucidating the significance of these properties, skinned fiber preparations with intact contractile machinery and defined contractile states should be performed.

14.9 Interconnections between Thick Filaments?

In theory, a more rigid muscle with high stretch resistance can also be reached if the contractile functional units become larger during the catch state (Rüegg, 1971, Sugi et al., 1999, Mukou et al., 2004). This conclusion is clarified by considering basic principles playing a role in muscle: The shortening of a muscle fiber is determined by the functional units in series; in contrast, force and rigidity of a muscle fiber is determined by only one functional unit, because all other units compensate each other. Larger functional units have higher outputs in force and rigidity. Therefore, muscle fibers with a few large functional units have higher force and rigidity than muscle fibers with many small functional units.

An increase in size of the functional units can be reached in catch muscles, if adjacent thick filaments which usually are longitudinally displaced are laterally interconnected. There are, in fact, some experimental suggestions for this: Besides the aggregations between thick filaments mentioned already above, also distinct interconnections between thick filaments were observed in the electron micrographs of catch muscles. Such interconnections were found occasionally in the relaxed state and more frequently in the catch state (Takahashi et al., 2003). Based on this circumstance, they may not be a simple correlate of myomesin interconnections of thick filaments within the sarcomeres of striated vertebrate muscle (Auerbach et al., 1999); they may rather have a function in the catch mechanism. In view of the relatively low number of tied myosin heads, these interconnections could increase the stretch resistance during the catch state. If this applies, two types of catch linkages would exist: interconnections between thick and thin filaments (tied myosin heads) and interconnections among thick filaments of unknown nature. The two types of catch linkages could correspond to the two regulatory mechanisms for catch: cAMP and alkalization.

A study using synchrotron radiation supports the idea that both thick and thin filaments and thick filaments among each other are interconnected during the catch state (Tajima et al., 2008). In this study periodicities along the thick and thin filaments were measured for detecting the lengths of these filaments. When the muscle was activated by acetylcholine, thick and thin filaments elongated because of the increase in tension. After removal of acetylcholine, when the tension slightly decreased while entering in the catch state, thick filaments reduced their elongation less than thick filaments. This circumstance probably suggests that the mechanical load is partially transferred from thin to thick filaments—as it is expected when thick filaments interconnect among each other for forming larger contractile units.

14.10   Additional Kinases and Phosphatases

Kinases and phosphatases play an important role in the regulation of catch. To date three kinases and one phosphatase are known. The following enzymes were already mentioned in this chapter:

(1) PKA-like kinase which is stimulated by cAMP (the second messenger of serotonin) and phosphorylates specific sites near the N-terminus (D1, DX) and the C-terminus (D2) of twitchin.

(2) Twitchin kinase which is a short segment of the long molecular chain of twitchin. It resembles the MLCK of vertebrate smooth muscle and is probably activated by stretch. Its phosphorylation targets are unclear. Candidates are the N-terminal domain of myorod and twitchin itself (autophosphorylation).

(3) Calcineurin-like phosphatase which is activated by Ca^{2+} and dephosphorylates twitchin.

None of these enzymes are specific for catch muscles. However, there exists, in fact, an enzyme which is specific for catch muscles, namely the so-called myosin-associated kinase (Castellani and Cohen, 1987; Sobieszek et al., 2010). This kinase is tightly bound to myosin. It phosphorylates a non-helical C-terminal tailpiece of myorod and myosin. Micromolar concentrations of Ca^{2+} inhibit the phosphorylation of myosin, but not the phosphorylation of myorod. The functional consequence of the tailpiece phosphorylation of myosin and myorod is unknown; however, it is imaginable that the attachment of myosin and myorod to the thick filament is changed after phosphorylation (Sobieszek et al., 2010). Such a change would have great impact on contraction; and therefore, detailed investigations are desirable.

It is worth to speculate, if the Ca^{2+}-stimulated calcineurin-like enzyme is the only phosphatase in catch muscles. Specific findings are easier to explain if an additional phosphatase existed. Thus, it is unclear why the phosphatase inhibitor NaF causes rapid relaxation of catch force (Castellani and Cohen, 1987). This finding namely indicates that a NaF-sensitive phosphatase is active during the catch state. The calcineurin-like phosphatase may not be the active enzyme, because this phosphatase is thought to be turned off by the basal levels of cytosolic Ca^{2+} concentration during the catch state.

14.11 Catch during Active Contraction

In the usual physiological context, the catch state becomes apparent after the end of muscle stimulation when cytosolic Ca^{2+} concentration is decreased to basal levels. Force which was generated during active contraction is partially or even totally maintained during the catch state. It is plausible to assume that the linkages maintaining the catch force are formed already during active contraction, at least in a preliminary state. Otherwise, the linkages would have to be formed very quickly exactly in the moment when the myosin heads detach from actin filaments.

The assumption that catch linkages are established already during active contraction is supported by the finding that application of myosin inhibitors like blebbistatin during active contraction is not able to depress force to zero (Butler et al., 2006; Andruchov et al., 2006). The remaining force seems to be catch force, because it can be relaxed by the catch releasing factors cAMP and alkalization. This catch force is observed during a period of constantly high (even maximally activating) Ca^{2+} concentration, and therefore, it may suggest that catch linkages are formed already during the phase of active contraction (Galler, 2008).

Interestingly, cAMP is able to depress Ca^{2+}-activated force also when no myosin inhibitor is applied. This fact may not necessarily be related to catch linkages; however, if it really is, then catch linkages must be under tension during the active state of muscle (Galler, 2008). This is because only structures under tension can cause a force decrease after detachment. Interestingly, the cAMP effect diminishes with increasing Ca^{2+} concentration and disappears at maximal Ca^{2+} stimulation (Butler et al., 1998, 2006).

As mentioned above, myosin heads which are bound to the actin filament by a segment of twitchin (tied myosin heads) are likely candidates for catch linkages although perhaps also additional types of catch linkages may exist. The concept of tied myosin heads offers a tentative explanation for the parallel existence of force-generating myosin heads and catch linkages. Thus, it is imaginable that the tied myosin heads are surpassed by the active myosin heads during Ca^{2+} activation because the number of active myosin heads is up to seven to eight times higher. Perhaps Ca^{2+} bound myosin heads which are normally cycling and Ca^{2+} free myosin heads which have a twitchin segment attached compete for the same binding place on actin. Since the number of Ca^{2+} free myosin heads decreases with increasing Ca^{2+} concentration the influence of these heads may fade away at maximal Ca^{2+} stimulation. If these Ca^{2+} free myosin heads with attached twitchin contribute to isometric force, the force depressing effect of cAMP which disappears at maximal Ca^{2+} stimulation would be explained. It is imaginable that myosin heads with a twitchin segment bound remain in a standby mode when Ca^{2+} binds; they may switch in the strongly bound catch mode when Ca^{2+} dissociates (Galler, 2008). This mechanism would explain why a catch state is established not only after submaximal but also after maximal Ca^{2+} activation.

Independent of such mechanistic details it can be stated that catch linkages may not impede muscle fiber shortening (Butler et al., 1998), if they are fully established already during active contraction. In other words, they should have ratchet-like properties. Thus, they may provide resistance only against muscle lengthening but not against shortening. This assumption was already made by Von Uexküll (1912) who observed that a muscle in catch has a high resistance to stretch but almost no resistance to shortening when a load in the shortening direction is applied.

14.12 Conclusion

The mechanism of the catch state of mollusk smooth muscle is still unclear. Twitchin has an important regulatory function in the catch phenomenon. Twitchin might also mechanically contribute to the force maintenance during catch. However, compared to myosin, the number of twitchin molecules is only low, and therefore, additional linkages which contribute for the maintenance of catch force may exist.

Catch muscles differ from other muscles by two specific proteins, namely myorod and myosin-associated kinase. Therefore, progress in research is especially expected from further studies on these specific proteins, particularly by using skinned fiber preparations with intact contractile structure. Insights are also expected from studies of the stretch sensitive twitchin kinase and from detailed morphological investigations of the contractile machinery. Finally, the investigation of the pH dependency of catch might also contribute to unraveling the mechanism of the puzzling catch phenomenon.

Acknowledgements

I am grateful to Caspar Rüegg (Heidelberg) and Apolinary Sobieszek (Innsbruck) for stimulating discussions during many years. Dorothee Günzel (Berlin) and Anita Holzinger (Salzburg) are acknowledged for correcting the manuscript. I am thankful to Haruo Sugi (Tokyo) for the invitation to write this book contribution.

References

Achazi, R. K., Dolling, B., and Haakshorst, R. (1974) 5-HT-induced relaxation and cyclic AMP in a molluscan smooth muscle. *Pflügers Archiv: European Journal of Physiology*, 349: 19–27.

Andruchov, O., Andruchova, O., and Galler, S. (2006) The catch state of molluscan catch muscle is established during activation: Experiments on skinned fibre preparations of the anterior byssus retractor muscle of *Mytilus edulis* L. using the myosin inhibitors orthovanadate and blebbistatin. *Journal of Experimental Biology*, 209: 4319–4328.

Andruchov, O., Andruchova, O., Wang, Y., and Galler, S. (2004) Kinetic properties of myosin heavy chain isoforms in mouse skeletal muscle: Comparison with rat, rabbit and human and correlation with amino acid sequence. *American Journal of Physiology*, 287: C1725–C1732.

Andruchova, O., Höpflinger, M. C., Andruchov, O., and Galler, S. (2005) No effect of twitchin phosphorylation on the rate of myosin head detachment in molluscan catch muscle: Are myosin heads involved in the catch state? *Pflügers Archiv: European Journal of Physiology*, 450: 326–334.

Auerbach, D., Bantle, S., Keller, S., Hinderling, V., Leu, M., Ehler, E., and Perriard, J. C. (1999) Different domains of the M-band protein myomesin are involved in myosin binding and M-band targeting. *Molecular Biology of the Cell*, 10: 1297–1308.

Avrova, S. V., Rysev, N. A., Matusovsky, O. S., Shelud'ko, N. S., and Borovikov, Y. S. (2012) Twitchin can regulate the ATPase cycle of actomyosin in a phosphorylation-dependent manner in skinned mammalian skeletal muscle fibres. *Archives of Biochemistry and Biophysics*, 521: 1–9.

Avrova, S. V., Shelud'ko, N. S., Borovikov, Y. S., and Galler, S. (2009) Twitchin of mollusc smooth muscles can induce "catch"-like properties in human skeletal muscle: Support for the assumption that the "catch" state involves twitchin linkages between myofilaments. *Journal of Comparative Physiology B*, 179: 945–950.

Baguet, F., and Gillis, J. M. (1968) Energy cost of tonic contraction in lamellibranch catch muscle. *Journal of Physiology (London)*, 198: 127–143.

Bejar, P., and Villamarin, J. A. (2006) Catalytic subunit of cAMP-dependent protein kinase from a catch muscle of the bivalve mollusk *Mytilus galloprovincialis*: Purification, characterization, and phosphorylation of muscle proteins. *Archives of Physiology and Biochemistry*, 450: 133–140.

Bennett, P. M., and Elliott, A. (1989) The "catch" mechanism in molluscan muscle: An electron microscopy study of freeze-substituted anterior byssus retractor muscle of *Mytilus edulis*. *Journal of Muscle Research and Cell Motility*, 10: 297–311.

Brecht, K., Utz, G., and Lutz, E. (1955) Über die Atmung quergestreifter und glatter Muskeln von Kaltblütlern in Ruhe, Dehnung, Kontraktion und Kontraktur. *Pflügers Archiv: European Journal of Physiology*, 260: 524–537.

Butler, T. M., Mooers, S. U., Li, C., Narayan, S., and Siegman, M. J. (1998) Regulation of catch muscle by twitchin phosphorylation: Effects on force, ATPase, and shortening. *Biophysical Journal*, 75: 1904–1914.

Butler, T. M., Mooers, S. U., Narayan. S. R., and Siegman, M. J. (2010) The N-terminal region of twitchin binds thick and thin contractile filaments: Redundant mechanisms of catch force maintenance. *Journal of Biological Chemistry*, 285: 40654–40665.

Butler, T. M., Mooers, S. U., Siegman, M. J. (2006) Catch force links and the low to high force transition of myosin. *Biophysical Journal*, 90: 3193–3202.

Butler, T. M., Narayan, S. R., Mooers, S. U., Hartshorne, D. J., and Siegman, M. J. (2001) The myosin cross-bridge cycle and its control by twitchin phosphorylation in catch muscle. *Biophysical Journal*, 80: 415–426.

Butler, T. M., and Siegman, M. J. (2010) Mechanism of catch force: Tethering of thick and thin filaments by twitchin. *Journal of Biomedicine & Biotechnology*, 2010: 725207.

Butler, T. M., and Siegman, M. J. (2011) A force-activated kinase in a catch smooth muscle. *Journal of Muscle Research and Cell Motility*, 31: 349–358.

Castellani, L., and Cohen, C. (1987) Myosin rod phosphorylation and the catch state of molluscan muscles. *Science*, 235: 334–337.

Castellani, L., and Cohen, C. (1992) A calcineurin-like phosphatase is required for catch contraction. *FEBS Letters*, 309: 321–326.

Cornelius, F. (1982) Tonic contraction and the control of relaxation in a chemically skinned molluscan smooth muscle. *Journal of General Physiology*, 79: 821 834.

Elfvin, M., Levine, R. J., and Dewey, M. M. (1976) Paramyosin in invertebrate muscles. I. Identification and localization. *Journal of Cell Biology*, 71: 261–272.

Elliott, G. F. (1982) Structure of paramyosin filaments. In *Basic Biology of Muscles: A Comparative Approach* (Twarog, B. M., Levine, R. J. C., and Dewey, M. M., eds.), New York: Raven Press, New York.

Flechter, C. M. (1937) The relation between the mechanical and electrical activity of a molluscan unstriated muscle. *Journal of Physiology (London)*, 91: 172–185.

Franke, A. S., Mooers, S. U., Narayan, S. R., Siegman, M. J., and Butler, T. M. (2007) Myosin cross-bridge kinetics and the mechanism of catch. *Biophysical Journal*, 93: 554–565.

Funabara, D., Hamamoto, C., Yamamoto, K., Inoue, A., Ueda, M., Osawa, R., Kanoh, S., Hartshorne, D. J., Suzuki, S. & Watabe, S. (2007) Unphosphorylated twitchin forms a complex with actin and myosin that may contribute to tension maintenance in catch. *Journal of Experimental Biology*, 210: 4399–4410.

Funabara, D., Kanoh, S., Siegman, M. J., Butler, T. M., Hartshorne, D. J., and Watabe, S. J. (2005) Twitchin as a regulator of catch contraction in molluscan smooth muscle. *Journal of Muscle Research and Cell Motility*, 26: 455–460.

Funabara, D., Kinoshita, S., Watabe, S., Siegman, M. J., Butler, T. M., and Hartshorne, D. J. (2001) Phosphorylation of molluscan twitchin by the cAMP-dependent protein kinase. *Biochemistry*, 40: 2087–2095.

Funabara, D., Osawa, R., Ueda, M., Kanoh, S., Hartshorne, D. J., and Watabe, S. (2009) Myosin loop 2 is involved in the formation of a trimeric complex of twitchin, actin, and myosin. *Journal of Biological Chemistry*, 284: 18015–18020.

Funabara, D., Osawa, R., Ueda, M., Kanoh, S., Hartshorne, D. J., and Watabe, S. (2010) Myosin Loop 2 is involved in the formation of a trimeric complex of twitchin, actin, and myosin. *Journal of Biological Chemistry*, 284: 18015–18020.

Funabara, D., Watabe, S., Mooers, S. U., Narayan, S., Dudas, C., Hartshorne, D. J., Siegman, M. J., and Butler, T. M. (2003) Twitchin from molluscan catch muscle: Primary structure and relationship between sitespecific phosphorylation and mechanical function. *Journal of Biological Chemistry*, 278: 29308–29316.

Galler, S. (2008) Molecular basis of the catch state in molluscan smooth muscles: A catchy challenge. *Journal of Muscle Research and Cell Motility*, 29: 73–99.

Galler, S., Höpflinger, M. C., Andruchov, O., Andruchova, O., and Grassberger, H. (2005) Effects of vanadate, phosphate and 2,3-butanedione monoxime (BDM) on skinned molluscan catch muscle. *Pflügers Archiv: European Journal of Physiology*, 449: 372–383.

Galler, S., Kögler, H., Ivemeyer, M., and Rüegg, J. C. (1999) Force responses of skinned molluscan catch muscle following photoliberation of ATP. *Pflügers Archiv: European Journal of Physiology*, 438: 525–530.

Galler, S., Litzlbauer, J., Kröss, M., and Grassberger, H. (2010) The highly efficient holding function of the mollusc "catch" muscle is not based on decelerated myosin head cross-bridge cycles. *Proceedings of the Royal Society B*, 277: 803–808.

Gilloteaux, J., and Baguet, F. (1977) Contractile filaments organization in functional states of the anterior bysssus retractor muscle (ABRM) of *Mytilus edulis* L. *European Journal of Cell Biology*, 15: 192–220.

Grützner, P. (1904) Die glatten Muskeln. *Ergebnisse der Physiologie*, 3: 12–188.

Hauck, R., and Achazi, R. K. (1987) The ultrastructure of a molluscan catch muscle during a contraction–relaxation cycle. *European Journal of Cell Biology*, 45: 30–35.

Heumann, H. G., and Zebe, E. (1968) Über die Funktionweise glatter Muskelfasern, Elektronenmikroskopische Untersuchungen am Byssusretraktor (ABRM) von *Mytilus edulis*. *Zeitschrift für Zellforschung und Mikroskopische Anatomie (Vienna, Austria: 1948)*, 85: 534–551.

Höpflinger, M. C., Andruchova, O., Andruchov, O., Grassberger, H., and Galler, S. (2006) Effect of pH on the rate of myosin head detachment in molluscan catch muscle: Are myosin heads involved in the catch state? *Journal of Experimental Biology*, 209: 668–676.

Hooper, S. L., Hobbs, K. H., and Thuma, J. B. (2008) Invertebrate muscles: Thin and thick filament structure; molecular basis of contraction and its regulation, catch and asynchronous muscle, *Progress in Neurobiology*, 86: 72–127.

Huxley, A. F. (1957) Muscle structure and theories of contraction. *Progress in Biophysics and Biophysical Chemistry*, 7: 255–318.

Huxley, H. E. (1969) The mechanism of muscular contraction. *Science*, 164: 1356–1365.

Ishii, N., Simpson, A. W., and Ashley, C. C. (1989) Free calcium at rest during 'catch' in single smooth muscle cells. *Science*, 243: 1367–1368.

Kobayashi, T., Ichikawa, C., and Sugi, H. (1985) Differential effects of sinusoidal vibrations on tension and stiffness in *Mytilus* smooth muscle during catch state. *Japanese Journal of Physiology*, 35: 689–692.

Kovacs, M., Toth, J., Hetenyi, C., Malnasi-Csizmadia, A., and Sellers, J. R. (2004) Mechanism of blebbistatin inhibition of myosin II. *Journal of Biological Chemistry*, 279: 35557–35563.

Lowy, J., and Millman, B. M. (1963) The contractile mechanism of the anterior byssus retractor muscle (ABRM) of *Mytilus edulis*. *Proceedings of the Royal Society B*, 246: 105–148.

Lowy, J., Millman, B. M., and Hanson, J. (1964) Structure and function in smooth tonic muscles of lamellibranch molluscans. *Proceedings of the Royal Society B*, 160: 525–536.

Matusovsky, O. S., Matusovskaya, G. G., Dyachuk, V. A., and Shelud'ko, N. S. (2011) Molluscan catch muscle myorod and its N-terminal peptide bind to F-actin and myosin in a phosphorylation-dependent manner. *Archives of Biochemistry and Biophysics*, 509: 59–65.

Matusovsky, O. S., Shelud'ko, N. S., Permyakova, T. V., Zukowska, M., and Sobieszek, A. (2010) Catch muscle of bivalve molluscs contains myosin- and twitchin-associated protein kinase phosphorylating myorod. *Biochimica et Biophysica Acta*, 1804: 884–890.

Matusovsky, O. S., Shevchenko, U. V., Matusovskaya, G. G., Sobieszek, A., Dobrzhanskaya, A. V., and Shelud'ko, N. S. (2015) Catch muscle myorod modulates ATPase activity of Myosin in a phosphorylation-dependent way. *PLoS One*, 10: e0125379.

Morita, F., and Kondo, S. (1982) Regulatory light chain contents and molecular species of myosin in catch muscle of scallop. *Journal of Biochemistry*, 92: 977–983.

Mukou, M., Kishi, H., Shirakawa, I., Kobayashi, T., Tominaga, K., Imanishi, H., and Sugi, H. (2004) Marked load-bearing ability of *Mytilus* smooth muscle in both active and catch states as revealed by quick increases in load. *Journal of Experimental Biology*, 207: 1675–1681.

Pellegrino, M. A., Canepari, M., Rossi, R., D'Antona, G., Reggiani, C., and Bottinelli, R. (2003) Orthologous myosin isoforms and scaling of shortening velocity with body size in mouse, rat, rabbit and human muscles. *Journal of Physiology (London)*, 546: 677–689.

Pfitzer, G., and Rüegg, J. C. (1982) Molluscan catch muscle: Regulation and mechanics in living and skinned anterior byssus retractor muscle of *Mytilus edulis*. *Journal of Comparative Physiology B*, 147: 137–142.

Rüegg, J. C. (1963) Actomyosin inactivation by thiourea and the nature of viscous tone in a molluscan smooth muscle. *Proceedings of the Royal Society B*, 158: 177–195.

Rüegg, J. C., and Strassner, E. (1963) Sperrtonus und Nukleosidtriphosphate. *Zeitschrift für Naturforschung. Teil B*, 18:133–138.

Rüegg, J. C. (1964) Tropomyosin paramyosin system and prolonged contraction in a molluscan smooth muscle. *Proceedings of the Royal Society B*, 160: 536–542.

Rüegg, J. C. (1965) Physiologie und Biochemie des Sperrtonus. *Helvetica Physiologica et Pharmacologica Acta. Supplementum*, 16: 1–76.

Rüegg, J. C. (1971) Smooth muscle tone. *Physiological Reviews*, 51: 201–248.

Shelud'ko, N. S., Matusovskaya, G. G., Permyakova, T. V., and Matusovsky, O. S. (2004) Twitchin, a thick-filament protein from molluscan catch muscle, interacts with F-actin in a phosphorylation-dependent way. *Archives of Biochemistry and Biophysics*, 432: 269–277.

Shelud'ko, N. S., Tuturova, K. F., Permyakova, T. V., Plotnikov, S. V., and Orlova, A. (1999) A novel thick filament protein in smooth muscles of bivalve molluscans. *Comparative Biochemistry and Physiology*, 122B: 277–285.

Siegman, M. J., Funabara, D., Kinoshita, S., Kinoshita, W. S., Hartshorne, D. J., and Butler, T. M. (1998) Phosphorylation of a twitchin-related protein controls catch and calcium sensitivity of force production in invertebrate smooth muscle. *Proceedings of the National Academy of Sciences of the United States of America*, 95: 5383–5388.

Siegman, M. J., Mooers, S. U., Li, C., Narayan, S., Trinkle-Mulcahy, L., Watabe, S., Hartshorne, D. J., and Butler, T. M. (1997) Phosphorylation of a high molecular weight (approximately 600 kDa) protein regulates catch in invertebrate smooth muscle. *Journal of Muscle Research and Cell Motility*, 18: 655–670.

Sobieszek, A. (1973) The fine structure of the contractile apparatus of the anterior byssus retractor muscle of *Mytilus edulis*. *Journal of Ultrastructure Research*, 43: 313–343.

Sobieszek, A., Matusovsky, O. S., Permyakova, T. V., Sarg, B., Lindner, H., and Shelud'ko, N. S. (2006) Phosphorylation of myorod (catchin) by kinases tightly associated to molluscan and vertebrate smooth muscle myosins. *Archives of Biochemistry and Biophysics*, 454: 197–205.

Sobieszek, A., Sarg, B., Lindner, H., Matusovsky, O. S., and Zukowska, M. (2010) Myosin kinase of molluscan smooth muscle. Regulation by binding of calcium to the substrate and inhibition of myorod and twitchin phosphorylation by myosin. *Biochemistry*, 49:4191–4199.

Sugi, H., Iwamoto, H., Shimo, M., and Shirakawa, I. (1999) Evidence for load-bearing structures specialized for the catch state in *Mytilus* smooth muscle. *Comparative Biochemistry and Physiology*, 122A: 347–353.

Tajima, Y., Takahashi, W., Ito, A. (2008) Small-angle X-ray diffraction studies of a molluscan smooth muscle in the catch state. *Journal of Muscle Research and Cell Motility*, 29: 57–68.

Takahashi, I., Shimada, M., Akimoto, T., Kishi, T., and Sugi, H. (2003) Electron microscopic evidence for the thick filament interconnections associated with the catch state in the anterior byssal retractor muscle of *Mytilus edulis*. *Comparative Biochemistry and Physiology*, 134A: 115–120.

Takahashi, M., Sohma, H., and Morita, F. (1988) The steady state intermediate of scallop smooth muscle myosin ATPase and effect of light chain phosphorylation. A molecular mechanism for catch contraction. *Journal of Biochemistry*, 104: 102–107.

Twarog, B. M. (1954) Responses of a molluscan smooth muscle to acetylcholine and 5-hydroxytryptamine. *Journal of Cellular Physiology*, 44: 141–163.

Twarog, B. M. (1967) The regulation of catch in molluscan muscle. *Journal of General Physiology*, 50: 157–169.

Twarog, B. M. (1976) Aspects of smooth muscle function in molluscan catch muscle. *Physiological Reviews*, 56: 829–838.

Twarog, B. M. (1979) The nature of catch and its control. In *Motility in Cell Function* (Pepe, F. A., Sanger, J. W., and Nachmias, V. T., ed.), Academic Press, New York, pp. 231–241.

Von Uexküll, J. (1912) Studien über den Tonus VI. Die Pilgermuschel. *Zeitschrift für Biologie*, 58: 305–322.

Watabe, S., and Hartshorne, D. J. (1990) Paramyosin and the catch mechanism. *Comparative Biochemistry and Physiology*, 96B: 639–646.

Weiss, A., Schiaffino, S., and Leinwand, L. A. (1999) Comparative sequence analysis of the complete human sarcomeric myosin heavy chain family: Implications for functional diversity. *Journal of Molecular Biology*, 290: 61–75.

Yamada, A., Yoshio, M., Kojima, H., and Oiwa, K. (2001) An in vitro assay reveals essential protein components for the "catch" state of invertebrate smooth muscle. *Proceedings of the National Academy of Sciences of the United States of America*, 98: 6635–6640.

Yamada, A., Yoshio, M., Nakamura, A., Kohama, K., and Oiwa, K. (2004) Protein phosphatase 2B dephosphorylates twitchin, initiating the catch state of invertebrate smooth muscle. *Journal of Biological Chemistry*, 279: 40762–40768.

Yamada, A., Yoshio, M., Oiwa, K., and Nyitray, L. (2000) Catchin, a novel protein in molluscan catch muscles, is produced by alternative splicing from the myosin heavy chain gene. *Journal of Molecular Biology*, 295: 169–178.

Part III
Cell Motility

Chapter 15

Regulation of Dynein Activity in Oscillatory Movement of Sperm Flagella

Chikako Shingyoji

Department of Biological Sciences, Graduate School of Science,
The University of Tokyo, Hongo, Tokyo 113-0033, Japan

chikako@bs.s.u-tokyo.ac.jp

A prominent feature of flagellar and ciliary motility is oscillation. Mechanical conditions influence the frequency of oscillation. Within a beating flagellum, regardless of the presence of ATP the activity of dynein arms except those of the doublet microtubules on the two sides of the central-pair is kept low, indicating oscillation requires several modes of dynein activity. Experiments by using dynein arms (21S dynein) isolated from sea urchin sperm flagellar axonemes suggest that the dynein molecule is not only responsible for the oscillatory force generation but also capable of changing its mode under mechanical signal of bending of the interacting microtubule. The motor activity of dynein can be modified by forced bending of the interacting doublet. A possible mode change from a forward to a backward also occurred in dyneins, still attached on the doublet, as a transient event upon an application of the mechanical signal of bending.

Muscle Contraction and Cell Motility: Fundamentals and Developments
Edited by Haruo Sugi
Copyright © 2017 Pan Stanford Publishing Pte. Ltd.
ISBN 978-981-4745-16-1 (Hardcover), 978-981-4745-17-8 (eBook)
www.panstanford.com

15.1 Introduction

Cell motility in living organisms is mainly achieved by two different mechanical systems: One is muscular motility and the other is ciliary and flagellar motility. Muscular motility is achieved by muscle contraction, while ciliary and flagellar motility are achieved by ciliary and flagellar beating generating currents of fluid. These two motile systems are caused, respectively, by different molecular motors myosin and dynein. Although their basic mechanisms at a molecular level seem to share a similar rule of "sliding" (Cooke, 2004; Gibbons, 1981), a self-oscillatory nature is unique in dyneins thereby produces continuous regular beating of cilia and flagella. In addition, cilia and flagella are able to respond to mechanical strain given to them.

This chapter overviews the regulation of dynein motile activity underlying the oscillatory bending movement of flagella, in particular focusing on the studies using sea urchin sperm.

15.2 Basic Features of Flagellar Movement

A sperm flagellum is a motile apparatus to bring genetic information to eggs. The movement of sperm flagella is characterized by oscillatory beating, or continuous beating due to cyclical bending (Fig. 15.1A). Bending formation is capable at any place along the flagellum, although usually bends are initiated at the base and propagated to the tip of the flagellum. In sea urchin sperm, principal bends (P bends, a larger bends) and reverse bends (R bends) are alternately formed and propagated in the plane perpendicular to the plane of central-pair microtubules (CP) (Fig. 15.1B) (Omoto et al., 1999). Dynein arms in the 9+2 structure (Fig. 15.1C) generate force that produces sliding between adjacent doublet microtubules, and the spatial control of microtubule sliding along the flagellum is important for the formation of bending. However, the mechanism responsible for oscillation of flagella and cilia is still a mystery.

Our present understanding of the mechanism of regulation in dynein function to induce flagellar oscillation involves at least following four levels (Fig. 15.2): (1) the first level of regulation involves the intrinsic ATP-driven oscillation of the dynein arms, (2) a second level is required to coordinate the activity of the dynein arms arranged along each doublet microtubule with the initiation and propagation of successive flagellar bends, (3) a third level, associated with the regulation of the pattern of active sliding around the axonemal axis, is a switching mechanism that alternates activity between dynein arms located on opposite sides of the axonemes, and (4) the fourth level of regulation is responsible for the overall initiation of flagellar beating, coupled with the mechanical force of bending and, in some cases, involves phosphorylation of axonemal polypeptides.

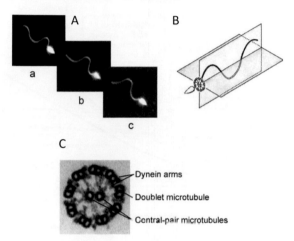

Figure 15.1 Basic characteristics of sperm flagella. A, Sequential images of a swimming sea urchin sperm. Dark-field images. Bending waves are alternately formed in both directions of the flagellum at its base (a, b) and propagating towards its tip (b, c). B, A diagram shows that the plane of bending (blue) is perpendicular to the plane of central-pair microtubules (pink) in a sea urchin sperm flagellum. An image of the 9+2 structure (not to scale) at the base of a flagellum shows the orientation of the central-pair microtubules. C, An electron micrograph showing the internal structure (axoneme) of a flagellum, called the "9+2" structure, consisting of nine doublet microtubules surrounding the central-pair microtubules, in transverse section. Dynein arms are projecting from the A-tubule of each doublet towards the B-tubule of the adjacent doublet.

To understand the mechanism of flagellar oscillation, it is an important way of studying the nature of dyneins from the view of these four levels, because the structural hierarchy is coupled with the expression of higher-order function to generate oscillation (Fig. 15.2). Dynein molecules seem to have multiple potencies as their intrinsic properties and are able to respond and behave appropriately in any higher-order functions. In beating flagella single dynein activity is regulated basically by a self-regulatory feedback system including the mechanical force of bending. Such a characteristic feature of single dynein activity in the mechanism of flagellar oscillation has become known by the early basic studies.

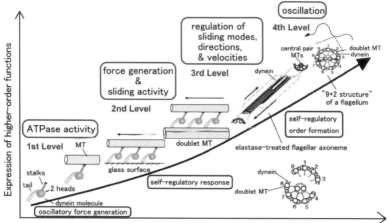

Structural hierarchy that supports the expression of self–regulatory functions of dynein, which is the basis of flagellar oscillation

Figure 15.2 A diagram illustrates the outline of mechanisms regulating active microtubule sliding to generate flagellar and ciliary beating. The mechanism of regulation in dynein function to induce flagellar oscillation involves at least four levels.

There is ample evidence that mechanical conditions influence the frequency of the flagellar oscillation (Shingyoji et al., 1991a, 1995). The sliding velocity in normally beating flagella depends not only on local variables, such as the ATP concentration, but also on other control mechanisms that are closely associated with the initiation of bending waves. Our developed procedure for changing the beat frequency by imposed vibration (Fig. 15.3) has an advantage as a reliable method for reversibly increasing or decreasing the beat frequency (Gibbons et al., 1987; Shingyoji et al., 1991a; Shingyoji et al., 1995; Omoto et al., 1999; Shingyoji, 2013) (Fig. 15.3C). Using this method we contributed to deepen our understanding the role of mechanical conditions regulating basic components of flagellar oscillation.

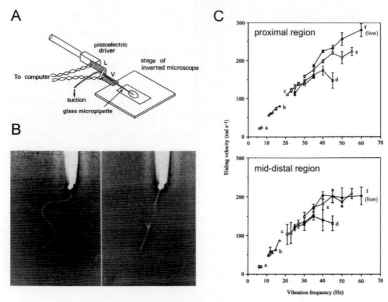

Figure 15.3 Measurement of sliding velocity under imposed vibration. (A) The device for vibrating and rotating the sperm head. (B) Photographs of beating, live flagella taken before (left) and after (right) rotation of 90° of the beat plane. (C) Effects of vibration frequency on the apparent sliding velocity in the proximal (upper panel) and the mid-distal region (lower panel). (a–e) reactivated sperm; (f) live sperm. (a) 10 µM ATP; (b) 20 µM ATP; (c) 50 µM ATP; (d) 100 µM ATP; (e) 2 mM ATP. Modified from Fig. 1 in Shingyoji et al. (1991b) and Fig. 5 in Shingyoji et al. (1995).

15.3 Dynein Force Generation and Microtubule Sliding in the Axoneme

Most of the early studies on the mechanism of flagellar oscillation are mainly considered at the third and fourth levels of Fig. 15.2 using intact axonemes (or whole sperm). The 9+2 structure is thought to be responsible for generating oscillation and the mechanism of flagellar oscillation has been thought to exist in the 9+2 structure itself, because there were no reports that showed oscillation is an inherent property of a single dynein or a row of dyneins on the doublets. Oscillatory movement is related to the force generation and sliding activity of dynein. Therefore, we should consider a possible contribution of dyneins. However, measuring the force of motor proteins, which has become popular recently, was not easy in the early days of the study.

An important pioneering work was done by Shinji Kamimura and Keiichi Takahashi, who succeeded in measuring the force of microtubule sliding in flagella as early as 1981 (Kamimura and Takahashi, 1981). They attached glass microneedles to an axoneme of sea urchin sperm flagella and measured the force generated by the dynein arms when sliding movement was induced between doublet microtubules (Fig. 15.4a). From the elastic deflection of the glass needle, they estimated the force of a single dynein molecule to be about 1 pN. Their work was the first successful attempt to measure the force of a motor protein directly, and provided an impetus for the development of force measurement of single motors. Furthermore, they found that the force of sliding sometimes oscillates under isometric conditions

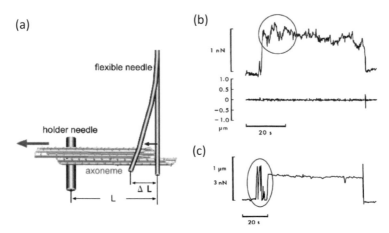

Figure 15.4   Measurement of microtubule sliding force in flagella. (a) Diagram illustrating the principle of experiment. The demembranated flagellar axoneme of sea urchin sperm was attached at two sites to a pair of microneedles—a stiff (holder) needle and a flexible needle. Application of ATP and elastase induced sliding disintegration of the axoneme. From the displacement of the flexible needle, the sliding force was calculated. (b and c) Isometric force recorded with a servo-controlled device (b) and a relatively stiff needle (c). The sliding force sometimes oscillates (red circles). Modified from Figs. 4 and 5 in Takahashi and Kamimura (1982).

(Fig. 15.4b,c) (Takahashi and Kamimura, 1982). The findings of Kamimura and Takahashi seemed to indicate that the "9+2" structure is responsible for generating oscillation.

Oscillation of the sliding force means periodic reversal of the direction of microtubule sliding. If we assume that the direction of force generation by dynein is kept constant, a plausible mechanism for the oscillation would be that the activity of dynein arms on one side of the axoneme alternates with the activity of those on the opposite side of the axoneme. This is the switching hypothesis and there is some evidence to believe that the switching of dynein activity is mediated by the central pair microtubules, which are the center of the axoneme (see the next section). However, the role of the 9+2 structure for generating flagellar oscillation does not exclude the possibility of involvement of the dynein arms themselves in the oscillation.

Using optical trap nanometry, measurement of force generated by dynein at a single molecular level was succeeded around 1998 (Shingyoji et al., 1998; Sakakibara et al., 1999; Hirakawa et al., 2000). Figure 15.5 illustrates an outline of our experiment. We obtained doublet microtubules with dynein arms by inducing sliding disintegration of elastase-treated sea urchin sperm flagella with ATP. A singlet microtubule derived from a bovine brain was brought close to the doublet by means of an optically trapped bead. For transient activation of dynein on the doublet, ATP was applied by UV-photolysis of caged ATP. Since dynein arms are arranged with a 24 nm periodicity, and the diameter of the microtubule is 25 nm, the force of a single dynein arm can be measured if the dynein can generate force when the microtubule interacts with the doublet at an angle larger than 45°. Measurement at 90° would be ideal. We found that the peak force was always 6 pN regardless of the angle of interaction. At 90° one or two of the dynein arms can interact with the microtubule and the maximum force of 15 pN was nearly twice as large as the peak force of 6 pN. The force record sometimes showed wavy tracings that seem to contain some regular components. Figure 15.6b is an example of regular fluctuations when the microtubule was interacted with a dynein at 90° (Fig. 15.6a) and it is clear that we are looking at an oscillatory response. These observations indicate that the force of a single dynein molecule oscillates.

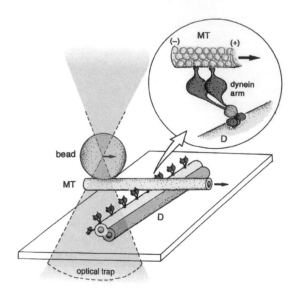

Figure 15.5 Force measurement of dynein arms. Schematic diagram of exposed dynein arms on a doublet microtubule (D) interacting with a singlet microtubule (MT). The singlet microtubule is manipulated with an optically trapped streptavidin-coated bead. Modified from Fig. 1 in Shingyoji et al. (1998).

Observations of oscillatory responses in dynein force reveal that dynein still attached on the doublet seems to have within itself a regulatory mechanism. How does dynein behave under various mechanical signals? Experiments by using dynein arms (21S dynein) isolated from sea urchin sperm flagellar axonemes suggest that the dynein molecule is capable of changing its mode (e.g., from an active forward mode to bidirectional mode) (Shingyoji et al., 2015). However, oscillation does not occur in isolated dyneins. Oscillation requires an intact arrangement of dynein on the doublet microtubule. A mode change from a forward to a backward occurred in dyneins, again still attached on the doublet, as a transient event upon an application of mechanical signal (Shingyoji et al., 2015). Figure 15.6c shows an example of generation of the backward force oscillation induced after an application of mechanical shifting of the doublet (toward the "pulling" direction in Fig. 15.6a is more effective than the opposite direction). These findings indicate that shifting of the doublet microtubule acts as a mechanical signal for the dynein-microtubule interaction, thereby inducing a structural change in the dynein and/or the microtubule. The axonemal dynein molecule is not simply a minus-end-directed force generator, but rather a direction-switching force generator (Shingyoji et al., 2015).

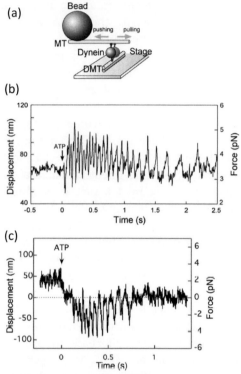

Figure 15.6 Measurement of force-generating interaction between a dynein arm on a doublet microtubule (DMT) and a singlet microtubule (MT) attached to an optically trapped bead. (a) Scheme of experiment (not to scale). ATP induces an active force to displace the MT towards its plus end (pulling or pushing). (b and c) Oscillation of the force following an application of ATP (arrows) in the normal (b) and in the reverse (c) directions. (b) The force of 3–5 pN was maintained in the normal direction during the oscillation and its frequency decreased gradually due to the decrease of ATP concentration. (c) Oscillation of the force in the reverse direction (indicated by the negative value). The force of 2–6 pN changed across the zero level from –5 to 2 pN. The frequency also decreased gradually.

15.4 Control of Microtubule Sliding and Bend Formation

To induce rhythmic oscillatory bending movement, microtubule sliding in the axoneme has to be regulated. In sea urchin sperm flagella, the bending movement occurs in one plane, the plane perpendicular to the plane of the central-pair microtubules (Figs. 15.1 and 15.7). How the sliding of the nine doublet microtubules can produce such planar bending is one of the important questions. Our early work contributed to the establishment of the sliding microtubule model of flagellar motility (Shingyoji et al., 1977; Shingyoji and Takahashi, 1995; Shingyoji, 2011). Demembranated sperm flagellum becomes motionless as long as there is no ATP. Local application of ATP by iontophoresis activated a small region of the flagellum and induced a localized bending near the ATP pipette. This experiment provided a strong evidence for the sliding microtubule model, that is, sliding movement of microtubules is the basis for bending.

Local bending can be induced in two opposite directions when a small amount of ATP is applied repetitively to any region of the flagellum (Shingyoji and Takahashi, 1995). Thus, the repetitive application of ATP alternated the direction of local bending. Peter Satir proposed a switching hypothesis for the oscillatory bending of cilia and flagella (Satir and Matsuoka, 1989), in which the activity of dynein arms on one side and that on the opposite side of the axoneme alternate to produce the sliding necessary for the principal and the reverse bends. The question is whether the dynein activity alternates always between the doublet no. 3 side and the opposite, doublet no. 7 side (Fig. 15.7).

In 1987 in collaboration with Ian Gibbons, we found an interesting behavior of the sea urchin sperm flagella, which showed a versatile property of dynein arms. When we caught a single sperm by holding its head by gentle suction with a micropipette and the micropipette was vibrated laterally, the flagellum began to beat in phase with the imposed vibration (Gibbons et al., 1987) (Fig. 15.3A). Furthermore, we found that when the plane of pipette vibration was rotated around the head axis without rotating the pipette itself induced a corresponding rotation of the plane of beating, in both live and reactivated sperm (Shingyoji et al., 1991b; Takahashi et al., 1991). We attached small

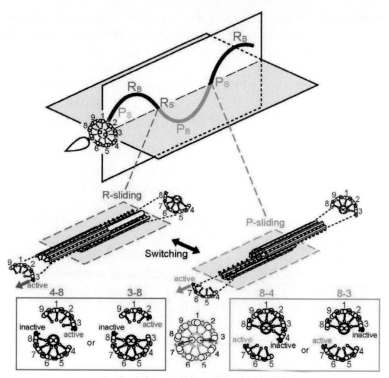

Bending induces switching of the dyein activity
through the CP/RS system

Figure 15.7 Diagrams showing the possible regulation of dynein activity in a switching model of beating flagella. In the sea urchin sperm flagellum, principal and reverse bends (P_B and R_B) are cyclically formed in the plane of beat (white), which is perpendicular to the plane of the central-pair (grey). The formation of P_B and R_B is due to P-sliding (P_S) and R-sliding (R_S), respectively, induced by the dynein arms of doublets 7 and 3 (or 2). The model is based on our previous studies (Shingyoji and Takahashi, 1995; Hayashi and Shingyoji, 2008). In the locally reactivated axonemes, microtubule sliding occurs only in the region (inter-bend region) between the pair of opposite bends and a reversal of the sliding direction between the two bends leads to alternation of the direction of these bends. Coupling of the mechanism regulating microtubule sliding in the paired bend formation with the regulation that induces a reversal of the sliding direction in the bending region may constitute a basis for the coordination of cyclical bend formation and propagation (Hayashi and Shingyoji, 2008; Shingyoji, 2011).

polystyrene beads as markers on the peripheral doublet microtubules and observed the positions of the beads relative to the axoneme during rotation of the beat plane (Takahashi et al., 1991). During rotation of the beat plane, the position of a bead on the axoneme always remains constant relative to the flagellum and appears unaffected by the rotation. This indicates that the axonemal framework of nine outer doublet microtubules does not twist but only the beating plane is rotated. It is possible that the rotation of the plane of imposed vibration causes rotation of the central pair, inducing rotation of the beat plane (Fig. 15.3B). These observations indicate that the regulation of dynein activity is not related to particular fixed dynein arms in the axoneme but is mediated by a rotatable component, probably the central pair microtubules.

Cyclical bending can be induced locally in demembranated flagellar axonemes by repetitive application of a minute amount of ATP (Shingyoji and Takahashi, 1995). In trypsin-treated axonemes we could not induce such a cyclical bending. This is what we explained by the later finding in the effects of trypsin on the dynein activity (Imai and Shingyoji, 2003). However, when the axoneme was treated with elastase instead of trypsin, cyclical bending was induced by local applications of ATP, although elastase can induce microtubule sliding in the axonemes (Shingyoji and Takahashi, 1995).

In addition to the ability of cyclical bending the elastase-treated axonemes showed unique characteristics in the sliding patterns depending on the concentrations of ATP and Ca^{2+} (Nakano et al., 2003). At lower concentrations of ATP and Ca^{2+}, the elastase-treated axonemes showed sliding disintegration into individual doublet microtubules. This is very similar to the sliding disintegration observed in the trypsin-treated axonemes. At higher concentrations of ATP (≥ 0.1 mM) and Ca^{2+} ($\sim 10^{-4}$ M), however, the sliding was restricted at one or two inter-doublet sites, which induced splitting of the axoneme into a thinner and a thicker doublet bundles. Now, the thinner bundles were always distinguished from the thicker bundles. Electron micrographs of thicker and thinner bundles showed fixed numbers of doublet microtubules within the bundles. The thicker bundles contained five or six doublet microtubules with the central pair microtubules, whereas the thinner bundles contained the remaining four or three doublet bundles without the central pair.

Furthermore the members of doublets within the bundles were also fixed. The thicker bundles showed mainly four patterns of doublet groups, that is, 8-4, 8-3, 4-8 and 3-8 patterns. By increasing the concentration of Ca^{2+}, the 8-4 and 8-3 patterns of the four became dominant. This indicates that the splitting of the axoneme may be related to some mechanism regulating the dynein activity, which is sensitive to Ca^{2+} (Nakano et al., 2003). Thus, the elastase-treated axonemes still retain a mechanism responding to Ca^{2+} as well as producing cyclical bending. We analyzed the direction of the movement of the thinner bundles relative to the microtubule polarity and found that the thinner bundles always move toward the minus end of the microtubule. This indicates that the dynein arms on the thinner bundles were dominantly active to induce splitting of the axonemes into paired bundles. The alternate activation of these dyneins on the thinner bundles on both sides of the central pair microtubules may be the basis of switching mechanism underlying flagellar oscillation.

15.5 Regulation of Dynein Activities by Mechanical Signal

We succeeded to demonstrate the switching of dynein sliding activity by bending the axonemal microtubules (Morita and Shingyoji, 2004; Hayashi and Shingyoji, 2008). As described above, elastase-treated flagellar axonemes split into two bundles at high ATP and high Ca^{2+} conditions. When we activated the axoneme by photolysis of caged ATP by a UV flash, the restricted sliding between a thinner and a thicker bundles occurred for a short distance. In order to investigate the effect of forced bending, we bent either the thinner or the thicker bundle with a glass microneedle and applied a UV flash to release ATP from caged ATP. As a result we found that bending switched the dynein activity on and off. In addition to the activation of the inactive dyneins, the forced bending of doublet microtubules sometimes induced a change of direction of sliding.

The dynein arms on the thinner bundle, but not those on the thicker bundles, are mainly active to induce the separation of elastase-treated axonemes. Thus, we conclude that switching of the dynein activity from no. 7 to no. 3 is induced by bending, although at the present time we cannot rule out the possibility that some dynein arms on doublet no. 7 behave as a plus end motor (Shingyoji et al., 2015). Our present understanding is that it is not simply the bending doublet microtubules but the direction of bending that determines the timing of the switching of dynein activity (Hayashi and Shingyoji, 2008).

This is a model for the self-regulatory feedback for flagellar oscillation. A difference in sliding amounts induces bending, and the bending triggers to activate the inactive dynein as well as switches the dynein activity on and off at doublets no. 7 and no. 3 through the central pair microtubules. Our findings show that the initial step of switching during each beat cycle primarily involves a mechanical effect of bending rather than chemical reactions as an essential element in the self-regulatory feedback system for oscillatory flagellar movement. And at this switching, the backward force generation that is, discussed in the previous section, induced by the bending and appears for limited period just after the bending (Shingyoji et al., 2015) may play an essential role for the activation of dyneins on the opposite side of the central pair microtubules.

15.6 Outlook

There are many unsolved problems about activation and regulation of dynein. Among those, the roles of the dynein force oscillation, the central-pair microtubules, the state of ATP binding and the mechanical bending of dynein-microtubule interacting system are important ones we should consider.

Acknowledgments

I thank Prof. Hideo Higuchi for valuable discussions and the members of my group, past and present, who have collaborated with me on dynein and sea urchin sperm flagella (M. Yoshimura, T. Kobayashi, H. Imai, I. Nakano, Y. Inoue, Y. Morita, R. Ishikawa, A. Yoshimura, S. Hayashi, Y. Kambara, H. Yoke, Y. Watanabe, I. Nishide, S. Umeyama, Y. Izawa, T. Fujiwara, and J. Jirakulsomchok).

References

Cooke, R. (2004). The sliding filament model: 1972–2004. *J. Gen. Physiol.*, **123**, 643–656.

Gibbons I. R. (1981). Cilia and flagella of eukaryotes. *J. Cell Biol.*, **91**, 107s–124s.

Gibbons, I. R., Shingyoji, C., Murakami, A., and Takahashi, K. (1987). Spontaneous recovery after experimental manipulation of the plane of beat in sperm flagella. *Nature*, **325**, 351–352.

Hayashi, S., and Shingyoji, C. (2008). Mechanism of flagellar oscillation—bending-induced switching of dynein activity in elastase-treated axonemes of sea urchin sperm. *J. Cell Sci.*, **121**, 2833–2843.

Hirakawa, E., Higuchi, H., and Toyoshima Y. Y. (2000). Processive movement of single 22S dynein molecules occurs only at low ATP concentrations. *Proc. Natl. Acad. Sci. U. S. A.*, **97**, 2533–2537.

Imai, H., and Shingyoji, C. (2003). Effects of trypsin-digested outer-arm dynein fragments on the velocity of microtubule sliding in elastase-digested flagellar axonemes. *Cell Struct. Funct.*, **28**, 71–86.

Kamimura, S., and Takahashi, K. (1981). Direct measurement of the force of microtubule sliding in flagella. *Nature*, **293**, 566–568.

Morita, Y., and Shingyoji, C. (2004). Effects of imposed bending on microtubule sliding in sperm flagella. *Curr. Biol.*, **14**, 2113–2118.

Nakano, I., Kobayashi, T., Yoshimura, M., and Shingyoji, C. (2003). Central-pair-linked regulation of microtubule sliding by calcium in flagellar axonemes. *J. Cell Sci.*, **116**, 1627–1636.

Omoto, C. K., Gibbons, I. R., Kamiya, R., Shingyoji, C., Takahashi, K., and Witman, G. B. (1999). Rotation of the central pair microtubules in eukaryotic flagella. *Mol. Biol. Cell,* **10**, 1–4.

Sakakibara, H., Kojima, H., Sakai, Y., Katayama, E., and Oiwa, K. (1999). Inner-arm dynein c of Chlamydomonas flagella is a single-headed processive motor. *Nature*, **400**, 586–590.

Satir, P., and Matsuoka, T. (1989). Splitting the ciliary axoneme: Implications for a "switch-point" model of dynein arm activity in ciliary motion. *Cell Motil. Cytoskeleton*, **14**, 345–358.

Shingyoji, C. (2011). Regulation of dynein in ciliary and flagellar movement. In *Dyneins –Structure, Biology and Disease* (King, S. M., ed.), UK: Elsevier, pp. 367–393.

Shingyoji, C. (2013). Measuring the regulation of dynein activity during flagellar motility. *Methods Enzymol.,* **524**, 147–169.

Shingyoji, C., Gibbons, I. R., Murakami, A., and Takahashi, K. (1991a). Effect of imposed head vibration on the stability and waveform of flagellar beating in sea urchin spermatozoa. *J. Exp. Biol.,* **156**, 63–80.

Shingyoji, C., Higuchi, H., Yoshimura, M., Katayama, E., and Yanagida, T. (1998). Dynein arms are oscillating force generators. *Nature,* **393**, 711–714.

Shingyoji, C., Katada, J., Takahashi, K., and Gibbons, I. R. (1991b). Rotating the plane of imposed vibration can rotate the plane of flagellar beating in sea-urchin sperm without twisting the axoneme. *J. Cell Sci.,* **98**, 175–181.

Shingyoji, C., Murakami, A., and Takahashi, K. (1977). Local reactivation of Triton-extracted flagella by iontophoretic application of ATP. *Nature,* **265**, 269–270.

Shingyoji, C., Nakano, I., Inoue, Y., and Higuchi, H. (2015). Dynein arms are strain-dependent direction-switching force generators. *Cytoskeleton (Hoboken),* **72**, 388–401.

Shingyoji, C., and Takahashi, K. (1995). Cyclical bending movements induced locally by successive iontophoretic application of ATP to an elastase-treated flagellar axoneme. *J. Cell Sci.,* **108**, 1359–1369.

Shingyoji, C., Yoshimura, K., Eshel, D., Takahashi, K., and Gibbons, I. R. (1995). Effect of beat frequency on the velocity of microtubule sliding in reactivated sea urchin sperm flagella under imposed head vibration. *J. Exp. Biol.,* **198**, 645–653.

Takahashi, K., and Kamimura, S. (1982). The dynamics of microtubule sliding in flagella. In *Biological Function of Microtubules and Related Structures* (Sakai, H., Mohri, H., and Borisy, G, G., eds.), Academic Press. N.Y., pp. 177–187.

Takahashi, K., Shingyoji, C., Katada, J., Eshel, D., and Gibbons, I. R. (1991). Polarity in spontaneous unwinding after prior rotation of the flagellar beat plane in sea-urchin spermatozoa. *J. Cell Sci.,* **98**, 183–189.

Chapter 16

The Biomechanics of Cell Migration

Yoshiaki Iwadate

Faculty of Science, Yamaguchi University, Yamaguchi, Japan

Iwadate@yamaguchi-u.ac.jp

Crawling migration plays an essential role in a range of biological phenomena, including development, wound healing, and immune system function. The current consensus is that in the typical mechanism of crawling cell migration, actin polymerization pushes the cell front, and actomyosin contraction pulls the rear, although there is a high degree of variability in the details according to cell type. These forces are translated into traction forces on the substratum, and, simultaneously, appear to regulate the tension of the actin cytoskeleton. Cells can generate anterior-posterior polarity to drive migration by mechanical interaction between the cells and the substratum. In relation to this polarity generation, it has been shown that cofilin and myosin II change their affinity to actin filaments dependent on their tension. How cell migration is regulated by mechanical signals via cofilin and myosin II-related processes remains an interesting future area of study.

Muscle Contraction and Cell Motility: Fundamentals and Developments
Edited by Haruo Sugi
Copyright © 2017 Pan Stanford Publishing Pte. Ltd.
ISBN 978-981-4745-16-1 (Hardcover), 978-981-4745-17-8 (eBook)
www.panstanford.com

16.1   Introduction

The bodies of multicellular organisms are built up of a vast range of cell types. The total number of cells constituting the human body surpasses 10 trillion (Bianconi et al., 2013), all of which originate in a fertilized egg, a single cell. During development into the adult form, the separate and differentiated cells must move to the appropriate sites to form new tissues and organs. Cells do not migrate only during the developmental stage: Many other cell types in adult organisms are not static but continuously move around. Swimming is a major form of movement, exhibited by spermatozoa and many types of protozoan, such as *Paramecium* sp. (Jennings, 1906). They swim by beating their flagella or cilia (Huitorel, 1988). The other major form of movement is crawling migration. It plays an essential role in various biological phenomena, including development (Lauffenburger and Horwitz, 1996; Ridley et al., 2003), wound healing (Martin, 1997), immune system function (Weninger et al., 2014), not to mention cancer metastasis (Stroka and Konstantopoulos, 2014). The movement of cells in developmental stages, as mentioned above, is realized by crawling migration. In adult organisms, if the skin is damaged, the surrounding healthy epithelial cells migrate to the wound site and cover it. In immune response and cancer metastasis, neutrophils and metastatic cancer cells adhere to the endothelium and begin intraluminal crawling migration, after being carried via the bloodstream. They then infiltrate new sites.

The details of the characteristics of crawling migration vary according to cell type. For instance, the migration velocity of fibroblasts is about one-tenth that of *Dictyostelium* cells and neutrophils (Iwadate et al., 2013; Shin et al., 2010; Wang et al., 2001). On the other hand, the traction force applied to the substratum by fibroblasts is 10 times greater than that by *Dictyostelium* cells and neutrophils (Iwadate and Yumura, 2008; Lombardi et al., 2007; Munevar et al., 2001a; Shin et al., 2010; Tanimoto and Sano, 2014). However, actin cytoskeleton-related common mechanisms are conserved in the migration of most cell types. Migrating cells adhere to the substratum by forming connections between the intracellular actin cytoskeleton and the extracellular matrix via focal adhesions throughout the cell bottom (Giannone et al., 2009). Actin polymerization takes place at the leading edge and pushes the front of the cell (Parent, 2004), whereas the detachment and retraction of the rear of the cell from the substratum is induced by contraction effected by myosin-II-dependent processes (Chen, 1981). Traction forces are exerted mainly at the front and the rear focal adhesions of the cell (Beningo et al., 2001).

The initial step to realizing crawling migration is the generation of anterior-posterior polarity within the cell. Most cell types can sense chemoattractants and migrate toward them (Parent, 2004). However, they will migrate even in the absence of chemoattractants. The cells seem to be able to sense not only chemical signals but also mechanical ones from their surrounding environment, such as the hardness and shape of the substratum (Lo et al., 2000; Miyoshi et al., 2010). Movement of all objects, including living cells, is caused by forces. Actin and actomyosin play an important role in migrating cells, in that they apply traction forces to the substratum. In this chapter, we will examine the mechanism of crawling cell migration, focusing particularly on its mechanical aspects.

16.2   The Cytoskeleton

A cell's cytoskeleton is a complex architecture of filaments that maintains its shape. It also has important roles in physiological functions, such as the generation of the mechanical forces required for intra- and extracellular motion and intracellular transport of organelles and vesicles. In most cell types, there are three main types of cytoskeleton that consist of different proteins, such as filamentous actin (F-actin), intermediate filaments, and microtubules. F-actin, with a diameter of ~7 nm, takes the form of a double helix made of two actin strands coiled around each other. Microtubules are hollow tubular fibers with a much thicker diameter of ~25 nm (internal diameter ~15 nm) made of the protein tubulin. The diameter of intermediate filaments is between those of F-actin and microtubules. There are several kinds of intermediate filaments that differ from each other depending on their constituent proteins: vimentin, keratin, lamin, and so on. These cytoskeleton filaments perform different functions within the cell. Interaction of these filaments with the other organelles and proteins, such as focal adhesions, bipolar myosin II filaments, kinesin, and dynein, enables a cell to deform its shape for migration and cell division, and to transport organelles.

16.2.1 Cytoskeleton and Cell Type

Migrating cells can be classified into two groups that differ from each other according to their migration speeds (Table 16.1). *Dictyostelium* cells, neutrophils, and fish epidermal keratocytes are fast-moving cell types. Their speed of migration (~10 µm/min) is much faster than that of slow-moving cell types, such as fibroblasts (<1 µm/min), suggesting that different cytoskeleton systems in each group realize migration at the optimal speed for that cell type.

The shape of slow-migrating cells, such as fibroblasts, is maintained by the tension of thick contractile bundles of F-actin termed "stress fibers." F-actin bundles in stress fibers are cross-linked with various proteins, such as α-actinin and bipolar myosin II filaments, to form a highly regulated actomyosin structure. The ends of stress fibers are bonded to the extracellular matrix via focal adhesions. On the other hand, in fast-migrating cells, such as *Dictyostelium* cells, a dense meshwork of thin F-actin bundles, rather than stress fibers, extends just under the cell membrane throughout the cell (Fig. 16.1). Interestingly, *Dictyostelium* cells seem to be able to adhere to a substrate without any extracellular matrix. Moreover, no macromolecular assemblies similar to focal adhesions that would adhere to the substratum have as yet been discovered. The adhesion strength to the substratum of *Dictyostelium* cells (Tarantola et al., 2014) is much smaller than that of fibroblasts (Engler et al., 2009).

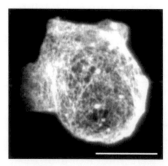

Figure 16.1   Actin meshwork in a *Dictyostelium* cell. Bar = 5 µm.

Table 16.1 The characteristics of migrating cells

Cell types	Slow	Fast		
	Fibroblasts	Dictyostelium discoideum	Neutrophils or HL-60 cells	Keratocytes
Size [μm]	~100[1]	~10[2]	~10[2]	~10 (longitudinal) ~50 (lateral)[2]
Velocity [μm/min]	0.827 ± 0.175[1]	7.1 ± 0.17[3]	9.95 ± 3.37[4]	10.1 ± 0.64[5]
Adhesion strength (Shear stress [Pa])	73 ± 3[6]	<1[7]	—	—
Stress fiber	Yes[8]	No[3]	—	Yes[9]
Peak traction force [kPa]	~10[10]	~0.66[11]	0.638 ± 0.102[12]	0.66 ± 0.29[5]
Passive mechanosensing Migration	—	Migration perpendicular to the periodic stretching of substratum[3]	Migration perpendicular to the periodic stretching of substratum[13]	Migration perpendicular or parallel to the periodic stretching of substratum[14]
Passive mechanosensing Cytoskeleton	Stress fiber alignment perpendicular to the periodic stretching of substratum[15]	Myosin II accumulation at both stretching side equally[3]	—	Stress fiber alignment perpendicular to the periodic stretching of substratum[14]
Active mechanosensing	Durotaxis[16]	—	—	—
Contact guidance	Aligned[17]	—	Aligned[18]	Turn[19]

[1]Wang et al., 2001, [2]Tsugiyama et al., 2013, [3]Iwadate et al., 2013, [4]Chen et al., 2012, [5]Nakashima et al., 2015, [6]Engler et al., 2009, [7]Tarantola et al., 2014, [8]Hotulainen and Lappalainen, 2006, [9]Svitkina et al., 1997, [10]Munevar et al., 2001, [11]Iwadate and Yumura, 2008, [12]Shin et al., 2010, [13]Okimura et al., in press, [14]Okimura and Iwadate, in press, [15]Zhao et al., 2011, [16]Lo et al., 2000, [17]Brunette, 1986, [18]Tan and Saltzman, 2002, [19]Miyoshi et al., 2010.

16.2.2 Fundamental Mechanism of Cell Migration Based on Actin Polymerization and Actomyosin Contraction

Analogous to the way that a car moves by only the rotation of its tires, *Paramecium* cells move without changing their architecture by swimming using their periodically beating cilia (Iwadate and Nakaoka, 2008; Narematsu et al., 2015). However, an adherent cell, which comprises soft matter, realizes movement by employing a highly orchestrated, continuous deformation of its architecture. To be able to move forward, an adherent cell extends its front and retracts its rear. The main cause of extension at the front is actin polymerization (Fig. 16.2). At the front, the plus end (or barbed end) of F-actin faces the cell membrane and the minus end (or pointed end) faces the cytoplasmic side. It has been thought that thermal undulations of F-actin create a gap between the cell membrane and its plus end. Actin monomer then fits into the gap and binds to the plus end of F-actin, generating the force that pushes the plasma membrane forward and F-actin rearward (Mogilner and Oster, 1996). This induces the simultaneous extension of the leading edge and of F-actin retrograde flow (ARF). Thus, the rate of extension of the leading

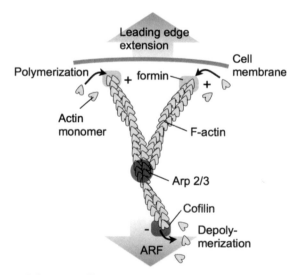

Figure 16.2 Schematic illustration of the mechanism for leading-edge extension by actin polymerization. +: plus end, –: minus end.

edge is the difference between the rate of actin polymerization and that of ARF. ARF and leading-edge extension can be measured simultaneously using fluorescence speckle microscopy (Fig. 16.3; Waterman-Storer et al., 1998), in which a limited proportion of actin monomer molecules are fluorescently labeled. The actin polymerization rate can be calculated by adding them together.

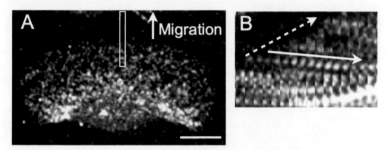

Figure 16.3 Measurement of the rates of ARF and leading-edge extension. (A) Fluorescence speckle observation of ARF in keratocytes. Bar = 5 μm. (B) Kymograph constructed from image strips, of width 1 μm (white rectangles in A) from consecutive images taken at 2 s intervals. The rates of ARF and leading-edge extension, respectively, can be calculated from the slopes of the arrow and the dotted arrow.

The actin polymerization rate and spatial organization are controlled by the complex regulation of regulatory proteins, including formin, Arp2/3, cofilin, and so on. Formin is localized at the plus end of the F-actin and enhances actin polymerization. Arp 2/3 binds to the side of preexisting F-actin and nucleates a new filament from it, resulting in a branched filamentous structure. On the other hand, cofilin binds to the minus end of F-actin and enhances its depolymerization.

The rate of leading-edge extension is affected not only by actin polymerization but also by ARF. F-actin is thought to be coupled to the cell adhesion molecules via a molecular clutch (Fig. 16.4) such as shootin (Shimada et al., 2008; Toriyama et al., 2013). When the molecular clutch is engaged, the rate of ARF is reduced (Chan and Odde, 2008; Koch et al., 2012). Myosin II motor molecules also have the role of regulating the rate of ARF: they accelerate it by pulling the F-actin network back from the leading edge (Lin et al., 1996) and by disassembling F-actin (Fuchs et al., 2014; Wilson et al., 2010).

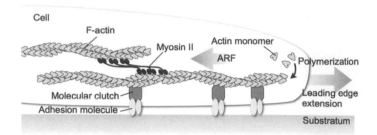

Figure 16.4 Schematic illustration of the relationship between actin polymerization, leading-edge extension, and ARF. The extension rate of the leading edge is the difference between the rate of actin polymerization and that of ARF. When the molecular clutch is engaged, focal adhesion molecules reduce the rate of ARF. Conversely, myosin II motor molecules accelerate the rate of ARF by pulling the F-actin network back from the leading edge.

It is now generally believed that retraction of the rear, induced by contraction through myosin II-dependent processes (Chen, 1981; Jay et al., 1995), is the driving force for crawling migration. In fibroblasts, the myosin II contractile force along the stress fibers induces the forward translocation of the cell body at the trailing edge (Vicente-Manzanares et al., 2009). In *Dictyostelium* cells, which have no stress fibers, myosin II accumulates at the rear actin meshwork and induces retraction there (Fig. 16.5).

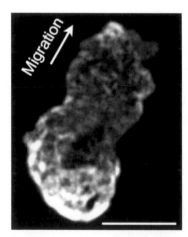

Figure 16.5 Myosin II distribution in a freely migrating *Dictyostelium* cell. Bar = 5 μm.

16.2.3   The Role of Microtubules in Maintaining Anterior-Posterior Polarity

In response to extracellular signals such as chemotactic stimuli, the cells generate anterior-posterior polarity. How do cells maintain the polarity needed for stable migration? Microtubules are thought to play an essential role in cell migration, due to observations that the application of colcemid, a microtubule depolymerizing drug, inhibits fibroblast migration (Vasiliev et al., 1970). In the front of the cell, the small Rho GTPase Rac1 is thought to coordinate Arp2/3 and formin-dependent actin nucleation (Brandt and Grosse, 2007); and in the rear, RhoA activates myosin II through Rho kinase (ROCK) activation and generates contractility (Amano et al., 2010). It is now generally believed that numerous processes essential to actin polymerization and actomyosin contraction are regulated by microtubules, and depend on distinct modes of microtubule dynamics (Fig. 16.6) (Akhshi et al., 2014; Kaverina and Straube, 2011). The microtubule plus end binding protein CLASP tethers microtubules to focal adhesions (Stehbens et al., 2014). Polymerized microtubules activate Rac1 (Waterman-Storer et al., 1999) through the activation of TIAM1 (Montenegro-Venegas et al., 2010) or STEF (TIAM2) (Rooney et al., 2010), while microtubule depolymerization promotes RhoA activation (Chang et al., 2008; Enomoto, 1996; Krendel et al., 2002) through the activation of GEF-H1 (Lfc) (Nalbant et al., 2009).

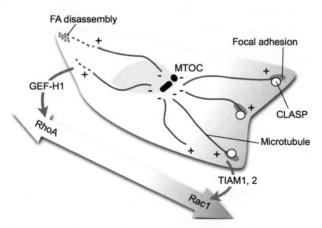

Figure 16.6   Schematic illustration of potential mechanisms by which a microtubule cytoskeleton might maintain cell polarization.

16.2.4   Variety of Cell Migration Mechanisms

It should be noted that a complete combination of the roles of actin polymerization at the front, actomyosin contraction at the rear, and microtubule dynamics in the middle is seen only rarely in migrating cells. For instance, *Amoeba proteus* appears to extend its leading edge without actin polymerization (Pomorski et al., 2007; Stockem et al., 1982), showing "bleb"-driven migration, in which actomyosin at the rear cortex first contracts to increase the hydrostatic pressure inside the cell; then the front cortex is broken and extension takes place there. Microtubules are not required by fish epidermal keratocytes for crawling migration: treatment with nocodazole, an inhibitor of microtubule polymerization, has no effect on either their migration velocity (Euteneuer and Schliwa, 1984) or the rate of ARF, the distributions of focal adhesions or myosin II, or traction force (Nakashima et al., 2015). Migrating keratocytes rarely generate cytoplasmic fragments spontaneously in culture (Euteneuer and Schliwa, 1984). Cytoplasmic fragments also maintain their migratory behavior in a pattern indistinguishable from that of intact cells, in spite of the fragments containing no microtubules or centrioles (Euteneuer and Schliwa, 1984; Verkhovsky et al., 1999).

16.3   Traction Forces

Actin meshworks and stress fibers are anchored to the substratum via focal adhesions. Their contraction, via a myosin II-dependent process, translates into traction forces on the substratum. Polymerizing F-actin at the leading edge is also anchored to the substratum via focal adhesions and molecular clutches. Actin polymerization also applies a traction force to the substratum just beneath the focal adhesions through molecular clutches. In general, the magnitude and spatial distribution of traction forces are related to cell shape and mode of movement (Oliver et al., 1999).

16.3.1   Traction Forces Exerted by Fibroblasts

"Traction force microscopy" is a technique for converting the measurement of elastic substratum deformation in migrating cells into traction stresses using the finite element method (Dembo and Wang, 1999; Dembo et al., 1996) and has been used to map forces in migrating cells (Beningo et al., 2001). Deformation of the elastic substratum can be measured as the displacement of fluorescent beads embedded in the surface of the substratum (Fig. 16.7). Traction force microscopy revealed dynamic traction stresses at the leading edges and the contractile tails of migrating fibroblasts (Munevar et al., 2001a, 2001b). At the leading edge, the force is exerted in the opposite direction to that of the leading-edge extension. On the other hand, at the rear of the cell, the force is exerted in the same direction as that of cell rear retraction. The magnitude of the force represented by the stress generated in the substrate at the leading edge and cell rear end of fibroblasts is about several tens of kPa (Table 16.1).

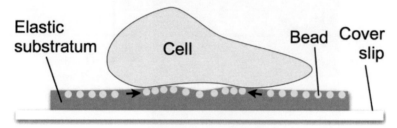

Figure 16.7 Schematic illustration of cell-traction force measurement by the displacement of fluorescent beads. Deformation of the elastic substratum is measured using the displacement of the fluorescent beads (arrows).

16.3.2 Traction Forces Exerted by *Dictyostelium* Cells and Neutrophils

Unlike fibroblasts, *Dictyostelium* cells have no stress fibers. Their migration velocity is about 10-fold that of fibroblasts, whereas their cell size is about 10 times smaller. The magnitude and spatial distribution of traction forces are dependent on cell type (Table 16.1). In *Dictyostelium* cells, rearward traction forces are exerted under spot-like regions in the front of cells where F-actin accumulates (Fig. 16.8; Iwadate and Yumura, 2008). Cells migrate in a direction counter to the sum of the force vectors at each spot. On the other hand, forward forces emerge at the posterior edge. The magnitude of the forward forces at the posterior edge is greater than those of the rearward forces at each spot in the anterior. The maximum traction force exerted by *Dictyostelium* cells is about 1 kPa, which is 10 times smaller than that exerted by fibroblasts. As expected from the analogy of cell size and cytoskeleton between *Dictyostelium* cells and neutrophils, the directionality and velocity of neutrophil migration are similar to those of *Dictyostelium* cells (Parent, 2004). The magnitude and spatial distribution of traction forces exerted by neutrophils are also similar to those of *Dictyostelium* cells (Shin et al., 2010).

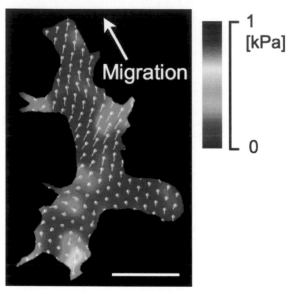

Figure 16.8 Traction force map of a *Dictyostelium* cell. The direction of strain in the substratum at each small white circle is indicated by a white bar. The length of the white bars is three times as long as the strain. Bar = 3 μm.

16.3.3   Traction Forces Applied by Keratocytes

Keratocytes are uniquely fast-moving cell types that maintain an overall fan shape while crawling (Lee et al., 1993; Fig. 16.9). In keratocytes, unlike in *Dictyostelium* cells and neutrophils, stress fibers composed of actomyosin are positioned so as to connect the rear left and right focal adhesions (Barnhart et al., 2011; Doyle and Lee, 2005; Lee and Jacobson, 1997). In fan-shaped keratocytes, traction forces are chiefly generated at the trailing left and right ends (Burton et al., 1999; Chen et al., 2013; Doyle et al., 2004; Fournier et al., 2010; Sonoda et al., 2016). The vector of the forces at the trailing left and right ends is not parallel to the direction of migration but directs to the center of the cell area. The magnitude of the forces measured at the trailing left and right ends is from several hundred Pa to several kPa (Jurado et al., 2005; Nakashima et al., 2015; Nakata et al., 2016). Although rearward traction forces might be exerted at the leading edge, their magnitude is much less than that at the trailing left and right ends. In keratocytes, contraction of the stress fibers is thought to be one of the sources of traction forces, in the same way as in fibroblasts.

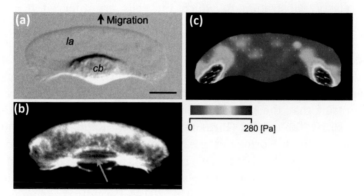

Figure 16.9 Actin cytoskeleton and traction force map of keratocytes. (a) DIC image. Bar = 5 μm. A keratocyte has a broad lamellipodium (*la*) in front of its cell body (*cb*). (b) F-actin staining. To be able to see the stress fibers in the cell body (green arrow), the brightness of the images has been increased. (c) Traction force map. The directions of the forces at each position are indicated by white arrows.

16.4   Mechanosensing and Cell Migration

As mentioned above, crawling cells exert traction forces on the substratum. The magnitude and spatial distribution of these traction forces depend on the cell type. Is traction force just the simple result of actin polymerization at the front or actomyosin contraction at the rear? Or does it trigger a further physiological step for migration via mechano-transduction? Crawling cells seem to sense mechanical forces by converting them to the tension of F-actin and utilizing them for migration. For instance, the rise in traction forces enhances the tension of the actin cytoskeleton and, at the same time, matures the focal adhesions (Galbraith et al., 2002), since crawling cells exert traction forces on the substratum via focal adhesions (Harris et al., 1980, 1981).

16.4.1 Passive Mechanosensing

Both the focal adhesions and the F-actin bundles sense these forces. To investigate the relationship between signals derived from mechanical stimuli and related cell functions, one effective technique for artificially applying mechanical stimuli is to stretch the elastic substratum to which the cells are adhered (Crosby et al., 2011; Desai et al., 2010; Iwadate and Yumura, 2009; Naruse et al., 1998a, 1998b). In response to periodic stretching of the elastic substratum, intracellular stress fibers in fibroblasts, and endothelial, osteosarcoma, and smooth muscle cells rearrange themselves perpendicular to the direction of stretching, with the result that the shape of the cells extends in that direction (Fig. 16.10; Table 16.1) (Birukov et al., 2003; Kaunas et al., 2005; Lee et al., 2010; Morioka et al., 2011; Sato et al., 2005; Tondon et al., 2011; Zhao et al., 2011). This reaction is thought to be caused by the higher affinity of cofilin to relaxed stress fibers than to extended stress fibers (Hayakawa et al., 2011, 2014). Periodic stretching and relaxation causes cofilin to bind to stress fibers that are parallel to the stretching direction at the moment when they are relaxed. The stress fibers then depolymerize, with the result that only the stress fibers perpendicular to the stretching direction remain.

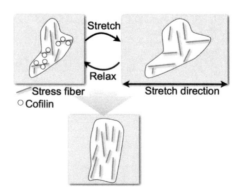

Figure 16.10 Schematic illustration of slow-migrating cells responding to periodic stretching of the substratum.

Both cofilin and myosin II seem to change their affinity to F-actin in relation to the tension of F-actin (Uyeda et al., 2011). In response to the stretching of the surface of a *Dictyostelium* cell, applied by suction using a micropipette, myosin II localizes at the tip of the sucked cell lobe (Fig. 16.11) (Merkel et al., 2000; Ren et al., 2009; Uyeda et al., 2011), indicating that the localization of myosin II is regulated by mechanical forces. Unlike cofilin, myosin II appears to bind to the extended F-actin.

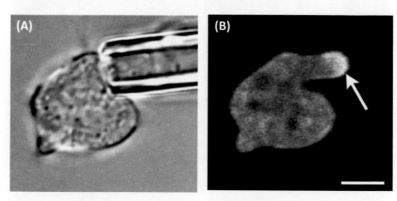

Figure 16.11 Myosin II localization by applying suction to a *Dictyostelium* cell with a micropipette. (A) Phase image. (B) Myosin II fluorescence. Myosin II localized at the tip of the sucked cell lobe (arrow).

The presence of these mechanosensing reactions of focal adhesion, cofilin and myosin II suggests that cells utilize mechanical signals for crawling migration. In actuality, periodic stretching of the substratum induces myosin II localization equally on both stretching sides in *Dictyostelium* cells (Fig. 16.12; Table 16.1), and the cells migrate perpendicular to the direction of stretch (Iwadate et al., 2013; Fig. 16.13). Neutrophil-like differentiated HL-60 cells also show the same directional migration (Okimura et al., in press). Interestingly, intact keratocytes migrate parallel to the stretching direction. They seem to rearrange their stress fibers perpendicular to the direction of stretching in the same way as fibroblasts. However, blebbistatin-treated stress fiber-less keratocytes migrate perpendicular to the stretching direction, in the same way as seen in HL-60 cells and *Dictyostelium* cells (Okimura and Iwadate, in press; Table 16.1). This reaction can be termed "passive mechanosensing," because artificial force was applied to crawling cells via the substratum in this experiment.

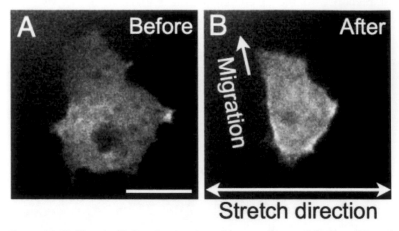

Figure 16.12 Myosin II distribution in a *Dictyostelium* cell before (A) and after four stretches of the substratum (B). Bar = 5 μm.

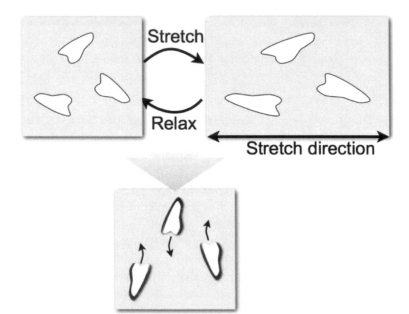

Figure 16.13 Schematic illustration of *Dictyostelium* cells in response to periodic stretching of the substratum. *Dictyostelium* cells migrate perpendicular to the direction of stretch. Red: myosin II. Arrows in the bottom panel: direction of migration.

16.4.2   Active Mechanosensing

It appears that the forces that the adherent cells experience from the substratum are reaction forces from the substratum caused by their own traction to it. Reaction forces from the substratum have the same amplitude but act in the opposite direction to the traction forces. The magnitude of the traction/reaction forces is likely to depend on the rigidity of the substratum. The traction forces generated by both fibroblasts (Wang et al., 2001) and neutrophils (Shin et al., 2010) are significantly greater on hard substrata than on soft substrata. The sensing of substratum rigidity via reaction of the traction forces has been proposed as "active touch" mechanosensing (Kobayashi and Sokabe, 2010). It is established that substrate rigidity plays a crucial role in differentiation (Engler et al., 2009). Mesenchymal stem cells differentiate into osteogenic-like cells on a rigid substrate that mimics bone tissue; on a soft gel that mimics neuronal tissue, the cells differentiate into neurons; and on an intermediately hard surface that mimics muscle tissue, mesenchymal stem cells differentiate into myogenic cells. Active touch also plays an important role in cell migration. At the boundary between rigid and soft areas of a substratum, fibroblasts move only from the soft part to the rigid part (a process called durotaxis) (Fig. 16.14; Table 16.1; Lo et al., 2000; Plotnikov et al., 2012), never in the other direction.

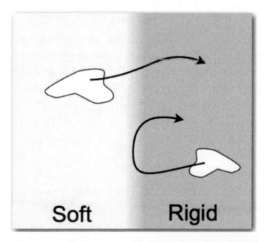

Figure 16.14 Schematic illustration of durotaxis of fibroblasts. Arrows: direction of migration.

16.4.3   Contact Guidance

Cells are able to sense both the rigidity and the topography of a substratum. The shape of cells is regulated by the topography of the substratum due to the restrictions upon the cytoskeleton rearrangements necessary for movement. This topographical effect is termed "contact guidance" (Weiss, 1934, 1945). Neurons (Rajnicek et al., 1997), fibroblasts (Brunette, 1986; Crouch et al., 2009) and neutrophils (Tan and Saltzman, 2002) placed in parallel micrometer- or 10 micrometer-sized grooves/ridges show extension parallel to the groove (Table 16.1). Not only the shape but also the migration of crawling cells is restricted by the topography of the substratum. When migrating keratocytes encounter a single linear groove whose width is 1.5 μm, they form a new leading lamella in the backward direction and reverse their migratory direction (Miyoshi et al., 2010). If the width of the groove is 20 μm, cells enter the groove, climb to the top of the opposite side and continue their migration without changing their direction of migration.

16.5   Conclusion and Future Perspectives

In this chapter, we have described the mechanisms of the extension of the cell front induced by actin polymerization at the leading edge and of the retraction of the cell rear by actomyosin contraction. These two mechanisms appear to be the basis of cell-crawling migration. However, the migration properties such as velocity, directionality and the shape of the leading edge vary greatly according to cell type. Moreover, many mechanisms are not very common. Numerous questions remain unanswered: For instance, why can *Dictyostelium* cells migrate 10 times faster than fibroblasts, even when both cell types extend the cell front by the same mechanism of actin polymerization? Their structural differences are due to the presence or absence of stress fibers and their binding strength to the substratum due to different types of focal adhesions. A theory that can explain the crawling migration of all cell types in relation to their structure remains to be established.

The signaling pathway of chemotaxis that generates anterior-posterior polarity has been extensively investigated. Since there is a great deal of literature on chemotaxis, we did not cover it in detail in this chapter. Neutrophil-like differentiated HL-60 cells localize F-actin to one portion of the cell and myosin IIA or RhoA to the opposite side in media containing a uniform concentration of their chemoattractant fMLP (Shin et al., 2010; Xu et al., 2003). In *Dictyostelium* cells, a mutant lacking the chemotaxis pathway can still migrate, albeit at low velocities (Hoeller and Kay, 2007). Polarity generation for migration that utilizes forces applied to the substratum may therefore be a deeply fundamental mechanism, with chemotaxis being a highly sophisticated version.

The mechanism of polarity generation by mechanosensing remains poorly understood. At the cellular level of mechano-transduction, fast-migrating cells such as *Dictyostelium* cells migrate perpendicular to the direction of stretch in response to periodic stretching of the substratum. Slow-migrating cells such as fibroblasts move only from the soft area to the rigid area at the boundary between the rigid and the soft area of a substratum. At the molecular level, cofilin binds to the relaxed actin filaments and myosin II seems to bind to stretched actin filaments. It is not known whether the molecular mechanisms behind durotaxis employ focal adhesions (Plotnikov et al., 2012) or stretch-activated calcium channels (Hayakawa et al., 2008; Kobayashi and Sokabe, 2010; Lee et al., 1999) that respond to mechanical stimuli.

The solutions to these questions will lead to a much broader understanding of crawling cell migration. Advanced understanding of the mechanics of cell migration, including mechanosensing reactions, may enable us to control the direction and speed of metastatic cancer cells and embryonic cells in developmental stages without the use of chemicals. This future technology, if established, would represent a major breakthrough in medical innovation.

Acknowledgement

Some photographs in figures were taken by Chika Okimura. I thank her for preparing figures and helpful discussions.

References

Akhshi, T. K., Wernike, D., and Piekny, A. (2014). Microtubules and actin crosstalk in cell migration and division. *Cytoskeleton*, 71, 1–23.

Amano, M., Nakayama, M., and Kaibuchi, K. (2010). Rho-kinase/ROCK: A key regulator of the cytoskeleton and cell polarity. *Cytoskeleton*, 67, 545–554.

Barnhart, E. L., Lee, K.-C., Keren, K., Mogilner, A., and Theriot, J. A. (2011). An adhesion-dependent switch between mechanisms that determine motile cell shape. *PLoS Biol.*, 9, e1001059.

Beningo, K. A., Dembo, M., Kaverina, I., Small, J. V., and Wang, Y. L. (2001). Nascent focal adhesions are responsible for the generation of strong propulsive forces in migrating fibroblasts. *J. Cell Biol.*, 153, 881–888.

Bianconi, E., Piovesan, A., Facchin, F., Beraudi, A., Casadei, R., Frabetti, F., Vitale, L., Pelleri, M. C., Tassani, S., Piva, F., et al. (2013). An estimation of the number of cells in the human body. *Ann. Human Biol.*, 40, 463–471.

Birukov, K. G., Jacobson, J. R., Flores, A. A., Ye, S. Q., Birukova, A. A., Verin, A. D., and Garcia, J. G. N. (2003). Magnitude-dependent regulation of pulmonary endothelial cell barrier function by cyclic stretch. *Am. J. Physiol. Lung Cell Mol. Physiol.*, 285, L785–L797.

Brandt, D. T., and Grosse, R. (2007). Get to grips: Steering local actin dynamics with IQGAPs. *EMBO Rep.*, 8, 1019–1023.

Brunette, D. M. (1986). Fibroblasts on micromachined substrata orient hierarchically to grooves of different dimensions. *Exp. Cell Res.*, 164, 11–26.

Burton, K., Park, J. H., and Taylor, D. L. (1999). Keratocytes generate traction forces in two phases. *Mol. Biol. Cell*, 10, 3745–3769.

Chan, C. E., and Odde, D. J. (2008). Traction dynamics of filopodia on compliant substrates. *Science*, 322, 1687–1691.

Chang, Y.-C., Nalbant, P., Birkenfeld, J., Chang, Z.-F., and Bokoch, G. M. (2008). GEF-H1 couples nocodazole-induced microtubule disassembly to cell contractility via RhoA. *Mol. Biol. Cell*, 19, 2147–2153.

Chen, W. T. (1981). Mechanism of retraction of the trailing edge during fibroblast movement. *J. Cell Biol.*, 90, 187–200.

Chen, Z., Lessey, E., Berginski, M. E., Cao, L., Li, J., Trepat, X., Itano, M., Gomez, S. M., Kapustina, M., Huang, C., et al. (2013). Gleevec, an Abl family inhibitor, produces a profound change in cell shape and migration. *PLoS ONE*, 8, e52233.

Chen, L., Vicente-Manzanares, M., Potvin-Trottier, L., Wiseman, P. W., and Horwitz, A. R. (2012). The integrin-ligand interaction regulates adhesion and migration through a molecular clutch. *PLoS ONE*, 7, e40202.

Crosby, L. M., Luellen, C., Zhang, Z., Tague, L. L., Sinclair, S. E., and Waters, C. M. (2011). Balance of life and death in alveolar epithelial type II cells: Proliferation, apoptosis, and the effects of cyclic stretch on wound healing. *Am. J. Physiol. Lung Cell Mol. Physiol.*, 301, L536–L546.

Crouch, A. S., Miller, D., Luebke, K. J., and Hu, W. (2009). Correlation of anisotropic cell behaviors with topographic aspect ratio. *Biomaterials*, 30, 1560–1567.

Dembo, M., Oliver, T., Ishihara, A., and Jacobson, K. (1996). Imaging the traction stresses exerted by locomoting cells with the elastic substratum method. *Biophys. J.*, 70, 2008–2022.

Dembo, M., and Wang, Y. L. (1999). Stresses at the cell-to-substrate interface during locomotion of fibroblasts. *Biophys. J.*, 76, 2307–2316.

Desai, L. P., White, S. R., and Waters, C. M. (2010). Cyclic mechanical stretch decreases cell migration by inhibiting phosphatidylinositol 3-kinase- and focal adhesion kinase-mediated JNK1 activation. *J. Biol. Chem.*, 285, 4511–4519.

Doyle, A. D., and Lee, J. (2005). Cyclic changes in keratocyte speed and traction stress arise from Ca^{2+}-dependent regulation of cell adhesiveness. *J. Cell. Sci.*, 118, 369–379.

Doyle, A., Marganski, W., and Lee, J. (2004). Calcium transients induce spatially coordinated increases in traction force during the movement of fish keratocytes. *J. Cell Sci.*, 117, 2203–2214.

Engler, A. J., Chan, M., Boettiger, D., and Schwarzbauer, J. E. (2009). A novel mode of cell detachment from fibrillar fibronectin matrix under shear. *J. Cell. Sci.*, 122, 1647–1653.

Enomoto, T. (1996). Microtubule disruption induces the formation of actin stress fibers and focal adhesions in cultured cells: Possible involvement of the rho signal cascade. *Cell Struct. Funct.*, 21, 317–326.

Euteneuer, U., and Schliwa, M. (1984). Persistent, directional motility of cells and cytoplasmic fragments in the absence of microtubules. *Nature*, 310, 58–61.

Fournier, M. F., Sauser, R., Ambrosi, D., Meister, J.-J., and Verkhovsky, A. B. (2010). Force transmission in migrating cells. *J. Cell Biol.*, 188, 287–297.

Fuhs, T., Goegler, M., Brunner, C. A., Wolgemuth, C. W., and Kaes, J. A. (2014). Causes of retrograde flow in fish keratocytes. *Cytoskeleton*, 71, 24–35.

Galbraith, C. G., Yamada, K. M., and Sheetz, M. P. (2002). The relationship between force and focal complex development. *J. Cell Biol.*, 159, 695–705.

Giannone, G., Mège, R.-M., and Thoumine, O. (2009). Multi-level molecular clutches in motile cell processes. *Trends Cell Biol.*, 19, 475–486.

Harris, A. K., Stopak, D., and Wild, P. (1981). Fibroblast traction as a mechanism for collagen morphogenesis. *Nature*, 290, 249–251.

Harris, A. K., Wild, P., and Stopak, D. (1980). Silicone rubber substrata: A new wrinkle in the study of cell locomotion. *Science*, 208, 177–179.

Hayakawa, K., Sakakibara, S., Sokabe, M., and Tatsumi, H. (2014). Single-molecule imaging and kinetic analysis of cooperative cofilin-actin filament interactions. *Proc. Natl. Acad. Sci. U. S. A.*, 111, 9810–9815.

Hayakawa, K., Tatsumi, H., and Sokabe, M. (2008). Actin stress fibers transmit and focus force to activate mechanosensitive channels. *J. Cell. Sci.*, 121, 496–503.

Hayakawa, K., Tatsumi, H., and Sokabe, M. (2011). Actin filaments function as a tension sensor by tension-dependent binding of cofilin to the filament. *J. Cell Biol.*, 195, 721–727.

Hoeller, O., and Kay, R. R. (2007). Chemotaxis in the absence of PIP3 gradients. *Curr. Biol.*, 17, 813–817.

Hotulainen, P., and Lappalainen, P. (2006). Stress fibers are generated by two distinct actin assembly mechanisms in motile cells. *J. Cell Biol.*, 173, 383–394.

Huitorel, P. (1988). From cilia and flagella to intracellular motility and back again: A review of a few aspects of microtubule-based motility. *Biol. Cell*, 63, 249–258.

Iwadate, Y., and Nakaoka, Y. (2008). Calcium regulates independently ciliary beat and cell contraction in *Paramecium* cells. *Cell Calcium*, 44, 169–179.

Iwadate, Y., Okimura, C., Sato, K., Nakashima, Y., Tsujioka, M., and Minami, K. (2013). Myosin-II-mediated directional migration of *Dictyostelium* cells in response to cyclic stretching of substratum. *Biophys. J.*, 104, 748–758.

Iwadate, Y., and Yumura, S. (2008). Actin-based propulsive forces and myosin-II-based contractile forces in migrating *Dictyostelium* cells. *J. Cell Sci.*, 121, 1314–1324.

Iwadate, Y., and Yumura, S. (2009). Cyclic stretch of the substratum using a shape-memory alloy induces directional migration in *Dictyostelium* cells. *BioTechniques*, 47, 757–767.

Jay, P. Y., Pham, P. A., Wong, S. A., and Elson, E. L. (1995). A mechanical function of myosin II in cell motility. *J. Cell Sci.*, 108, 387–393.

Jennings, H. S. (1906). *Behavior of the Lower Organisms*. (New York: Columbia University Press), pp. 47–54.

Jurado, C., Haserick, J. R., and Lee, J. (2005). Slipping or gripping? Fluorescent speckle microscopy in fish keratocytes reveals two different mechanisms for generating a retrograde flow of actin. *Mol. Biol. Cell*, 16, 507–518.

Kaunas, R., Nguyen, P., Usami, S., and Chien, S. (2005). Cooperative effects of Rho and mechanical stretch on stress fiber organization. *Proc. Natl. Acad. Sci. U. S. A.*, 102, 15895–15900.

Kaverina, I., and Straube, A. (2011). Regulation of cell migration by dynamic microtubules. *Semin. Cell Dev. Biol.*, 22, 968–974.

Kobayashi, T., and Sokabe, M. (2010). Sensing substrate rigidity by mechanosensitive ion channels with stress fibers and focal adhesions. *Curr. Opin. Cell Biol.*, 22, 669–676.

Koch, D., Rosoff, W. J., Jiang, J., Geller, H. M., and Urbach, J. S. (2012). Strength in the periphery: Growth cone biomechanics and substrate rigidity response in peripheral and central nervous system neurons. *Biophys. J.*, 102, 452–460.

Krendel, M., Zenke, F. T., and Bokoch, G. M. (2002). Nucleotide exchange factor GEF-H1 mediates cross-talk between microtubules and the actin cytoskeleton. *Nat. Cell Biol.*, 4, 294–301.

Lauffenburger, D. A., and Horwitz, A. F. (1996). Cell migration: A physically integrated molecular process. *Cell*, 84, 359–369.

Lee, C.-F., Haase, C., Deguchi, S., and Kaunas, R. (2010). Cyclic stretch-induced stress fiber dynamics–dependence on strain rate, Rho-kinase and MLCK. *Biochem. Biophys. Res. Commun.*, 401, 344–349.

Lee, J., Ishihara, A., Oxford, G., Johnson, B., and Jacobson, K. (1999). Regulation of cell movement is mediated by stretch-activated calcium channels. *Nature*, 400, 382–386.

Lee, J., Ishihara, A., Theriot, J. A., and Jacobson, K. (1993). Principles of locomotion for simple-shaped cells. *Nature*, 362, 167–171.

Lee, J., and Jacobson, K. (1997). The composition and dynamics of cell-substratum adhesions in locomoting fish keratocytes. *J. Cell Sci.*, 110, 2833–2844.

Lin, C. H., Espreafico, E. M., Mooseker, M. S., and Forscher, P. (1996). Myosin drives retrograde F-actin flow in neuronal growth cones. *Neuron*, 16, 769–782.

Lo, C. M., Wang, H. B., Dembo, M., and Wang, Y. L. (2000). Cell movement is guided by the rigidity of the substrate. *Biophys. J.*, 79, 144–152.

Lombardi, M. L., Knecht, D. A., Dembo, M., and Lee, J. (2007). Traction force microscopy in *Dictyostelium* reveals distinct roles for myosin II motor and actin-crosslinking activity in polarized cell movement. *J. Cell Sci.*, 120, 1624–1634.

Martin, P. (1997). Wound healing--aiming for perfect skin regeneration. *Science*, 276, 75–81.

Merkel, R., Simson, R., Simson, D. A., Hohenadl, M., Boulbitch, A., Wallraff, E., and Sackmann, E. (2000). A micromechanic study of cell polarity and plasma membrane cell body coupling in *Dictyostelium*. *Biophys. J.*, 79, 707–719.

Miyoshi, H., Ju, J., Lee, S. M., Cho, D. J., Ko, J. S., Yamagata, Y., and Adachi, T. (2010). Control of highly migratory cells by microstructured surface based on transient change in cell behavior. *Biomaterials*, 31, 8539–8545.

Mogilner, A., and Oster, G. (1996). Cell motility driven by actin polymerization. *Biophys. J.*, 71, 3030–3045.

Montenegro-Venegas, C., Tortosa, E., Rosso, S., Peretti, D., Bollati, F., Bisbal, M., Jausoro, I., Avila, J., Cáceres, A., and Gonzalez-Billault, C. (2010). MAP1B regulates axonal development by modulating Rho-GTPase Rac1 activity. *Mol. Biol. Cell*, 21, 3518–3528.

Morioka, M., Parameswaran, H., Naruse, K., Kondo, M., Sokabe, M., Hasegawa, Y., Suki, B., and Ito, S. (2011). Microtubule dynamics regulate cyclic stretch-induced cell alignment in human airway smooth muscle cells. *PLoS ONE*, 6, e26384.

Munevar, S., Wang, Y., and Dembo, M. (2001a). Traction force microscopy of migrating normal and H-ras transformed 3T3 fibroblasts. *Biophys. J.*, 80, 1744–1757.

Munevar, S., Wang, Y. L., and Dembo, M. (2001b). Distinct roles of frontal and rear cell-substrate adhesions in fibroblast migration. *Mol. Biol. Cell*, 12, 3947–3954.

Nakashima, H., Okimura, C., and Iwadate, Y. (2015). The molecular dynamics of crawling migration in microtubule-disrupted keratocytes. *Biophys. Physicobiol.*, 12, 21–29.

Nakata, T., Okimura, C., Mizuno, T., and Iwadate, Y. (2016). The role of stress fibers in the shape determination mechanism of fish keratocytes. *Biophys. J.*, 110, 481–492.

Nalbant, P., Chang, Y.-C., Birkenfeld, J., Chang, Z.-F., and Bokoch, G. M. (2009). Guanine nucleotide exchange factor-H1 regulates cell migration via localized activation of RhoA at the leading edge. *Mol. Biol. Cell*, 20, 4070–4082.

Narematsu, N., Quek, R., Chiam, K.-H., and Iwadate, Y. (2015). Ciliary metachronal wave propagation on the compliant surface of *Paramecium* cells. *Cytoskeleton*, 72, 633–646.

Naruse, K., Yamada, T., Sai, X. R., Hamaguchi, M., and Sokabe, M. (1998a). Pp125FAK is required for stretch dependent morphological response of endothelial cells. *Oncogene*, 17, 455–463.

Naruse, K., Yamada, T., and Sokabe, M. (1998b). Involvement of SA channels in orienting response of cultured endothelial cells to cyclic stretch. *Am. J. Physiol. Heart Circ. Physiol.*, 274, H1532–H1538.

Okimura, C., Ueda, K., Sakumura, Y., and Iwadate, Y. (in press). Fast-crawling cell types migrate to avoid the direction of periodic substratum stretching. *Cell Adhes. Migration*. doi: 10.1080/19336918.2015.1129 482.

Okimura, C., and Iwadate, Y. (in press). Hybrid mechanosensing system to generate the polarity needed for migration in fish keratocytes. *Cell Adhes. Migration*, doi: 10.1080/19336918.2016.1170268.

Oliver, T., Dembo, M., and Jacobson, K. (1999). Separation of propulsive and adhesive traction stresses in locomoting keratocytes. *J. Cell Biol.*, 145, 589–604.

Parent, C. A. (2004). Making all the right moves: Chemotaxis in neutrophils and *Dictyostelium*. *Curr. Opin. Cell Biol.*, 16, 4–13.

Plotnikov, S. V., Pasapera, A. M., Sabass, B., and Waterman, C. M. (2012). Force fluctuations within focal adhesions mediate ECM-rigidity sensing to guide directed cell migration. *Cell*, 151, 1513–1527.

Pomorski, P., Krzemiński, P., Wasik, A., Wierzbicka, K., Barańska, J., and Kłopocka, W. (2007). Actin dynamics in *Amoeba proteus* motility. *Protoplasma*, 231, 31–41.

Rajnicek, A., Britland, S., and McCaig, C. (1997). Contact guidance of CNS neurites on grooved quartz: Influence of groove dimensions, neuronal age and cell type. *J. Cell Sci.*, 110, 2905–2913.

Ren, Y., Effler, J. C., Norstrom, M., Luo, T., Firtel, R. A., Iglesias, P. A., Rock, R. S., and Robinson, D. N. (2009). Mechanosensing through Cooperative Interactions between Myosin-II and the Actin Crosslinker Cortexillin-I. *Curr. Biol.*, 19, 1421–1428.

Ridley, A. J., Schwartz, M. A., Burridge, K., Firtel, R. A., Ginsberg, M. H., Borisy, G., Parsons, J. T., and Horwitz, A. R. (2003). Cell migration: Integrating signals from front to back. *Science*, 302, 1704–1709.

Rooney, C., White, G., Nazgiewicz, A., Woodcock, S. A., Anderson, K. I., Ballestrem, C., and Malliri, A. (2010). The Rac activator STEF (Tiam2) regulates cell migration by microtubule-mediated focal adhesion disassembly. *EMBO Rep.*, 11, 292–298.

Sato, K., Adachi, T., Matsuo, M., and Tomita, Y. (2005). Quantitative evaluation of threshold fiber strain that induces reorganization of cytoskeletal actin fiber structure in osteoblastic cells. *J. Biomech.*, 38, 1895–1901.

Shimada, T., Toriyama, M., Uemura, K., Kamiguchi, H., Sugiura, T., Watanabe, N., and Inagaki, N. (2008). Shootin1 interacts with actin retrograde flow and L1-CAM to promote axon outgrowth. *J. Cell Biol.*, 181, 817–829.

Shin, M. E., He, Y., Li, D., Na, S., Chowdhury, F., Poh, Y.-C., Collin, O., Su, P., de Lanerolle, P., Schwartz, M. A., et al. (2010). Spatiotemporal organization, regulation, and functions of tractions during neutrophil chemotaxis. *Blood*, 116, 3297–3310.

Sonoda, A., Okimura, C., and Iwadate, Y. (2016). Shape and area of keratocytes are related to the distribution and magnitude of their traction forces. *Cell Struct. Funct.*, 41, 33–43.

Stehbens, S. J., Paszek, M., Pemble, H., Ettinger, A., Gierke, S., and Wittmann, T. (2014). CLASPs link focal-adhesion-associated microtubule capture to localized exocytosis and adhesion site turnover. *Nat. Cell Biol.*, 16, 558–570.

Stockem, W., Hoffmann, H. U., and Gawlitta, W. (1982). Spatial organization and fine structure of the cortical filament layer in normal locomoting Amoeba proteus. *Cell Tissue Res.*, 221, 505–519.

Stroka, K. M., and Konstantopoulos, K. (2014). Physical biology in cancer. 4. Physical cues guide tumor cell adhesion and migration. *Am. J. Physiol. - Cell Physiol.*, 306, C98–C109.

Svitkina, T. M., Verkhovsky, A. B., McQuade, K. M., and Borisy, G. G. (1997). Analysis of the actin-myosin II system in fish epidermal keratocytes: Mechanism of cell body translocation. *J. Cell Biol.*, 139, 397–415.

Tan, J., and Saltzman, W. M. (2002). Topographical control of human neutrophil motility on micropatterned materials with various surface chemistry. *Biomaterials*, 23, 3215–3225.

Tanimoto, H., and Sano, M. (2014). A simple force-motion relation for migrating cells revealed by multipole analysis of traction stress. *Biophys. J.*, 106, 16–25.

Tarantola, M., Bae, A., Fuller, D., Bodenschatz, E., Rappel, W.-J., and Loomis, W. F. (2014). Cell substratum adhesion during early development of *Dictyostelium discoideum*. *PLoS ONE*, 9, e106574.

Tondon, A., Hsu, H.-J., and Kaunas, R. (2011). Dependence of cyclic stretch-induced stress fiber reorientation on stretch waveform. *J. Biomech.*, 45, 728–735.

Toriyama, M., Kozawa, S., Sakumura, Y., and Inagaki, N. (2013). Conversion of a signal into forces for axon outgrowth through Pak1-mediated shootin1 phosphorylation. *Curr. Biol.*, 23, 529–534.

Tsugiyama, H., Okimura, C., Mizuno, T., and Iwadate, Y. (2013). Electroporation of adherent cells with low sample volumes on a microscope stage. *J. Exp. Biol.*, 216, 3591–3598.

Uyeda, T. Q. P., Iwadate, Y., Umeki, N., Nagasaki, A., and Yumura, S. (2011). Stretching actin filaments within cells enhances their affinity for the myosin II motor domain. *PLoS One*, 6, e26200.

Vasiliev, J. M., Gelfand, I. M., Domnina, L. V., Ivanova, O. Y., Komm, S. G., and Olshevskaja, L. V. (1970). Effect of colcemid on the locomotory behaviour of fibroblasts. *J. Embryol. Exp. Morphol.*, 24, 625–640.

Verkhovsky, A. B., Svitkina, T. M., and Borisy, G. G. (1999). Self-polarization and directional motility of cytoplasm. *Curr. Biol.*, 9, 11–20.

Vicente-Manzanares, M., Ma, X., Adelstein, R. S., and Horwitz, A. R. (2009). Non-muscle myosin II takes centre stage in cell adhesion and migration. *Nat. Rev. Mol. Cell Biol.*, 10, 778–790.

Wang, H. B., Dembo, M., Hanks, S. K., and Wang, Y. (2001). Focal adhesion kinase is involved in mechanosensing during fibroblast migration. *Proc. Natl. Acad. Sci. U. S. A.*, 98, 11295–11300.

Waterman-Storer, C. M., Desai, A., Chloe Bulinski, J., and Salmon, E. D. (1998). Fluorescent speckle microscopy, a method to visualize the dynamics of protein assemblies in living cells. *Curr. Biol.*, 8, 1227–1230.

Waterman-Storer, C. M., Worthylake, R. A., Liu, B. P., Burridge, K., and Salmon, E. D. (1999). Microtubule growth activates Rac1 to promote lamellipodial protrusion in fibroblasts. *Nat. Cell Biol.*, 1, 45–50.

Weiss, P. (1934). In vitro experiments on the factors determining the course of the outgrowing nerve fiber. *J. Exp. Zool.*, 68, 393–448.

Weiss, P. (1945). Experiments on cell and axon orientation in vitro: The role of colloidal exudates in tissue organization. *J. Exp. Zool.*, 100, 353–386.

Weninger, W., Biro, M., and Jain, R. (2014). Leukocyte migration in the interstitial space of non-lymphoid organs. *Nat. Rev. Immunol.*, 14, 232–246.

Wilson, C. A., Tsuchida, M. A., Allen, G. M., Barnhart, E. L., Applegate, K. T., Yam, P. T., Ji, L., Keren, K., Danuser, G., and Theriot, J. A. (2010). Myosin II contributes to cell-scale actin network treadmilling through network disassembly. *Nature*, 465, 373–377.

Xu, J., Wang, F., Van Keymeulen, A., Herzmark, P., Straight, A., Kelly, K., Takuwa, Y., Sugimoto, N., Mitchison, T., and Bourne, H. R. (2003). Divergent signals and cytoskeletal assemblies regulate self-organizing polarity in neutrophils. *Cell*, 114, 201–214.

Zhao, L., Sang, C., Yang, C., and Zhuang, F. (2011). Effects of stress fiber contractility on uniaxial stretch guiding mitosis orientation and stress fiber alignment. *J. Biomech.*, 44, 2388–2394.

Chapter 17

Role of Dynamic and Cooperative Conformational Changes in Actin Filaments

Taro Q. P. Uyeda

Department of Physics, Faculty of Advanced Science and Engineering,
Waseda University, 3-4-1 Okubo, Shinjuku, Tokyo 169-8555, Japan
School of Materials Sciences and Engineering
and Institute for Biomedical Engineering and Nano Science,
Tongji University, Shanghai 200092, China

t-uyeda@waseda.jp

The amino acid sequence of actin is highly conserved in most eukaryotic organisms, and each cell typically expresses only one or two homologous actin isoforms. Yet, actin filaments play important roles in a number of diverse activities in each cell. Each of those actin functions depends on interactions with a specific set of actin binding proteins. In a number of cases, local biochemical signaling to regulate the activities of actin binding proteins has been shown to contribute to the spatial and temporal regulation of those interactions, but not all aspects of the local regulations of the interactions can be explained by local biochemical signaling.

Muscle Contraction and Cell Motility: Fundamentals and Developments
Edited by Haruo Sugi
Copyright © 2017 Pan Stanford Publishing Pte. Ltd.
ISBN 978-981-4745-16-1 (Hardcover), 978-981-4745-17-8 (eBook)
www.panstanford.com

Interactions with actin binding proteins or mechanical force induce conformational changes to actin protomers in the filaments, and the resultant conformational changes often propagate to neighboring actin protomers in the filament. We propose that those cooperative conformational changes in actin filaments may contribute to functional differentiation of actin filaments by specifying which actin binding protein to bind, enabling each actin filament to perform different functions within a common cytoplasm. In addition, we discuss the possible involvement of such cooperative conformational changes of actin filaments in interaction with myosin to generate force.

17.1 An Exceptionally Conservative and Multifunctional Protein: Actin

Actin was originally isolated from skeletal muscle more than 70 years ago (Straub, 1942), but it was later found in virtually all eukaryotic organisms, ranging from slime molds (Hatano and Oosawa, 1966) and yeasts to higher plants and animals. Homologues have been discovered in bacteria and archaea. Thus, muscle contraction is a highly specialized function of actin that appeared at a later stage of eukaryotic evolution, and to understand the structure and function of actin, it is necessary to study actin in non-muscle cells. This may in turn lead to a deeper understanding of actin's roles in muscle contraction.

Eukaryotic actin, hereafter simply actin, is a very abundant protein, not only in muscle but also in a variety of cells. Yet, it is an exceptional protein for two reasons. First, it is a very conservative protein. The amino acid sequence is identical between chicken and human skeletal muscle α actins, and the human β cytoplasmic actin sequence is 90% identical to yeast actin. Lower eukaryotes such as yeasts and cellular slime molds express a single species of actin, while cells of higher plants and animals, except for muscle cells, typically express two isoforms. For instance, mammalian non-muscle cells express β and γ cytoplasmic actins, which differ by only four amino acid residues near the N-terminus. Reproductive cells of *Arabidopsis* also express two actin isoforms, ACT1 and ACT11. Vegetative *Arabidopsis* cells express three isoforms, ACT2, ACT7, and ACT8, but ACT2 and ACT8 are virtually identical with only one conservative substitution (McDowell et al., 1996).

Second, actin is highly multifunctional. Just to name a few functions, actin filaments serve as tracks for myosin to move, which drives muscle contraction, cytokinesis, intracellular transport, and protoplasmic streaming, among others. Actin filaments also play mechanical roles by underlining the cell membrane, and also support thin projections such as filopodia, stereocilia, and microvilli. Polymerization of actin itself produces a pushing force to drive processes such as lamellipodial extension and intracellular bacterial motility. Nuclear actin has been implicated in a number of structural and regulatory processes. Intriguingly, γ cytoplasmic actin is dispensable for normal mouse development (Belyantseva et al., 2009). Thus, all those diverse functions, except for muscle contraction, are supported by a single species of actin molecules in yeast, cellular slime molds, and mutant mouse cells that lack γ cytoplasmic actin. A notable exception is muscle cells that additionally express α muscle actin, which is not functionally interchangeable with cytoplasmic actin (Brault et al., 1999; McKane et al., 2006). The situation in higher plants is also somewhat different, because the two major isoforms of *Arabidopsis* vegetative actin, ACT2 and ACT7, differ by 7% in primary structure and are functionally non-equivalent (Kandasamy et al., 2009) with distinct biochemical properties (Kijima et al., 2015).

Nonetheless, we can generalize that eukaryotic actins are extremely conservative and, with the exception of α muscle actin, highly multifunctional. This is in sharp contrast to bacterial actins, which have highly divergent amino acid sequences, but each has one or a limited number of functions. In other words, bacteria changed actin's amino acid sequence to meet diverse functional needs, which is typical of evolution. In contrast, eukaryotes took a very different approach and realized multiple functions of actin without changing the amino acid sequence. The question is, how?

Each function of actin is believed to depend on an interaction with a specific set of actin binding proteins (ABPs). Thus, actin filaments in a cell have to interact with a large number of ABPs in a spatially and temporally regulated manner (dos Remedios et al., 2003; Pak et al., 2008; Pollard and Cooper, 2009). The extraordinary level of actin's sequence conservation has been traditionally attributed to the need to interact with a large number of ABPs. However, this only explains why residues on the surface of the molecule are difficult to change, while sequence conservation is equally high inside the actin molecule. Galkin et al. (2011, 2012, 2015) suggested that cooperative polymorphism in the filament structure is the basis of actin's multiple functions, and for actin protomers in filaments to undergo multiple distinct forms of cooperative conformational changes, an intricate network of allosteric conformational changes is essential, which precluded amino acid changes inside the actin molecule. In this chapter, we will review the roles of actin filaments from this point of view.

17.2   Structural Polymorphism of Actin Filaments

Crystallization of actin filaments has not been realized, and hence, high-resolution structural information of actin filaments has been obtained using diffraction studies of oriented filaments (Holmes et al., 1990; Oda et al., 2009) and electron microscopic analyses of rapidly frozen specimens (Fujii et al., 2010; Galkin et al., 2010, 2015; Murakami et al., 2010; von der Ecken et al., 2015). Of particular relevance to this review is the discovery by Galkin et al. (2010) that actin filaments have at least six different conformations that can be distinguished by electron microscopic analyses of a medium level (10~16 Å) of resolution. In that study, Galkin et al. (2010) took a single-particle analysis approach, treating 17 consecutive protomers in filaments as a single particle, implying that at least 17 consecutive protomers in filaments are likely to take the same one of six conformations (cooperative structural polymorphism).

When G actin carrying ATP polymerizes in vitro, it preferentially binds to the barbed or plus ends of growing filaments, which stimulates ATP hydrolysis and the subsequent release of Pi. Therefore, actin protomers near the barbed ends carry ATP, which abut a section of the filament rich with protomers carrying ADP and Pi, while protomers far from the barbed ends only carry ADP. Protomers carrying different nucleotides (ATP or ADP + Pi vs. ADP only) have different structures (Orlova and

Egelman, 1992; Murakami et al., 2010). This structural difference has profound functional consequences, including different critical concentrations for polymerization between the two ends (Carlier, 1990), and a different affinity for ABPs, as has been demonstrated for cofilin binding (Carlier et al., 1997; Blanchoin and Pollard, 1999; Umeki et al., 2012). Judging from the way in which actin filaments were prepared, however, it is unlikely that ATP hydrolysis or Pi release contributed to the structural polymorphism of pure actin filaments reported by Galkin et al. (2010).

In negatively stained electron micrographs, nascent actin filaments, within 20 min after the initiation of polymerization, appeared more irregular than the aged or mature filaments with the standard double-helical appearance (Steinmetz et al., 1997; Galinska-Rakoczy et al., 2009; Huang et al., 2010). This ageing process or "maturation" is apparently unrelated to ATP hydrolysis or Pi release (Huang et al., 2010), and may correspond to the "tilted-state (T-state)" that Egelman and his colleagues observed in nascent actin filaments by cryoelectron microscopy (Orlova et al., 2004; Galkin et al., 2010, 2015). Furthermore, nascent filaments are functionally different from mature filaments, in that they have a higher affinity for the Arp2/3 complex than mature filaments (Jensen et al., 2012). Although more needs to be elucidated regarding the exact nature of this nascent state, this state may be physiologically relevant, since actin filaments in non-muscle cells turnover very rapidly, with a half life in the order of seconds to minutes (Wang, 1985; Amato and Taylor, 1986; Theriot and Mitchison, 1991; Murthy and Wadsworth, 2005; Yumura et al., 2013), so that the majority of actin filaments in non-muscle cells may be of this nascent type, rather than the mature filaments that have been investigated extensively.

17.3 Cooperative Conformational Changes of Actin Filaments Induced by ABPs

It is well established that bound ABPs affect the conformation of actin filaments, often in a cooperative manner, providing an additional level of complexity to the structural polymorphism of actin filaments. Discussed below are examples of those cases.

17.3.1 Interaction with Cofilin

When a substoichiometric amount of cofilin is mixed with actin filaments, cofilin molecules bind to the filaments cooperatively and form tight clusters along the filament while leaving other sections of the filament bare, rather than binding uniformly sparsely along the filament (Hawkins et al., 1993; Hayden et al., 1993; McGough et al., 1997; De La Cruz, 2005; Hayakawa et al., 2014; Ngo et al., 2015). The two domains within each actin protomer in cofilin clusters are more twisted relative to each other than those in control filaments (Galkin et al., 2011), accompanying ~25% supertwisting of the filament helix (McGough et al., 1997; Galkin et al., 2001; Ngo et al., 2015).

Strikingly, this supertwisted structure propagates to neighboring bare zones (cooperative conformational change) (Galkin et al., 2001), and recent atomic force microscopy (AFM) observation (Fig. 17.1) (Ngo et al., 2015) demonstrated that these cooperative conformational changes propagate only to the pointed or minus end direction (unidirectional cooperative conformational change) (Fig. 17.1C,E). Furthermore, real-time high-speed AFM showed that cofilin clusters grow unidirectionally toward the pointed ends (Fig. 17.1F). This is presumably because cofilin prefers to bind to the bare section of the filaments with a supertwisted structure (Fig. 17.1G), as had been proposed earlier (Galkin et al., 2001).

The ADF/cofilin family of proteins severs actin filaments and plays essential roles in the dynamic regulation of actin filaments in vivo (Pollard et al., 2000; Bernstein and Bamburg, 2010). Real-time high-speed AFM found that severing occurs near, although not necessarily at, the boundary between cofilin clusters and the neighboring bare zones (Ngo et al., 2015). This is consistent with an earlier low-resolution fluorescence microscopy observation (Suarez et al., 2011), and supports the idea that structural discontinuity induced by cofilin clusters weakens bonds between actin protomers, leading to filament severing. Interestingly, recent fluorescence microscopy demonstrated that severing occurs preferentially on the pointed end side of cofilin clusters (Gressin et al., 2015), even though structural discontinuities are likely to be present on both sides of the cofilin clusters.

Each cofilin cluster induces 10~20 neighboring actin subunits on the pointed end side to undergo supertwisting (Ngo et al., 2015). When sparsely bound to actin filaments, each cofilin molecule has been shown to affect ~100 actin protomers, in terms of destabilizing the tertiary structure (Bobkov et al., 2006) or increasing microsecond rotational motion (Prochniewicz et al., 2005). The relationship between the two distinct modes of cooperative conformational changes is not understood.

17.3.2 Interaction with Myosin

There is no question that class II myosin is the ABP with the longest and most extensive research history. Pioneering work by Oosawa and his colleagues (1973) reported more than 40 years ago that binding of fragments of skeletal muscle myosin II changes the structure of fluorescently labeled actin filaments. In that experiment, each motor domain, called subfragment 1 or S1, of myosin II affected the structure of ~5 actin protomers around the actin protomer that it is bound to in the rigor conformation (i.e., in the absence of nucleotides), indicating that this is another case of ABP-induced cooperative conformational change. Similar myosin-induced cooperative changes in actin filaments were detected using other structural probes (Miki et al., 1982; Prochniewicz and Thomas, 1997; Prochniewicz et al., 2004; Siddique et al., 2005). Under certain conditions, the motor domain of myosin II binds cooperatively to actin filaments, forming either tight (Orlova and Egelman, 1997) or loose clusters (Tokuraku et al., 2009), which may be correlated with the myosin-induced cooperative conformational changes mentioned above. The helical pitch of actin filaments is slightly (0.3%) stretched when fully bound with rigor crossbridges (Holmes et al., 2003). Although this is a small change, it is noteworthy that this structural change (untwisting) is opposite to that induced by cofilin (supertwisting). The details of myosin-induced conformational changes of actin filaments are yet to be investigated. It is also unknown if the myosin motor domains induce different conformational changes to actin filaments under different chemical conditions (i.e., in the presence of ATP, ADP or ADP+Pi).

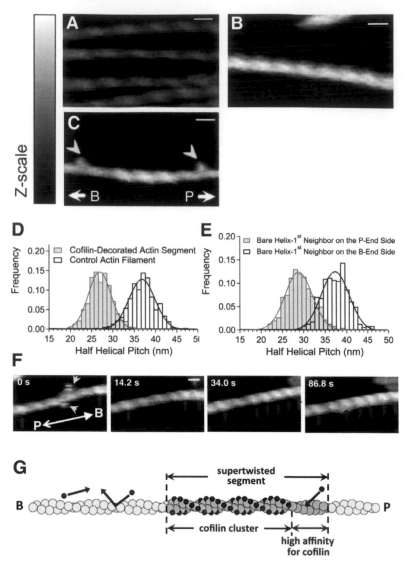

Figure 17.1 High-speed AFM observation of actin filaments interacting with cofilin. (A) and (B): AFM images of control actin filaments (A) and actin filaments fully bound with cofilin (B). The cofilin-bound filament is thicker and has a shorter half helical pitch (D), which is the distance between the crossover points. The crossover points are positions of the filament where the two protofilaments are aligned

vertically so that the filament is taller, and are shown in lighter colors in the AFM image. The filament shown in (C) has a short cofilin cluster in the middle. Pitches of half helices immediately neighboring cofilin clusters in images similar to this were measured to construct the histograms shown in (E). Measurements were made separately for the barbed-end neighbors and the pointed-end neighbors, based on the polarity of the filament determined from transiently binding S1 molecules (yellow arrowheads). The half helical pitch on the pointed end side is as short as that in cofilin clusters (~27 nm), while that on the barbed end side is similar to that of the control filaments (~37 nm). Growth of the cofilin cluster was also unidirectional toward the pointed end direction, as shown in the time-lapse sequence shown in (F). The cluster on the right grew to the left (toward the pointed end, as identified by the tilted transient binding of S1 indicated by yellow arrowheads), while that on the left did not grow to the right. Red arrows show positions of the crossover points within cofilin clusters, identified on the basis of greater height. These observations are summarized schematically in (G). Modified from Ngo et al. (2015).

17.3.3 Interaction with Drebrin

Drebrin is a neuron-specific ABP, and binds cooperatively to actin filaments, forming tight clusters as cofilin does. Unlike cofilin, however, drebrin binding untwists the actin helix by ~10% (Sharma et al., 2011). AFM observation demonstrated that this conformational change is also propagated to neighboring bare sections of the filament (Sharma et al., 2011), although the directionality of this cooperative conformational change is unknown.

17.3.4 Interactions with End-Binding ABPs

Cofilin, myosin II and drebrin all bind along the sides of actin filaments. There are classes of ABPs that bind to the ends of filaments, working as either cappers or polymerization promoting factors. Formins are a large family of proteins that bind to the barbed ends of filaments and catalyze polymerization, and filaments polymerized by the activity of formin have a structure different from those that polymerized spontaneously, even though there is only one dimeric formin complex at the barbed end of each filament (Papp et al., 2006). Gelsolin is a Ca^{2+}-dependent filament severing/barbed end capping protein, and in this case also, the structure of the entire filament is altered by gelsolin bound at the barbed end (Orlova et al., 1995). These are cases of extremely long-ranged unidirectional cooperative conformational changes, since hundreds of actin protomers are affected by a single formin dimer or a single gelsolin molecule at the barbed end.

Thus, many ABPs are known to change the structure of actin filaments, but this is expected, as it is natural that both proteins undergo certain conformation changes when two proteins bind to each other. Although only five ABPs, i.e., cofilin, myosin II, drebrin, formin and gelsolin, are mentioned above, many other ABPs, including tropomyosin (Butters et al., 1993), caldesmon (Collins et al., 2011; Jensen et al., 2012), fimbrin (Hanein et al., 1997; Galkin et al., 2008), α catenin (Hansen et al., 2013) and α actinin (Craig-Schmidt et al., 1981), have also been shown or suggested to change the structure of actin filaments. It is also not surprising that an ABP-induced conformational change that occurred in one actin protomer affects the neighboring protomer because a conformational change that occurred in one actin protomer is expected to exert a certain impact on the structure of a neighboring protomer with which it is in direct contact (cooperative conformational change). Subunits in hemoglobin complexes are well known to undergo cooperative conformational changes to alter the affinity of each subunit for oxygen. What is different from the case of hemoglobin is the number of subunits within a complex; hundreds to thousands of actin protomers in one actin filament versus four in a hemoglobin tetramer, which results in a significant difference in the geometric range that cooperative conformational changes can affect.

Another characteristic feature of cooperative polymorphism of actin filaments is the large repertoire of the polymorphism. As discussed above, cofilin, myosin II and drebrin induce different cooperative conformational changes to actin filaments, as characterized by different helical pitches. Other ABPs presumably evoke other conformational changes to actin filaments, and it is unknown how many distinct conformations actin filaments are able to assume, and how many distinct types of propagating conformational changes there are. It is also unknown if those ABP-induced cooperative polymorphisms are related to the cooperative polymorphism Galkin et al. (2010) observed in pure actin filaments, except that all six distinct structures that Galkin et al. (2010) observed did not have altered helical twists, and hence, they are probably different from those induced by cofilin, myosin II or drebrin.

17.4 Physiological Roles of Cooperative Polymorphism of Actin Filaments

As real-time high-speed AFM demonstrated (Ngo et al., 2015), cofilin-induced cooperative conformational changes that propagate into the neighboring bare zone are presumably the driving force for cooperative binding by promoting the growth of cofilin clusters. This scenario would probably apply to cooperative binding of other ABPs.

Furthermore, it has been demonstrated in a number of combinations that prior binding of one ABP inhibits binding of another ABP to that actin filament. Obviously, direct competition for binding sites on actin is a plausible mechanism by which one ABP inhibits the binding of another. As discussed above, however, both cofilin and myosin II motor domains bind cooperatively to actin filaments, and we have preliminary but compelling evidence to show that two distinct modes of cooperative conformational changes in actin filaments mediate mutually inhibitory effects of cooperative binding of cofilin and myosin II (Fig. 17.2) (Kijima et al., 2011; Umeki and Uyeda, 2012). Influencing the binding affinities of neighboring actin protomers by cooperative conformational changes that propagate along the filament may have a far greater impact than simple competition for binding sites, in the following three ways.

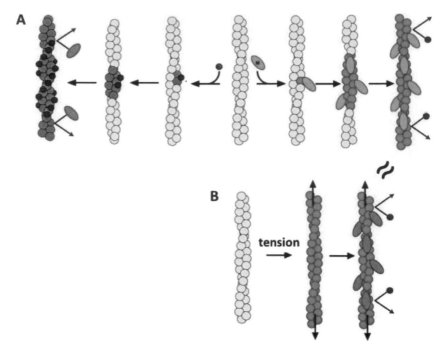

Figure 17.2 Schematic drawing of conformational changes of actin filaments induced by cofilin, myosin II motor domain (S1) (A), and mechanical stretching (B). Binding of cofilin (red spheres) induces specific cooperative conformational changes to actin filaments that accompany supertwisting of the helix (orange). This attracts more cofilin molecules to bind to form a cluster (Ngo et al., 2015), while potentially repelling the motor domain of myosin II (blue ellipses) in the presence of ATP. The motor domain of myosin II induces different conformational changes to actin filaments that accompany untwisting of the helix (green). This leads to formation of a loose cluster in the presence of low concentrations of ATP (Tokuraku et al., 2009), while repelling cofilin. Mechanical stretching also untwists the helix of actin filaments accompanying the attraction of myosin II motor domains (Uyeda et al., 2011) and repelling of cofilin (Hayakawa et al., 2011), such that the stretched state of actin filaments is similar to the myosin II-bound state. For the simplicity, cooperative conformational changes are depicted to propagate to both directions. In the case of cofilin, cooperative conformational changes have been shown to propagate only to the pointed ends (Ngo et al., 2015).

17.4.1   Segregation of ABPs along Actin Filaments

Direct competition for binding sites would allow many different ABP molecules to bind randomly on each actin protomer. In contrast, mutually exclusive binding of two or more ABPs that is mediated by cooperative conformational changes in actin filaments allows segregated binding of the ABPs. If a section of the actin filament is bound with one ABP or a specific group of ABPs, this would be more effective in specifying the function of the actin filament than when different ABP molecules are randomly distributed along the filament in a mixed manner (Fig. 17.3).

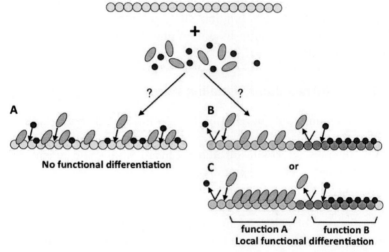

Figure 17.3   Schematic representation of two modes of mutually exclusive binding of two ABPs. Two ABPs (blue ellipses and red spheres) are bound to an actin filament (yellow), which is shown as a single strand for simplicity. In the direct competition model (left), each of the two ABP molecules binds to different actin protomers because they compete for a binding site on actin. Their distribution is random, however, and no functional differentiation is achieved. If both ABPs induce distinct cooperative conformational changes to actin protomers that propagate along the filament, and the conformational change attracts binding of the same ABP but repels the other ABP, the two ABPs would form segregated clusters, leading to local functional differentiation of the filament sections (right).

17.4.2 Amplification of the Inhibitory Effect

Mutually exclusive binding of two ABPs that is mediated by cooperative conformational changes in actin filaments allows a relatively small number of one type of ABP molecules to regulate the binding of the second ABP. For example, knocking down the expression of cofilin in HeLa cells led to hyper-activation of myosin II (Wiggan et al., 2012), and these authors interpreted their results to derive from direct competition for binding sites on actin between cofilin and myosin II. However, the normal expression level of cofilin is too low to occupy the majority of actin protomers in filaments (Pollard et al., 2000). This makes cooperative conformational change of the filaments, rather than direct competition for binding sites, a more likely mechanism of cofilin-mediated suppression of the activity of myosin II in untreated cells.

17.4.3 Intracellular Signaling Wire

Actin filaments polymerized by formin have a structure different from those formed by spontaneous nucleation/polymerization, in a manner detectable by fluorescence-lifetime and anisotropy-decay experiments (Papp et al., 2006). If this structural difference positively and negatively regulates binding of various other ABPs, those formin-dependent actin filaments are predisposed to bind specific sets of ABPs, again leading to functional differentiation. Although this is the case of a polymerization factor predisposing the function of the entire filament (thus, "identity acquired at birth" (Michelot and Drubin, 2011)), other side-binding ABPs could also influence the property of the filament over a long distance, allowing the actin filaments to act as an intracellular signaling wire.

17.5 Actin Filaments as Mechanosensors

It was experimentally demonstrated that actin filaments become slightly longer and untwisted when mechanical tension is applied (Wakabayashi et al., 1994). Similar results were obtained by molecular dynamics simulations (Matsushita et al., 2011). The structure of stretched actin filaments is at least qualitatively similar to that bound with myosin II in that they are both slightly untwisted, and is distinct from the supertwisted structure bound with cofilin. This suggests the possibility that stretching actin filaments would inhibit the interaction with cofilin, while promoting interactions with myosin II. This naïve prediction has been supported by experimental data, as described below.

Stress fibers are contractile structures found in many types of animal cells in culture. They contain actin filaments and myosin II filaments and are stable when mechanically stretched, but disassemble when tension is reduced either by inhibiting the activity of myosin II or by allowing the cell to contract on a compliant substrate. Hayakawa et al. (2011) demonstrated that stretched actin filaments are protected from the severing activity of cofilin in vitro, and that this is involved in the tension-dependent stability of stress fibers in vivo.

Actin filaments in the anterior lamellipodia of migrating amoeboid cells are mechanically compressed because polymerizing actin filaments push the cell membrane forward. In contrast, actin filaments in the posterior cortex are stretched through interaction with myosin II filaments, which drives retraction of the cell rear. Thus, distinct mechanical conditions of actin filaments in the front and rear of amoeboid cells would allow them to function as mechanosensors. Indeed, cofilin is excluded from the posterior actin cortex (Bravo-Cordero et al., 2013), and the stretching of posterior actin filaments may contribute to this intracellular localization of cofilin. Furthermore, the structural similarity between stretched and myosin II S1-bound actin filaments suggests that S1 may have a higher affinity for stretched actin filaments than for relaxed filaments. If so, because myosin II filaments are tension generators, increased tension would untwist the helical structure, which recruits more myosin II binding, resulting in increased tension, potentially forming a positive feedback loop in the posterior cortex. To test this possibility, we took advantage of a mutant form of *Dictyostelium* myosin II S1 that binds to actin filaments with high affinity even in the presence of ATP. In *Dictyostelium* cells, the mutant S1 fused with GFP was enriched along mechanically stretched actin filaments, including those in the posterior cortex in migrating cells and contractile rings in dividing cells, as well as those stretched by externally-applied force (Uyeda et al., 2011). Because S1 of myosin II is not biochemically regulated, selective binding of the mutant S1 to stretched actin filaments is likely dependent on the stretch-induced conformational change of actin filaments. Although in vitro studies are needed to examine the possible involvement of other factors in this process, the positive feedback loop that depends on the mechanosensitivity of actin filaments would stabilize the locally contractile state of the cortex, leading to stabilization of the front-rear axis of a cell. It would also confer the cells responsiveness to external mechanical stimuli, such as that observed in fragments of fish keratocytes (Verkhovsky et al., 1999).

The Arp2/3 complex, which binds to mother actin filaments and initiates polymerization of a daughter filament at a 70° degree branching angle, is involved in the formation of a dendric meshwork of actin filaments typically seen in lamellipodia. Pushing of the cell membrane by growing actin filaments in lamellipodia would not only compress but also bend the filaments. To maintain proper geometry of the dendric actin meshwork for efficient pushing of the cell membrane forward, it is necessary to generate and grow a daughter actin filament on the convex side of the bent mother filament, as this side of the filament faces the front of the cell. Consistent with this theoretical consideration, Risca et al. demonstrated that Arp2/3 preferentially binds to the convex side of bent actin filaments in vitro (Risca et al., 2012), and this is presumably related to the fact that bending applies asymmetric mechanical stress to the two protofilaments on the concave and convex sides of actin filaments. This is another intriguing case of mechanosensitivity of actin filaments playing physiologically important roles.

17.6 Why Is Actin So Conservative?

As discussed above, binding of an ABP due to local biochemical activation of that ABP, mechanical stretching, or even thermal fluctuation, changes the structure of actin protomers in filaments. Regardless of the trigger, certain types of conformational changes are propagated to neighboring protomers in the same filament, and this plays important roles in specifying what ABPs bind to those neighboring protomers. This depends on an allosteric regulation of actin protomers, involving a global conformational change of actin protomers, since each protomer receives a trigger from one side of the molecule and transmits the same signal to the next protomer on the other side of the molecule. Furthermore, because there appear to be multiple different types of cooperative polymorphism, actin protomers need the capacity to undergo multiple and different types of allosteric regulation, each involving a specific global conformational change. These functional restrictions are most likely the reason why the primary structure of actin, including that inside the molecule, is so difficult to change, as has been pointed out by Galkin et al. (2010, 2012, 2015).

It is still a mystery why bacterial actin took a standard approach of duplicating the parent gene and changing the sequence of the two genes independently to perform different multiple functions. This difference may be related to the fact that the number of bacterial actin functions in each cell type is relatively limited, whereas actin performs a much larger number of functions in one eukaryotic cell. Thus, one or a small number of divergent actin genes are sufficient for each bacterial cell, and even when a bacterial cell contains multiple actins, it is possible to regulate the assembly of that small number of specialized actins to form separate structures in one cell. In contrast, it was presumably easier for an ancestral eukaryotic cell to use a single species of actin filaments and functionally differentiate them by cooperative conformational changes to perform different functions, than to prepare a large number of specialized actins for each of the many actin functions and separately regulate that large number of mono-functional actins proteins that are structurally similar to each other.

Recently, a genome of archaea that contains many of the features expected for the most recent common ancestor of eukaryotes and archaea was published, and this genome contained a single copy of actin gene that is very similar to eukaryotic actin (Spang et al., 2015). The striking homology between eukaryotic actins and this archaeal actin includes Gly residues, which generally play important roles in conformational changes in proteins. For example, Gly146, located at the bridge between the two domains in the actin molecule, presumably function as a pivot of movements between the two domains. This Gly residue is absolutely conserved among all eukaryotic actins with known sequences, as well as in the archaeal actin (Spang et al., 2015), but not in the bacterial actin MreB (van den Ent et al., 2001). Characterization of this archaeal actin may give us clues as to why eukaryotic actin is so highly conservative.

17.7 Possible Dynamic Roles of Actin Filaments in Muscle Contraction

In the prevailing swinging lever arm model (Cooke et al., 1984; Uyeda et al., 1996; Holmes, 1997) that assumes the rotation of the lever arm domain around the catalytic domain within S1, or that assumes rotation of the lever arm domain around the S1-S2 junction (Sugi et al., 2003, 2015), actin filaments are presumed to play two, relatively passive roles in actin-myosin force production. The first role is to provide a foothold for myosin to transmit force during the force-generating and tension-bearing states in the actin-myosin crossbridges, and the second is to stimulate the release of Pi from myosin motors carrying ADP and Pi, which subsequently triggers the force-generating conformational change in the myosin motor (Huxley, 2000). However, it is clear that this simple view is insufficient and needs a revision. This is because actins modified with certain crosslinkers or protease, as well as certain mutant actins, are able to bind strongly to skeletal myosin motor, and stimulate its Pi release, and yet, are unable to move at all (Prochniewicz and Yanagida, 1990) or move only at a very slow speed (Schwyter et al., 1990; Prochniewicz et al., 1993; Kim et al., 1998; Noguchi et al., 2012). Treatment of actin filaments with crosslinking reagents (Prochniewicz et al., 1993; Kim et al., 1998) would restrict intramolecular and intermolecular motions in actin filaments. Thus, the above results suggest that dynamic conformational changes in or between actin protomers are essential for myosin to move.

This view is consistent with intramolecular Förster energy transfer (FRET) analysis of actin filaments along which processive myosin V moved in the presence of ATP (Kozuka et al., 2006).

In control actin filaments, there were two major peaks of FRET intensities, consistent with the polymorphic nature of actin filaments. When myosin V was allowed to move along those filaments, fractions of higher FRET intensity significantly increased, while in the immobile glutaraldehyde-crosslinked actin filaments, a low FRET state predominated. This intriguing observation suggested that myosin movement is associated with, and dependent on, conformational changes in actin protomers.

Analyses of an immobile mutant actin also provided further insight into the roles of actin dynamics in myosin motility. As mentioned above, Gly146 is located at the pivot between the two domains in the actin molecule, and a G146V mutant actin was isolated on the basis of dominant lethality in yeast, i.e., yeast cells carrying a wild-type and a G146V mutant copy of the actin gene cannot grow (Noguchi et al., 2010). This G146V mutant actin exhibited a number of interesting phenotypes in vitro. First, filaments of G146V actin were unable to bind cofilin. Furthermore, a copolymer of wild-type and mutant actins even at a 10:1 mixing ratio were also unable to bind cofilin. These results indicated that the G146V mutant actin either produced long-range cooperative conformational changes to neighboring wild-type actin protomers and rendered them inaccessible to cofilin, or inhibited propagation of cofilin-induced cooperative conformational changes that are necessary for cooperative cofilin binding (Noguchi et al., 2010).

In in vitro myosin motility assays, the G146V mutant actin filaments hardly moved on surfaces coated with skeletal muscle myosin II, which was accompanied by strong inhibition of force generation. Nonetheless, the mutant actin filaments moved at normal speeds on myosin V-coated surfaces, and produced normal force with myosin V (Noguchi et al., 2012). A similar myosin class-specific inhibition of motility was observed with another mutant actin carrying a mutation in the DNase binding loop (Kubota et al., 2009).

In recent intramolecular FRET analyses in the presence of ATP, Noguchi et al. (2015) discovered that control actin filaments undergo different conformational changes when interacting with myosin II or myosin V. Wild-type and G146V mutant filaments underwent similar conformational changes when interacting with myosin V in the presence of ATP, but the two types of filaments behaved differently when interacting with myosin II in the presence of ATP. These results are consistent with the previous transient phosphorescence anisotropy measurement that showed that myosin II and V induce different conformational changes to actin filaments (Prochniewicz et al., 2010). Furthermore, these results suggest that the requirements for conformational changes in actin protomers are different between the fast and non-processive myosin II and the slow and processive myosin V.

It is not known, however, in what way conformational changes in actin protomers are needed for the motility with myosin II. One extreme scenario is that actin protomers undergo cyclic conformational changes accompanying transient interaction with the myosin motor, and that this is needed for the fast and non-processive movement. This possibility cannot be excluded but is probably unrelated to the FRET data mentioned above, since the temporal resolution of the FRET measurements is too slow to detect short-lived intermediate conformations that might occur during the ATPase cycle. The other extreme scenario is that actin protomers need to assume a conformation that is suitable for productive interaction with myosin II, which itself is induced by binding with myosin II, and is long-lived such that it can be detected by relatively slow FRET measurements.

This latter possibility may be related to cooperative conformational changes of actin protomers induced by myosin II binding (Oosawa et al., 1973; Miki et al., 1982; Prochniewicz and Thomas, 1997; Prochniewicz et al., 2004; Siddique et al., 2005). It may also be correlated with the cooperative binding of myosin II motors to actin filaments (Orlova and Egelman, 1997; Tokuraku et al., 2009). If so, the myosin II-induced conformation of the actin protomers, which is necessary for productive force generation by myosin II, has a higher affinity for the myosin II motors. In particular, if the myosin-induced cooperative conformational changes of actin protomers to increase the affinity for myosin II are unidirectional, as was the case for cofilin-induced conformational changes (Ngo et al., 2015), this could drive unidirectional sliding or stepping motion of the motor II domain in a manner independent of the swinging motion of the lever arm (Fig. 17.4). Indeed, it was observed that a motor domain of skeletal muscle myosin II moves a long distance by making multiple 5.5 nm steps in one ATP hydrolysis cycle (Kitamura et al., 1999). A similar behavior was reproduced by a molecular dynamics simulation (Takano et al., 2010). To test this possibility, it is critically important to determine if the cluster of the myosin motor grows unidirectionally or bidirectionally along actin filaments.

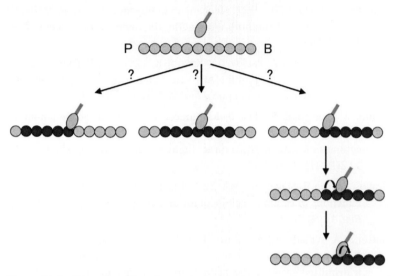

Figure 17.4 Directionality of myosin-induced cooperative conformational changes in an actin filament. For simplicity, the actin filament is shown as a single strand. If the cooperative conformational change (from pink to red) propagates only to the barbed-end direction as shown on the right, this could drive sliding or stepping of the myosin motor (blue ellipse) along the filament in a manner independent of the lever arm swing.

Acknowledgement

We thank Drs. Kiyotaka Tokuraku and Taro Noguchi for critical comments on the manuscript.

References

Amato PA, Taylor DL (1986) Probing the mechanism of incorporation of fluorescently labeled actin into stress fibers. *J Cell Biol,* **102**: 1074–1084.

Belyantseva IA, Perrin BJ, Sonnemann KJ, Zhu M, Stepanyan R, McGee J, Frolenkov GI, Walsh EJ, Friderici KH, Friedman TB, Ervasti

JM (2009) γ-Actin is required for cytoskeletal maintenance but not development. *Proc Natl Acad Sci U S A,* **106**: 9703–9708.

Bernstein BW, Bamburg JR (2010) ADF/cofilin: A functional node in cell biology. *Trends Cell Biol,* **20**: 187–195.

Blanchoin L, Pollard TD (1999) Mechanism of interaction of Acanthamoeba actophorin (ADF/Cofilin) with actin filaments. *J Biol Chem,* **274**: 15538–15546.

Bobkov AA, Muhlrad A, Pavlov DA, Kokabi K, Yilmaz A, Reisler E (2006) Cooperative effects of cofilin (ADF) on actin structure suggest allosteric mechanism of cofilin function. *J Mol Biol,* **356**: 325–334.

Brault V, Reedy MC, Sauder U, Kammerer RA, Aebi U, Schoenenberger C (1999) Substitution of flight muscle-specific actin by human β-cytoplasmic actin in the indirect flight muscle of Drosophila. *J Cell Sci,* **112**(Pt 21): 3627–3639.

Bravo-Cordero JJ, Magalhaes MA, Eddy RJ, Hodgson L, Condeelis J (2013) Functions of cofilin in cell locomotion and invasion. *Nat Rev Mol Cell Biol,* **14**: 405–415.

Butters CA, Willadsen KA, Tobacman LS (1993) Cooperative interactions between adjacent troponin-tropomyosin complexes may be transmitted through the actin filament. *J Biol Chem,* **268**: 15565–15570.

Carlier MF (1990) Actin polymerization and ATP hydrolysis. *Adv Biophys,* **26**: 51–73

Carlier MF, Laurent V, Santolini J, Melki R, Didry D, Xia GX, Hong Y, Chua NH, Pantaloni D (1997) Actin depolymerizing factor (ADF/cofilin) enhances the rate of filament turnover: Implication in actin-based motility. *J Cell Biol,* **136**: 1307–1322.

Collins A, Huang R, Jensen MH, Moore JR, Lehman W, Wang CL (2011) Structural studies on maturing actin filaments. *Bioarchitecture,* **1**: 127–133.

Cooke R, Crowder MS, Wendt CH, Barnett VA, Thomas DD (1984) Muscle cross-bridges: Do they rotate? *Adv Exp Med Biol,* **170**: 413–427.

Craig-Schmidt MC, Robson RM, Goll DE, Stromer MH (1981) Effect of alpha-actinin on actin structure. Release of bound nucleotide. *Biochim Biophys Acta,* **670**: 9–16.

De La Cruz EM (2005) Cofilin binding to muscle and non-muscle actin filaments: Isoform-dependent cooperative interactions. *J Mol Biol,* **346**: 557–564.

dos Remedios CG, Chhabra D, Kekic M, Dedova IV, Tsubakihara M, Berry DA, Nosworthy NJ (2003) Actin binding proteins: Regulation of cytoskeletal microfilaments. *Physiol Rev,* **83**: 433–473.

Fujii T, Iwane AH, Yanagida T, Namba K (2010) Direct visualization of secondary structures of F-actin by electron cryomicroscopy. *Nature,* **467**: 724–728.

Galinska-Rakoczy A, Wawro B, Strzelecka-Golaszewska H (2009) New aspects of the spontaneous polymerization of actin in the presence of salts. *J Mol Biol,* **387**: 869–882.

Galkin VE, Orlova A, Cherepanova O, Lebart MC, Egelman EH (2008) High-resolution cryo-EM structure of the F-actin-fimbrin/plastin ABD2 complex. *Proc Natl Acad Sci USA,* **105**: 1494–1498.

Galkin VE, Orlova A, Egelman EH (2012) Actin filaments as tension sensors. *Curr Biol,* **22**: R96–R101.

Galkin VE, Orlova A, Kudryashov DS, Solodukhin A, Reisler E, Schroder GF, Egelman EH (2011) Remodeling of actin filaments by ADF/cofilin proteins. *Proc Natl Acad Sci USA,* **108**: 20568–20572.

Galkin VE, Orlova A, Lukoyanova N, Wriggers W, Egelman EH (2001) Actin depolymerizing factor stabilizes an existing state of F-actin and can change the tilt of F-actin subunits. *J Cell Biol,* **153**: 75–86.

Galkin VE, Orlova A, Schroder GF, Egelman EH (2010) Structural polymorphism in F-actin. *Nat Struct Mol Biol,* **17**: 1318–1323.

Galkin VE, Orlova A, Vos MR, Schroder GF, Egelman EH (2015) Near-atomic resolution for one state of F-actin. *Structure,* **23**: 173–182.

Gressin L, Guillotin A, Guerin C, Blanchoin L, Michelot A (2015) Architecture dependence of actin filament network disassembly. *Curr Biol,* **25**: 1437–1447.

Hanein D, Matsudaira P, DeRosier DJ (1997) Evidence for a conformational change in actin induced by fimbrin (N375) binding. *J Cell Biol,* **139**: 387–396.

Hansen SD, Kwiatkowski AV, Ouyang CY, Liu H, Pokutta S, Watkins SC, Volkmann N, Hanein D, Weis WI, Mullins RD, Nelson WJ (2013) αE-catenin actin-binding domain alters actin filament conformation and regulates binding of nucleation and disassembly factors. *Mol Biol Cell,* **24**: 3710–3720.

Hatano S, Oosawa F (1966) Extraction of an actin-like protein from the plasmodium of a myxomycete and its interaction with myosin A from rabbit striated muscle. *J Cell Physiol,* **68**: 197–202.

Hawkins M, Pope B, Maciver SK, Weeds AG (1993) Human actin depolymerizing factor mediates a pH-sensitive destruction of actin filaments. *Biochemistry,* **32**: 9985–9993.

Hayakawa K, Sakakibara S, Sokabe M, Tatsumi H (2014) Single-molecule imaging and kinetic analysis of cooperative cofilin-actin filament interactions. *Proc Natl Acad Sci U S A,* **111**: 9810–9815.

Hayakawa K, Tatsumi H, Sokabe M (2011) Actin filaments function as a tension sensor by tension dependent binding of cofilin to the filament. *J Cell Biol,* **195**: 721–727.

Hayden SM, Miller PS, Brauweiler A, Bamburg JR (1993) Analysis of the interactions of actin depolymerizing factor with G-and F-actin. *Biochemistry,* **32**: 9994–10004.

Holmes KC (1997) The swinging lever-arm hypothesis of muscle contraction. *Curr Biol,* **7**: 112–118.

Holmes KC, Angert I, Kull FJ, Jahn W, Schroder RR (2003) Electron cryo-microscopy shows how strong binding of myosin to actin releases nucleotide. *Nature,* **425**: 423–427.

Holmes KC, Popp D, Gebhard W, Kabsch W (1990) Atomic model of the actin filament. *Nature,* **347**: 44–49

Huang R, Grabarek Z, Wang CL (2010) Differential effects of caldesmon on the intermediate conformational states of polymerizing actin. *J Biol Chem,* **285**: 71–79.

Huxley AF (2000) Mechanics and models of the myosin motor. *Philos Trans R Soc Lond B Biol Sci,* **355**: 433–440.

Jensen MH, Morris EJ, Huang R, Rebowski G, Dominguez R, Weitz DA, Moore JR, Wang CL (2012) The conformational state of actin filaments regulates branching by actin-related protein 2/3 (Arp2/3) complex. *J Biol Chem,* **287**: 31447–31453.

Kandasamy MK, McKinney EC, Meagher RB (2009) A single vegetative actin isovariant overexpressed under the control of multiple regulatory sequences is sufficient for normal Arabidopsis development. *Plant Cell,* **21**: 701–718.

Kijima ST, Hirose K, Kong SG, Wada M, Uyeda TQP (2016) Distinct biochemical properties of arabidopsis thaliana actin isoforms. *Plant Cell Physiol,* **57**: 46–56.

Kijima S, Noguchi TQP, Uyeda TQP, Tokuraku K (2011) Exclusive binding of myosin and cofilin to F-actin. *Seibutsu Butsuri,* **51**(Supplement 1): S85.

Kim E, Bobkova E, Miller CJ, Orlova A, Hegyi G, Egelman EH, Muhlrad A, Reisler E (1998) Intrastrand cross-linked actin between Gln-41 and Cys-374. III. Inhibition of motion and force generation with myosin. *Biochemistry,* **37**: 17801–17809.

Kitamura K, Tokunaga M, Iwane AH, Yanagida T (1999) A single myosin head moves along an actin filament with regular steps of 5.3 nanometres. *Nature,* **397**: 129–134.

Kozuka J, Yokota H, Arai Y, Ishii Y, Yanagida T (2006) Dynamic polymorphism of single actin molecules in the actin filament. *Nat Chem Biol,* **2**: 83–86.

Kubota H, Mikhailenko SV, Okabe H, Taguchi H, Ishiwata S (2009) D-loop of actin differently regulates the motor function of myosins II and V. *J Biol Chem,* **284**: 35251–35258.

Matsushita S, Inoue Y, Hojo M, Sokabe M, Adachi T (2011) Effect of tensile force on the mechanical behavior of actin filaments. *J Biomech,* **44**: 1776–1781.

McDowell JM, Huang S, McKinney EC, An YQ, Meagher RB (1996) Structure and evolution of the actin gene family in Arabidopsis thaliana. *Genetics,* **142**: 587–602.

McGough A, Pope B, Chiu W, Weeds A (1997) Cofilin changes the twist of F-actin: Implications for actin filament dynamics and cellular function. *J Cell Biol,* **138**: 771–781.

McKane M, Wen KK, Meyer A, Rubenstein PA (2006) Effect of the substitution of muscle actin-specific subdomain 1 and 2 residues in yeast actin on actin function. *J Biol Chem,* **281**: 29916–29928.

Michelot A, Drubin DG (2011) Building distinct actin filament networks in a common cytoplasm. *Curr Biol,* **21**: R560–R569.

Miki M, Wahl P, Auchet JC (1982) Fluorescence anisotropy of labeled F-actin: Influence of divalent cations on the interaction between F-actin and myosin heads. *Biochemistry,* **21**: 3661–3665.

Murakami K, Yasunaga T, Noguchi TQP, Gomibuchi Y, Ngo KX, Uyeda TQP, Wakabayashi T (2010) Structural basis for actin assembly, activation of ATP hydrolysis, and delayed phosphate release. *Cell,* **143**: 275–287.

Murthy K, Wadsworth P (2005) Myosin-II-dependent localization and dynamics of F-actin during cytokinesis. *Curr Biol,* **15**: 724–731.

Ngo KX, Kodera N, Katayama E, Ando T, Uyeda TQP (2015) Cofilin-induced unidirectional cooperative conformational changes in actin filaments revealed by high-speed atomic force microscopy. *Elife,* **4**, e04806.

Noguchi TQP, Komori T, Umeki N, Demizu N, Ito K, Iwane AH, Tokuraku K, Yanagida T, Uyeda TQP (2012) G146V mutation at the hinge region of actin reveals a myosin class-specific requirement of actin conformations for motility. *J Biol Chem,* **287**: 24339–24345.

Noguchi TQP, Morimatsu M, Iwane AH, Yanagida T, Uyeda TQP (2015) The role of structural dynamics of actin in class-specific myosin motility. *PLoS One,* **10**: e0126262.

Noguchi TQP, Toya R, Ueno H, Tokuraku K, Uyeda TQP (2010) Screening of novel dominant negative mutant actins using glycine targeted scanning identifies G146V actin that cooperatively inhibits cofilin binding. *Biochem Biophys Res Commun,* **396**: 1006–1011.

Oda T, Iwasa M, Aihara T, Maeda Y, Narita A (2009) The nature of the globular-to fibrous-actin transition. *Nature,* **457**: 441–445.

Oosawa F, Fujime S, Ishiwata S, Mihashi K (1973) Dynamic properties of F-actin and thin filament. *Cold Spring Harbor Symp Quant Biol,* **37**: 277–285.

Orlova A, Egelman EH (1992) Structural basis for the destabilization of F-actin by phosphate release following ATP hydrolysis. *J Mol Biol,* **227**: 1043–1053.

Orlova A, Egelman EH (1997) Cooperative rigor binding of myosin to actin is a function of F-actin structure. *J Mol Biol,***265**: 469–474.

Orlova A, Prochniewicz E, Egelman EH (1995) Structural dynamics of F-actin: II. Cooperativity in structural transitions. *J Mol Biol,* **245**: 598–607.

Orlova A, Shvetsov A, Galkin VE, Kudryashov DS, Rubenstein PA, Egelman EH, Reisler E (2004) Actin-destabilizing factors disrupt filaments by means of a time reversal of polymerization. *Proc Natl Acad Sci U S A,* **101**: 17664–17668.

Pak CW, Flynn KC, Bamburg JR (2008) Actin-binding proteins take the reins in growth cones. *Nat Rev Neurosci,* **9**: 136–147.

Papp G, Bugyi B, Ujfalusi Z, Barko S, Hild G, Somogyi B, Nyitrai M (2006) Conformational changes in actin filaments induced by formin binding to the barbed end. *Biophys J,* **91**: 2564–2572.

Pollard TD, Blanchoin L, Mullins RD (2000) Molecular mechanisms controlling actin filament dynamics in nonmuscle cells. *Annu Rev Biophys Biomol Struct,* **29**: 545–576.

Pollard TD, Cooper JA (2009) Actin, a central player in cell shape and movement. *Science,* **326**: 1208–1212.

Prochniewicz E, Chin HF, Henn A, Hannemann DE, Olivares AO, Thomas DD, De La Cruz EM (2010) Myosin isoform determines the conformational dynamics and cooperativity of actin filaments in the strongly bound actomyosin complex. *J Mol Biol*, **396**: 501–509.

Prochniewicz E, Janson N, Thomas DD, De La Cruz EM (2005) Cofilin increases the torsional flexibility and dynamics of actin filaments. *J Mol Biol*, **353**: 990–1000.

Prochniewicz E, Katayama E, Yanagida T, Thomas DD (1993) Cooperativity in F-actin: Chemical modifications of actin monomers affect the functional interactions of myosin with unmodified monomers in the same actin filament. *Biophys J*, **65**: 113–123.

Prochniewicz E, Thomas DD (1997) Perturbations of functional interactions with myosin induce long-range allosteric and cooperative structural changes in actin. *Biochemistry*, **36**: 12845–12853.

Prochniewicz E, Walseth TF, Thomas DD (2004) Structural dynamics of actin during active interaction with myosin: Different effects of weakly and strongly bound myosin heads. *Biochemistry*, **43**: 10642–10652.

Prochniewicz E, Yanagida T (1990) Inhibition of sliding movement of F-actin by crosslinking emphasizes the role of actin structure in the mechanism of motility. *J Mol Biol*, **216**: 761–772.

Risca VI, Wang FR, Chaudhuri O, Chia JJ, Geissler PL, Fletcher DA (2012) Actin filament curvature biases branching direction. *Proc Natl Acad Sci U S A*, **109**: 2913–2918.

Schwyter DH, Kron SJ, Toyoshima YY, Spudich JA, Reisler E (1990) Subtilisin cleavage of actin inhibits in vitro sliding movement of actin filaments over myosin. *J Cell Biol*, **111**: 465–470.

Sharma S, Grintsevich EE, Phillips ML, Reisler E, Gimzewski JK (2011) Atomic force microscopy reveals drebrin induced remodeling of f-actin with subnanometer resolution. *Nano Lett*, **11**: 825–827.

Siddique MS, Mogami G, Miyazaki T, Katayama E, Uyeda TQP, Suzuki M (2005) Cooperative structural change of actin filaments interacting with activated myosin motor domain, detected with copolymers of pyrene-labeled actin and acto-S1 chimera protein. *Biochem Biophys Res Commun*, **337**: 1185–1191.

Spang A, Saw JH, Jorgensen SL, Zaremba-Niedzwiedzka K, Martijn J, Lind AE, van Eijk R, Schleper C, Guy L, Ettema TJ (2015) Complex archaea that bridge the gap between prokaryotes and eukaryotes. *Nature*, **521**: 173–179.

Steinmetz MO, Goldie KN, Aebi U (1997) A correlative analysis of actin filament assembly, structure, and dynamics. *J Cell Biol,* **138**: 559–574.

Straub FB (1942) Actin. *Stud. Inst. Med. Chem. Univ. Szeged,* **2**: 3–15

Suarez C, Roland J, Boujemaa-Paterski R, Kang H, McCullough BR, Reymann AC, Guerin C, Martiel JL, De La Cruz EM, Blanchoin L (2011) Cofilin tunes the nucleotide state of actin filaments and severs at bare and decorated segment boundaries. *Curr Biol,* **21**: 862–868.

Sugi H, Akimoto T, Kobayashi T (2003) Evidence for the involvement of myosin subfragment 2 in muscle contraction. *Adv Exp Med Biol,* **538**: 317–332.

Sugi H, Chaen S, Akimoto T, Minoda H, Miyakawa T, Miyauchi Y, Tanokura M, Sugiura S (2015) Electron microscopic recording of myosin head power stroke in hydrated myosin filaments. *Sci Rep,* **5**: 15700.

Takano M, Terada TP, Sasai M (2010) Unidirectional Brownian motion observed in an in silico single molecule experiment of an actomyosin motor. *Proc Natl Acad Sci U S A,* **107**: 7769–7774.

Theriot JA, Mitchison TJ (1991) Actin microfilament dynamics in locomoting cells. *Nature,* **352**: 126–131.

Tokuraku K, Kurogi R, Toya R, Uyeda TQP (2009) Novel mode of cooperative binding between myosin and Mg^{2+}-actin filaments in the presence of low concentrations of ATP. *J Mol Biol,* **386**: 149–162.

Umeki N, Nakajima J, Noguchi TQP, Tokuraku K, Nagasaki A, Ito K, Hirose K, Uyeda TQP (2012) Rapid nucleotide exchange renders Asp-11 mutant actins resistant to depolymerizing activity of cofilin, leading to dominant toxicity in vivo. *J Biol Chem,* **288**: 1739–1749.

Umeki N, Uyeda TQP (2012) Cooperative conformational changes of actin filaments drive mutually exclusive cooperative binding of cofilin and HMM to the filaments. *Seibutsu Butsuri,* **52**(Supplement 1): S113.

Uyeda TQP, Abramson PD, Spudich JA (1996) The neck region of the myosin motor domain acts as a lever arm to generate movement. *Proc Natl Acad Sci U S A,* **93**: 4459–4464.

Uyeda TQP, Iwadate Y, Umeki N, Nagasaki A, Yumura S (2011) Stretching actin filaments within cells enhances their affinity for the myosin II motor domain. *PLoS One,* **6**: e26200.

van den Ent F, Amos LA, Löwe J (2001) Prokaryotic origin of the actin cytoskeleton. *Nature,* **413**: 39–44.

Verkhovsky AB, Svitkina TM, Borisy GG (1999) Self-polarization and directional motility of cytoplasm. *Curr Biol,* **9**: 11–20.

von der Ecken J, Muller M, Lehman W, Manstein DJ, Penczek PA, Raunser S (2015) Structure of the F-actin-tropomyosin complex. *Nature,* **519**: 114–117.

Wakabayashi K, Sugimoto Y, Tanaka H, Ueno Y, Takezawa Y, Amemiya Y (1994) X-ray diffraction evidence for the extensibility of actin and myosin filaments during muscle contraction. *Biophys J,* **67**: 2422–2435.

Wang YL (1985) Exchange of actin subunits at the leading edge of living fibroblasts: Possible role of treadmilling. *J Cell Biol,* **101**: 597–602.

Wiggan O, Shaw AE, DeLuca JG, Bamburg JR (2012) ADF/cofilin regulates actomyosin assembly through competitive inhibition of myosin II binding to F-actin. *Dev Cell,* **22**: 530–543.

Yumura S, Itoh G, Kikuta Y, Kikuchi T, Kitanishi-Yumura T, Tsujioka M (2013) Cell-scale dynamic recycling and cortical flow of the actin-myosin cytoskeleton for rapid cell migration. *Biol Open,* **2**: 200–209.

Index